Advances in Mobile Commerce Technologies

Ee-Peng Lim
Nanyang Technological University, Singapore

Keng Siau
University of Nebraska-Lincoln, USA

IGP
IDEA GROUP PUBLISHING
Hershey • London • Melbourne • Singapore • Beijing

Acquisition Editor:	Mehdi Khosrow-Pour
Senior Managing Editor:	Jan Travers
Managing Editor:	Amanda Appicello
Development Editor:	Michele Rossi
Copy Editor:	Amy Bingham
Typesetter:	Amanda Lutz
Cover Design:	Weston Pritts
Printed at:	Integrated Book Technology

Published in the United States of America by
Idea Group Publishing (an imprint of Idea Group Inc.)
701 E. Chocolate Avenue, Suite 200
Hershey PA 17033
Tel: 717-533-8845
Fax: 717-533-8661
E-mail: cust@idea-group.com
Web site: http://www.idea-group.com

and in the United Kingdom by
Idea Group Publishing (an imprint of Idea Group Inc.)
3 Henrietta Street
Covent Garden
London WC2E 8LU
Tel: 44 20 7240 0856
Fax: 44 20 7379 3313
Web site: http://www.eurospan.co.uk

Library of Congress Cataloging-in-Publication Data

Lim, Ee-Peng.
Advances in mobile commerce technologies / Ee-Peng Lim and Keng Siau.
p. cm.
Includes bibliographical references and index.
ISBN 1-59140-052-X (hardcover) -- ISBN 1-59140-089-9 (ebook)
1. Mobile communication systems. I. Siau, Keng, 1964- II. Title.
TK6570.M6L46 2003
621.382--dc21

2003000051

British Cataloguing in Publication Data
A Cataloguing in Publication record for this book is available from the British Library.

Advances in Mobile Commerce Technologies

Table of Contents

Preface

With the number of mobile device users exceeding that of PC users, conducting business and services over these mobile devices, which is known as mobile commerce, is becoming real and attractive. Although mobile commerce shares many similarities with traditional electronic commerce, it extends the latter by offering a wide range of personalized and location-aware services to users by integrating a myriad of technologies together. Some of these technologies are required to realize new mobile business opportunities, while others are needed to overcome the operating constraints within the mobile environment, such as limited screen size, less reliable and smaller bandwidth communication channel, shorter battery lifespan, and keyboardless input.

This book discusses mobile commerce with emphasis on both theory and application and serves as a good introductory guide for both researchers and practitioners. It consists of a collection of chapters on mobile commerce addressing a wide spectrum of technology and application issues. These chapters are essential to understanding the current state of mobile commerce applications and services. The book is structured into three parts.

Part I reviews the current trends and future development in mobile commerce applications and technology.

The article "Mobile Commerce: Current States and Future Trends," by Keng Siau, Ee-Peng Lim, and Zixing Shen presents an overview of mobile commerce development by examining the features of mobile commerce, the value-added applications, the enabling technologies, the business implications, and the challenges in implementing mobile commerce. The paper also provides an agenda for future research to enhance mobile commerce. The article provides the necessary background knowledge for readers to understand the rest of the book.

Part II focuses on the technological challenges facing mobile commerce.

In the chapter "Mobile E-Commerce on Mobile Phones," Do van Thanh describes the protocol and security issues involved in using mobile phones to conduct business-to-consumer transactions. The author explains the fundamental differences between mobile commerce and e-commerce in B2C transactions and identifies the limitations of Wireless Application Protocol (WAP) in mobile commerce. The author also proposes a solution known as Mobile ePay to provide authentication and micropayment services using mobile phones. A mobile commerce receipt system enabling instantaneous delivery is also described in detail.

The chapter on "Transactional Database Accesses for M-Commerce Clients," by Hong Va Leong, discusses the required generic architecture and appropriate mechanisms to be supported by database servers in the mobile environment. In particular, it focuses on the transaction processing component of the database server that ensures the atomicity and other desirable correctness criteria of the database accessing activities. The concept of transaction processing is generalized to encompass accessing multiple databases, while staying within the context of a mobile computing platform. Relevant issues on the broadcast database and the disconnected processing of transactions are also considered.

To overcome bandwidth and energy limitations resulting from short battery life of mobile devices, it is necessary to provide an energy-efficient wireless data dissemination architecture that supports broadcasting applications. The chapter "Techniques to Facilitate Information Exchange in Mobile Commerce," by Aslihan Celik and Anindya Datta, presents such an architecture. The chapter discusses the energy cost and access time of some proposed data broadcast and access protocols. It finally describes how secure data broadcasts can be achieved by incorporating encryption into the proposed protocols.

In the chapter "Digital Rights Management for Mobile Multimedia," Sai Ho Kwok proposes a digital rights management framework for mobile commerce. In the proposed framework, operations on digital rights, security, and payment are addressed. The framework can be adopted for the current 2.5G and 3G mobile technologies and even for 4G technologies.

For the chapter "Predicate Based Caching for Large Scale Mobile Distributed On-line Applications," the three authors, Abhinav Vora, Zahir Tari, and Peter Bertok, describe their experience in designing a predicate-based caching technique for mobile object-based middleware that optimizes the performance of the mobile medium by better utilizing the available bandwidth.

Part III covers the application studies and information systems issues in mobile commerce.

In the chapter "Modeling Static Aspects of Mobile Electronic Commerce Environments," by Jari Veijalainen and Mathias Weske, an object model that describes the fundamental static aspects of the mobile commerce environment and their relationships is presented. It distinguishes four spheres of concern: Regulatory Frameworks, Business Models, Enabling Technologies, and the Global Infrastructure. The spheres provide us a mean to understand and classify the development of mobile commerce applications and environment.

With location information about users and business entities, new kinds of mobile commerce applications can be developed. Stuart J. Barnes, in his chapter

"Known By the Network: The Emergence of Location-Based Mobile Commerce," examines the technologies, applications, and strategic issues associated with the commercialization of location based services.

The chapter "Usable M-Commerce Systems: The Need for Model-Based Approaches," by John Krogstie, Petter Bae Brandtzæg, Jan Heim, and Andreas L. Opdahl, discusses new challenges and possible solutions for developing and evolving usable mCommerce systems. The chapter focuses on model-based approaches. The authors summarize the main challenges on using model-based approaches to support the development of usable mCommerce systems and highlight research issues in this very dynamic area.

In the chapter "Managing the Interactions between Handheld Devices, Mobile Applications, and Users," by Maristella Agosti and Nicola Ferro, several issues related to managing the interactions between handheld devices, mobile applications, and users are discussed. The chapter suggests some approaches to overcome the constraints imposed by the mobile environment and to enhance the interactions between handheld devices and mobile applications.

Susy Chan and Xiaowen Fang, in the chapter "Mobile Commerce and Usability," analyse the usability issues that have great impact on the interface design, development, deployment and adoption of m-commerce applications. The chapter also highlights some usability topics for future research.

The chapter "Using Continuous Voice Activation Applications in Telemedicine to Transform Mobile Commerce" by James Rodger describes the use of mobile technologies in telemedicine efforts in defense. A strategy for implementing mobile telemedicine is given.

Finally, the chapter "Mobile Applications for Adaptive Supply Chains: A Landscape Analysis" by Ravi Kalakota, Marcia Robinson and Pavan Gundepudi examines the changes to supply chains brought about by mobile technologies.

The above collection of chapters provides a good mix of views on the technological and application aspects of mobile commerce. Mobile commerce is still in its infancy. More developments in this area are expected to take place in the near future. There will certainly be new technologies that will render some of the existing ideas obsolete. Nevertheless, this book will provide the necessary foundation for readers to understand the mobile commerce area and inspire more research work on mobile commerce-related technologies.

Part I:

Overview of Mobile Commerce

Chapter I

Mobile Commerce: Current States and Future Trends

Keng Siau
University of Nebraska-Lincoln, USA

Ee-Peng Lim
Nanyang Technological University, Republic of Singapore

Zixing Shen
University of Nebraska-Lincoln, USA

ABSTRACT

Advances in wireless technology increase the number of mobile device users and give pace to the rapid development of e-commerce using these devices. The new type of e-commerce, conducting transactions via mobile terminals, is called mobile commerce. Due to its inherent characteristics such as ubiquity, personalization, flexibility, and dissemination, mobile commerce promises business unprecedented market potential, great productivity, and high profitability. This paper presents an overview of mobile commerce development by examining the enabling technologies, the impact of mobile commerce on the business world, and the implications to mobile commerce providers. The paper also provides an agenda for future research in the area.

INTRODUCTION

Advances in wireless technology increase the number of mobile device users and give pace to the rapid development of e-commerce conducted with these devices. The new type of e-commerce transactions, conducted through mobile devices using wireless telecommunications network and other wired e-commerce technologies, is called mobile commerce (increasingly known as mobile e-commerce or m-commerce). Mobile commerce enables a new mode of information exchange and purchases, and it presents an unexplored domain. To consumers, it represents convenience; merchants associate it with a huge earning potential; service providers view it as a large unexplored market; governments look it as a viable and highly productive connection with their constituents. In short, mobile commerce promises many more alluring market opportunities than traditional e-commerce and the global mobile commerce market is expected to be worth a staggering US$200 billion by 2004 (Guy Singh, 2000). Because of the characteristics and constraints of mobile devices and wireless network, the emerging mobile commerce operates in an environment very different from e-commerce conducted over the wired Internet. Although mobile commerce will emerge as a major focus of the business world and telecommunication industry in the immediate future, the marriage of mobile devices and the Internet is filled with challenges as well.

The article is structured as follows. We first summarize the features of mobile commerce. Next, value-added applications of mobile commerce and an overview of mobile commerce technology are presented, and the business implications are discussed. We then highlight the challenges in implementing mobile commerce. Finally, we suggest possible directions for future mobile commerce research.

FEATURES OF MOBILE COMMERCE

The essence of mobile commerce revolves around the idea of reaching customers, suppliers, and employees regardless of where they are located. Mobile commerce is about delivering the right information to the right place at the right time. It gives users the ability to access the Internet from any location at any time, the capability to pinpoint an individual mobile terminal user's location, the functionality to access information at the point of need, and a need-based data/information update capability. Mobile commerce has features not available to traditional e-commerce, some of which we discuss next:

Ubiquity

Ubiquity is the primary advantage of mobile commerce. Users can get any information that they are interested in, whenever they want regardless of their location, through Internet-enabled mobile devices. In mobile commerce applications, users may be engaged in activities, such as meeting people or traveling, while conducting transactions or receiving information. In this sense, mobile commerce makes a service or an application available wherever and whenever such a need arises.

Reachability

Through mobile devices, business entities are able to reach customers anywhere anytime. With a mobile terminal, on the other hand, a user can be in touch with and available for other people anywhere anytime. Moreover, the user might also limit his/her reachability to particular persons or at particular times.

Localization

The knowledge of the user's physical location at a particular moment also adds significant value to mobile commerce. With location information available, many location-based applications can be provided. For example, with the knowledge of the user's location, the mobile service will quickly alert him/her when his or her friend or colleague is nearby. It will also help the user locate the nearest restaurant or ATM.

Personalization

An enormous number of information, services, and applications are currently available on the Internet, and the relevance of information users receive is of great importance. Since owners of mobile devices often require different sets of applications and services, mobile commerce applications can be personalized to represent information or provide services in ways appropriate to a specific user.

Dissemination

Some wireless infrastructures support simultaneous delivery of data to all mobile users within a specific geographical region. This functionality offers an efficient means to disseminate information to a large consumer population.

VALUE-ADDED APPLICATIONS

As mobile commerce extends current Internet sales channel into the more immediate and personalized mobile environment, it also revolutionizes the business world by presenting it tremendous opportunities to provide additional value to hard-to-reach end customers. These value-added services include:

- *Easy, timely access to information (e.g., the latest availability of flights).* Delivering a service that not only reaches more people but also is available all of the time, mobile commerce enables consumers to make purchases from wherever they are whenever they are ready. This will result in an increase in revenue to the company providing the mobile services.
- *Immediate purchase opportunity (e.g., last-minute purchases of tickets or gifts).* Provided with a personalized, immediate opportunity to purchase,

Figure 1: Applications of mobile technology

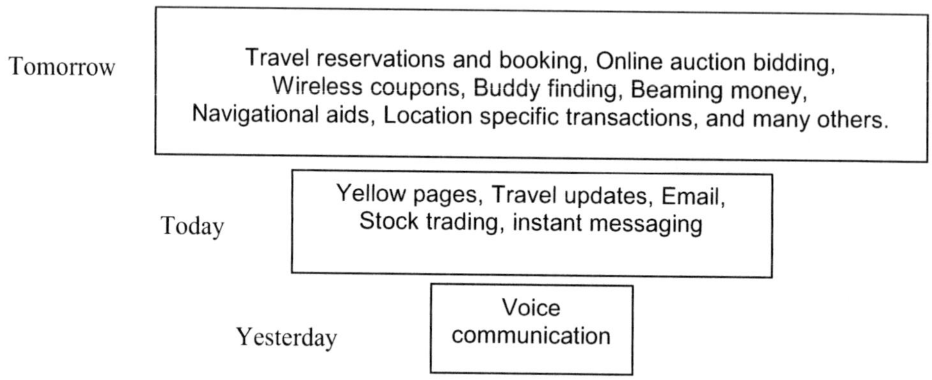

customer will make the purchasing decision on the spot and not go to an alternate source.

- *Wireless coupon based on user profiles.* Since a mobile device's location can be determined precisely, the stores around the mobile device user can transmit user-specific information, such as current sales or specials, and alert the user about similar upcoming events. Wireless coupons, which enable an advertiser to deliver geographically targeted and time-sensitive message to a willing consumer directly with a promotional offer virtually anytime and anywhere, will increase acquisition efficiency and allow direct offers suited to user profiles or stated users' preferences.
- *Beaming money.* Some bank transactions such as withdrawals and deposits will be conducted via mobile terminals in the near future. Electronic money can even be transferred to mobile devices allowing the latter to be used for electronic payments.

The only limit on the number and types of mobile commerce applications is our imagination. Varshney and Vetter (2001) identified a few important classes of applications such as mobile finance applications, mobile advertising, mobile inventory management, and product location shopping. As wireless technology further evolves, its application in business will only be broadened by more and more innovative mobile commerce possibilities (see Figure 1).

MOBILE COMMERCE TECHNOLOGY: AN OVERVIEW

Mobile commerce is enabled by a combination of technologies such as networking, embedded systems, databases, and security. Mobile hardware, software, and wireless networks enable mobile commerce systems to transmit data more quickly, locate users' positions more accurately, and conduct business with better security and reliability. In this section, we introduce the key technologies that make mobile commerce a reality and that will improve its performance and functionality in the near future.

Communication Technology

GSM

Global System for Mobile Communication (GSM) is so-called the second-generation (2G) digital network, operating in the 900 MHz and the 1800 MHz (1900 MHz in the US) frequency band. A circuit-switched service, where users must dial in to maintain a connection when data communications are desired, is the prevailing mobile standard in Europe and most of the Asia-Pacific region.

GPRS and EDGE

GPRS (General Packet Radio Service) and EDGE (Enhanced Data GSM Environment) are so-called 2.5G technologies. GPRS uses the existing network infrastructure but is being marketed as delivering ISDN-type speeds. Rather than sending a continuous stream of data over a permanent connection, GPRS's packet switching system only uses the network when there is data to be sent. Users can send and receive data at speeds of up to 115 kbits/second with GPRS. EDGE, a faster version of GSM, is designed to enable the delivery to multimedia and other broadband applications. It will use new modulation techniques to enable data rates of up to 384 kbits/second over the existing GSM infrastructure.

UMTS

Universal Mobile Telecommunications System (UMTS), the so-called third-generation (3G) technology, aims to offer higher-bandwidth, packet-based transmission of text, voice, video, and multimedia needed to support data-intensive applications. Once UMTS is fully implemented, computer and phone users can be constantly connected to the Internet and have access to a consistent set of services worldwide. Integrating the functions of a whole range of different equipments, the new 3G-enabled mobile phone can be used as a phone, a computer, a television, a paper, a video conferencing center, a newspaper, a diary, and even a credit card.

Fourth-Generation Technologies

Although 3G technologies are just emerging, research has commenced on fourth-generation (4G) technologies. These research initiatives encompass a variety of radio interfaces and even entirely new wireless access infrastructure. Better modulation methods and smart antenna technology are two of the main research areas that enable fourth-generation wireless systems to outperform third-generation wireless networks (PriceWaterHouseCoopers, 2001).

Figure 2: Evolution of wireless communication technology

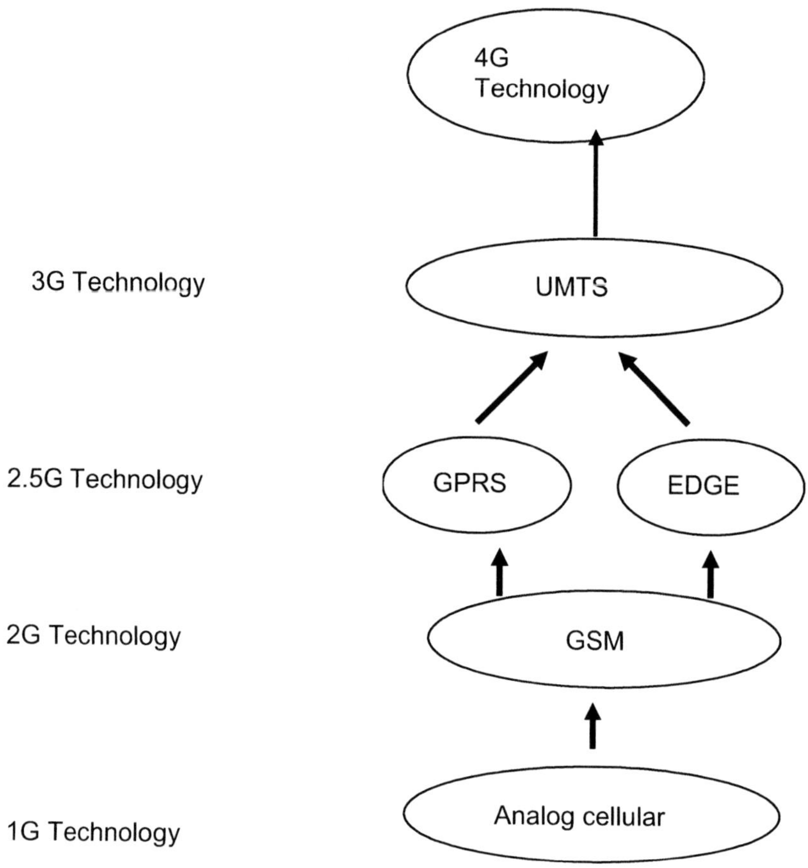

Bluetooth

Bluetooth, named after a tenth-century Danish king who conquered Scandinavia, is a low-power radio technology for communication and data exchange. Using a single chip with built-in radio-transmission circuitry, Bluetooth is an inexpensive short-range wireless standard supporting local area networks (LANs). It was developed to replace the cables and infrared links within a ten-meter diameter. Bluetooth can be used to link electronic devices, such as PCs, printers, mobile devices, and PDAs, to wireless data networks.

As depicted in Figure 3, the 1st generation wireless technology was the cellular phone. The 2nd generation wireless technology, which includes digital cellular phones, is narrow-band and currently in use worldwide. The 3rd generation wireless technology offers high bandwidth to support data-intensive applications.

Figure 3: Evolution of wireless technology

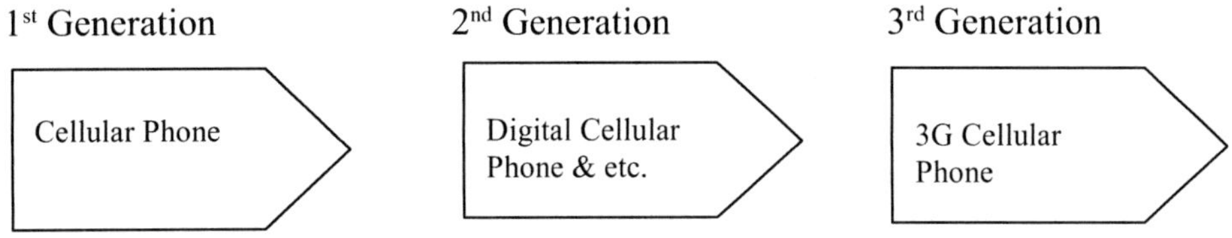

Figure 4. WAP Operation System

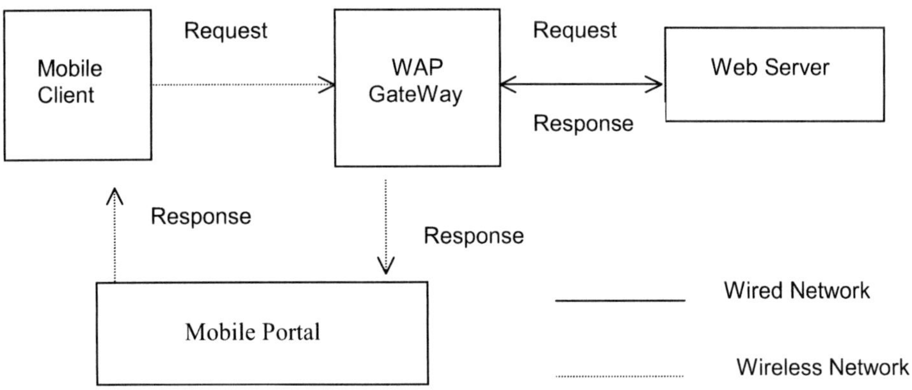

WAP

Wireless Application Protocol (WAP) is an open and global standard for mobile solutions, designed specifically to deliver Web information to mobile terminals (as shown in Figure 4). As an end-to-end application protocol, it attempts to provide solutions to the challenges in developing mobile applications, such as connecting mobile terminals to the Internet and making mobile terminals become communication devices capable of communicating with other devices over a wireless network. It also permits the design of interactive and real-time mobile services.

Information Exchange Technology

HTML

HTML (Hyper-Text Markup Language) is widely adopted by the Internet community as the document format for browsing. The availability of authoring tools and browsers make it easy for users to create HTML documents incorporating multimedia objects. Although HTML is not a suitable format for information exchange in the wireless domain, compact version of HTML, known as cHTML, has been used in the NTT DoCoMo's iMode services.

XML

eXtensible Markup Language (XML) is a meta-language, designed to communicate the meaning of the data through a self-describing mechanism. It tags data and puts content into context, therefore enabling content providers to encode semantics into their documents. For XML compliant information systems, data can be exchanged directly even between organizations with different operation systems and data models, as long as the organizations agree on the meaning of the data they exchange.

WML

Wireless Markup Language (WML), which has been derived from XML, has been developed especially for WAP. It allows information to be represented as cards suitable for display on mobile devices. So WML is basically to WAP what HTML is to the Internet.

SMS

Short Message Service (SMS) enables sending and receiving text messages to and from mobile phones. Up to 160 alphanumeric characters can be exchanged in each SMS message. Widely used in Europe, SMS messages are mainly voice mail notification and simple person-to-person messaging. It also provides mobile information services, such as news, stock quotes, sports, and weather. Recently, SMS chat and downloading of ringing tones have also been offered.

Location Identification Technology

In mobile communication, knowledge of the physical location of a user at any particular moment is central to offering relevant service. Location identification technologies are important to certain types of mobile commerce application, particularly those whose content is varied depending on location. Global Positioning System (GPS), a useful location technology, uses a system of satellites orbiting the earth. Because the satellites continuously broadcast their own positions and directions, GPS receivers can calculate the exact geographic locations with great accuracy. Originally developed in the US for military use, GPS is now also used for civilian purposes. For example, GPS can now be used in car navigation systems.

BUSINESS IMPLICATIONS

Like e-commerce, mobile commerce is a complex process involving a chain of operations. Transactions in mobile commerce typically involve customers, merchants and often banks, mobile network operators, and possibly other entities. In this section, we describe the business value chain and its primary participants, categorize its vast range of services, and illustrate its market segments.

Mobile Commerce Value Chain

As described by Barnett (2000), transport, basic enabling service, transaction support, presentation service, personalization support, user application, and content aggregators are the seven links in the mobile business value chain (illustrated in Table 1). Based on the value chain, we can identify seven players in mobile commerce as shown in Figure 5.

Table 1: Mobile commerce business value chain

Link	Name	Function
1	Transport	To maintain and operate the infrastructure and equipment to guarantee data communication between mobile users and application providers.
2	Basic enabling service	To provide services such as server hosting, data backup, and system integration.
3	Transaction support	To provide the mechanism for assisting transactions, for security, and for billing users.
4	Presentation service	To convert the content of Internet-based applications to a wireless standard suitable for the screens of mobile devices.
5	Personalization support	To gather users' personal information, which enables personalized applications for individual users.
6	Content aggregators	To provide information in a category or search facilities to help users find their way around the Internet.
7	User applications	To carry out mobile commerce transactions for mobile consumers.

Figure 5: Players in mobile commerce value chain

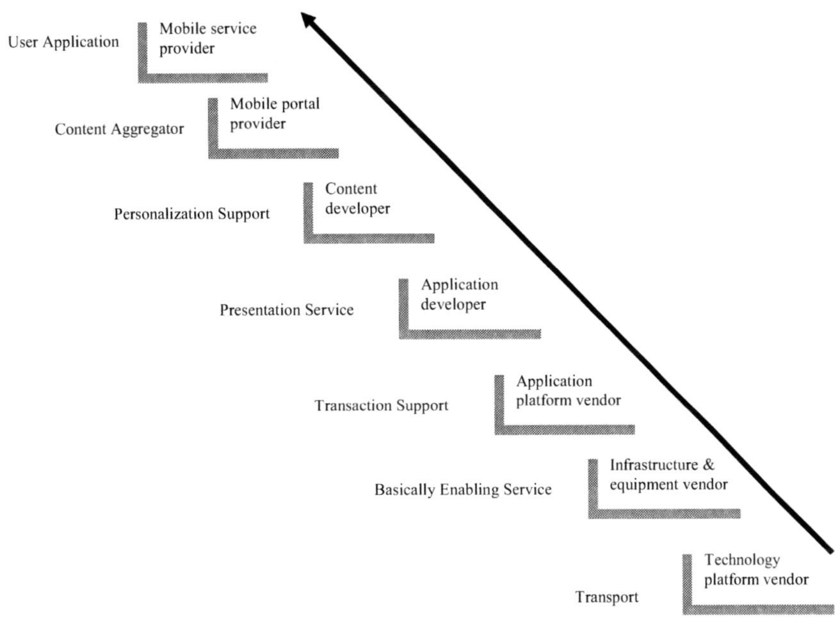

From the perspective of a transaction, the following entities are the main participants in mobile commerce: (1) *Customer*. He or she can initiate a transaction in one place, receive the service in another place, and complete the transaction in a third place. The places can be in different cities, states, and countries. (2) *Content provider*. It provides customers specific content, which can be transmitted through a WAP Gateway or through a portal. (3) *Mobile portal*. Different from an Internet portal, it offers customers services with a greater degree of personalization and localization. (4) *Mobile network provider*. It plays different roles in mobile commerce varying from a simple mobile network provider to an intermediary, portal or trusted third party, depending on where it stands in the mobile commerce chain.

Types of Mobile Commerce Markets

As an amalgamation of several different market segments, mobile commerce can be divided into three basic areas (Liebmann, 2000). They are internal business operations (also called internal mobility, probably the most common form of mobility currently in use), business-to-business applications using an extranet, and Web-based consumer services (illustrated in Figure 6).

Figure 6: Three market segments in mobile commerce

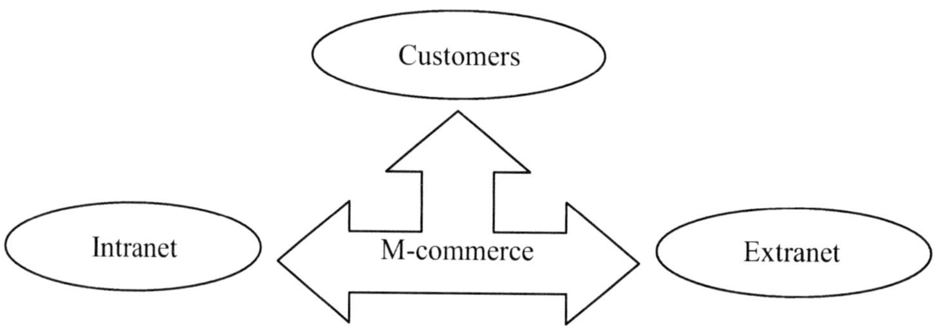

CHALLENGES AND CONCERNS

The prospect and advantages of mobile commerce may appear obvious to many of us, but the path to success using mobile commerce is not necessarily so plain. Technical restrictions of mobile devices and wireless communication, business concerns, and legal constraints complicate the practical use of mobile commerce. In this section, we focus on the obstacles confronted by mobile commerce applications.

Application Challenges

Absence of Killer Application(s)

A killer application for a computing platform is "an application compelling enough to motivate purchases of that platform" (PriceWaterHouseCoopers, 2001, p. 9). For example, access to the Internet is the killer application that spurs PC purchases in the second half of the 1990s. For the mobile commerce to succeed, one or more killer applications must be developed to compel organizations to purchase and use mobile devices in their daily operations.

Mobile Devices Limitations

Current wireless devices include phones, hand-held or palm-sized computers, laptops, and vehicle-mounted interfaces. Whereas mobile terminals demonstrate a greater extent of mobility and flexibility, they are inferior in several respects when compared to personal computers. The screen is small and the display resolution is low. The small and multifunction keypad complicates user input. Because of the need to be physically small and light, these input and output mechanisms impede the development of user-friendly interfaces and graphical applications for mobile devices. Mobile handsets are also limited in computational power, memory and disk capacity, battery life, and surfability. These drawbacks in mobile devices do not support complex applications and transactions, and consequently limit usage of mobile commerce.

User Distrust

In every transaction, each party involved needs to be able to authenticate its counterparts, to make sure that received messages are not tampered with, to keep the communication content confidential, and to believe that the received messages come from the correct senders. Due to the inherent vulnerability of the mobile environment, users in mobile commerce are more concerned about security issues involved with mobile transactions. Mobile commerce users need to be assured that their financial information is secure and that wireless transactions are safe. The mass adoption of mobile commerce will not be realized until users begin to trust mobile commerce (Siau & Shen, 2003).

Strategy Changes

To stay competitive and realize genuine productivity benefits from mobile commerce, many organizations actually need to be redesigned. They will have to make fundamental changes in organizational behavior, develop new business models, and eliminate the inefficiencies of the old organizational structures. The process of rethinking and reengineering is a demanding task. For example, implementing mobile government is more than developing a website on the mobile Internet. Actually, it is about rethinking and reengineering the way government does its business. It requires rethinking how government to be organized from the perspective of its citizens and reengineering how government to perform its functions according to the needs of its citizens rather than to the requirements of bureaucracies.

Investment Risk

A major problem faced by mobile commerce is the huge investment required to implement and operate it. Engineering massive organizational and system changes to reposition the organization strategically is complicated as well as expensive. For example, a company has to build a mobile infrastructure in order to better manage its supply chain. But the mobile technology itself does not guarantee the true benefits of mobile commerce. Expertise in fields other than technology is also prerequisites for successful applications of mobile commerce. How can organizations obtain a payoff from their investment in wireless technology? Understanding the costs and benefits of mobile commerce is difficult.

Network Obstacles

Incompatible Networks

Multiple, complex and competing protocols exist in today's cellular network standards. As previously mentioned, GSM is a single standard used by the network operators in Europe and Pacific Asian region. But TDMA (Time-division multiple access) and CDMA (Code-division multiple access) are widely used in the US. These different standards have resulted in the global incompatibility of cellular handsets. The network incompatibility poses problems for organizations to communicate and cooperate with their suppliers, distributors, retailers, and customers.

Bandwidth Access

The Federal Communications Commission (FCC) has established several frequency bands for use by cellular network operators across the country. In order to encourage competition, the FCC prohibits cellular operators from owning more than 45 MHz of radio spectrum in a given geographic region. Known as the "spectrum cap," this regulation imposes barrier for US cellular network operators who are attempting to implement the new high-bandwidth, next-generation networks.

Security Concerns

Compared with the wired counterpart, wireless communications are more vulnerable. Although most wireless data networks today provide reasonable levels of encryption and security, the technology does not ensure transmission security in the network infrastructure. Data can be lost due to mobile terminal malfunctions. Worse, these terminals can be stolen and ongoing transactions can be altered. In short, the mobility enjoyed by mobile commerce also raises many more challenging security tasks. Serious consideration must be given to the issue of security as mobile commerce applications play an increasingly significant role in our daily business and personal life.

Infrastructure Problems

Competing Web Languages

Today's mobile devices utilize a broad range of often incompatible standards, making the process of creating a successful m-commerce application even more difficult. Newer mobile phones will incorporate WAP and its WML. NTT DoCoMo's iMode, on the other hand, uses condensed HTML. The fact that incompatible standards are utilized in mobile devices today makes the process of creating successful m-commerce applications even more difficult.

Seamless Integration

The integration between network operators and businesses is a key issue for mobile commerce. In addition, to conduct business via mobile devices, companies must be capable of managing and supporting a large base of mobile customers or employees. This poses a challenge to the traditional helpdesk and customer care function. On one hand, companies must deal with the logistics, procurement, and asset management issues surrounding large numbers of devices and software. On the other hand, the broad range of mobile devices makes customer care far more complex and harder to manage.

Legal Concerns

Apart from its technical and business obstacles, the implementation of mobile commerce has its legal concerns, too. The application of traditional law to the mobile Internet is not always a straightforward process. Legal issues plaguing mobile commerce are similar to those facing e-commerce. Some of them are how to maintain privacy, how to deal with defamation, how to protect intellectual property, and how to treat Internet taxation (Dietel, Dietel & Steinbuhler, 2001). Like the wired Internet, the wireless Internet also poses significant challenges to our legal structure.

FUTURE RESEARCH AGENDAS

Research plays a vital role in solving problems in current mobile commerce applications and directing its future development. In this section, we delineate the research we feel should be carried out in mobile commerce in order to correct or address the challenges and tasks we discussed in the previous section.

Application-Level Issues

Seek Killer Application(s)

For the mobile commerce to succeed, one or more killer applications must be developed to compel individuals to purchase and use mobile devices in their daily commercial activities. The killer application(s) for mobile commerce should make full use of mobility, provide services directly relevant to mobile needs and benefit users in immediacy and efficiency.

Enhance Usability of Mobile Devices

As previously discussed, the usability of mobile devices is poor due to the various limitations of mobile terminals. Future mobile devices are expected to be smaller and more wearable, but they will also possess larger processing and storage resources. Screens for cellular phones also will be made larger, making them easier to use and more visually appealing. Meanwhile, offline methods without direct connection of mobile devices to the network can also help to minimize the technical limitations. Future mobile devices will also be able to integrate Bluetooth technology, allowing them to access nearby appliances such as vending machines and televisions using very low-cost, short-range moderate bandwidth connection. With such capabilities, mobile devices will have a combination of different communication connections to conduct mobile commerce.

Design User-Friendly Interface

Unlike the wired computing environment where large screens are available, mobile commerce applications have to operate on small and often wearable mobile devices that can only include small screens. Scientists are now developing voice-based and pen-based interaction to replace the present keyboard and mouse interaction. Pen-based interaction on touch screens may replace the mouse; voice-based interaction may be used for activation and control of functions like voice dialing. Some studies on the user interface for mobile devices have been reported in the Workshop series on Human Computer Interaction with Mobile Devices.

Build Business Models for Mobile Commerce

Although mobile commerce has the potential to improve the performance of organization, business models unique to mobile environment need to be built. Business models introduced within the e-commerce environment require further refinement to suit the mobile environment. It is vital to ensure that all the related applications and services can be accessed with ease and minimal cost.

User Infrastructure Issues

Consolidate Network Infrastructure

Bandwidth and coverage are major issues for the network infrastructure (Varshney, Vetter & Kalakota, 2000). The former allows more data to be exchanged between servers and the mobile devices, thus supporting multimedia content delivery. The latter minimizes the complications of connection losses when a mobile device moves beyond a network boundary or crosses from one network to another. These two issues directly affect the quality of mobile data transfer and, therefore, are critical for the further development and future deployment of mobile commerce applications.

Address Security Issues

Research on how to improve security in mobile commerce must be carried out due to the vulnerability of mobile devices and wireless networks. To meet security requirements including authentication, integrity, confidentiality, message authentication, and nonrepudiation in mobile commerce, additional security software and information (e.g., certificate, private, and public keys) will have to be installed on mobile devices. Nevertheless, due to the limited computing resource of mobile devices, at some point it will be necessary to establish additional servers to store information, perform security checking, and conduct electronic payment on behalf of mobile devices (Thanh, 2000).

Middleware Issues

Develop Content Delivery and a Format for Mobile Commerce

At present, much of the attention has been given to providing visual access to Web content. As a result, WML and compact HTML (cHTML) are widely used now. Voice access can also be employed to enable Web content to be displayed on mobile devices. VoiceXML (VoiceXML 1.0 Specification, 2000) is a new markup language for creating voice-user interfaces to Web applications or content using normal telephones. Since most mobile devices can be equipped with voice capabilities, it is therefore important to study how a combined voice, screen, and keyboard (or button) access to the Web can be realized by integrating the features in VoiceXML with wireless markup language

Improve Mobile Access to Databases

To allow users to run applications on their mobile devices without having to maintain constant connection with the servers and pay expensive connection time, at least part of the database systems must be able to reside on the mobile devices. These mobile database systems require little memory and are able to transfer their databases to the centralized database systems or to synchronize their databases with those at the centralized database systems. In some cases, a mobile database system may only manage a portion of a large central database, pulling in additional data on demand and pushing back data that are not required. In a mobile environment where databases are on the move and little computing resources are available, the database location, query processing, and data recovery capabilities of the mobile database systems will have to be further improved.

Explore Agent Technologies

The relatively high cost of connection time and data exchange for mobile devices discourages the adoption of mobile commerce by cost-sensitive organizations. Agent technologies can alleviate this problem. Mobile commerce users can contact agents to look for products and services, to locate merchants, to negotiate prices, and to make payments. All of these activities can be performed without having the mobile devices constantly connected to the network. In an agent-based mobile commerce framework, agents can be envisioned as merchants, consumers, and other brokering services, interacting with one another to enable electronic transactions.

CONCLUSIONS

Despite the skepticism around mobile commerce (e.g., Dugan, 2000), we share the industrial and academic communities' optimism. Though there remain a great number of technical, regulatory, and social challenges to overcome for their further development, we believe that mobile devices will continue to develop and incorporate additional functionality in the coming years and that the end result will be a global marketplace of mobile commerce.

For example, Japan and Europe are already witnessing early successes in mobile commerce. In Japan, NTT DoCoMo's iMode phone has emerged as a great success highlighting the application of wireless technology to a business environment. Introduced in February 1999, NTT DoCoMo iMode provides a continuous Internet connection via mobile phones and connects users to a wide range of online services, many of which are interactive. iMode has already attracted more than 13 million Japanese consumers, particularly youth. Connected continuously to the Internet, these 13 million users can send e-mail, get stock quotes, and play online games. Europe has also embraced a simple mobile data service wholeheartedly. SMS technology makes wireless email a reality, and the new WAP facilitates Web browsing and other Web-based transactions on mobile phones. Bluetooth, another European data initiative, further establishes a common standard for a wide range of appliances and industrial devices to communicate wirelessly.

North America, where people tend to have a PC-centric view of the Internet, has lagged behind in applications of mobile technology. But companies here have started to realize that they might miss business opportunities if they don't get a share of the current mobile commerce market, and they are attempting to catch up. Cellular operators such as Sprint PCS and Verizon now offer customers wireless access to news, the weather, sports, and financial information. MasterCard International and Motorola announced they would collaborate on mobile commerce projects.

As wireless technologies evolve, the coming mobile revolution will bring dramatic and fundamental changes to business strategies, enterprise resource planning, supply chains, and customer relations. This revolution has already begun, but it is still in its infancy. When complete, the revolution will impact numerous facets of everyday life. It will provide important data in real time to assist decision makers, exert great influence on the ways businesses communicate and develop relationships with consumers, and ultimately transform the way we do business.

REFERENCES

Barnett, N., Hodges, S., & Wilshire, M. J. (2000). M-Commerce: An operator's manual. McKinsey Quarterly, 3, 162-173. Retrieved from the World Wide Web: http://www.libfind.unl.edu:2020/journals/iris/busis.html.

Bluetooth White Paper. The Official Bluetooth Websites. Retrieved from the World Wide Web: http://www.bluetooth.com/developer/whitepaper/whitepaper.asp

Deitel, H. M., Deitel, P. J., & Steinbuhler, K. (2001). *e-Business and e-Commerce for Managers.* Upper Saddle River, New Jersey: Prentice Hall.

Dugan, S. M. (2000). Oh the horror, the horror: The new world of wireless commerce runs amok. Info World, 22 (25), 92. Retrieved from the World Wide Web: http://www.libfind.unl.edu:2020/journals/iris/busis.html.

Gosh, A. K., & Swaminatha, T. M. (2001). Software security and privacy risks in mobile e-commerce. *Communications of the ACM, 44* (2), 51-57.

Liebmann, L. (2000). Preparing for m-commerce. *Communication News, 37* (9), 132. Retrieved from the World Wide Web: http://www.libfind.unl.edu:2020/journals/iris/busis.html.

Muller-Veerse, F. (2000). Mobile commerce report. Durlacher Corporation. Retrieved from the World Wide Web: http://www.durlacher.com/downloads/mcomreport.pdf.

PriceWaterHouseCoopers (2001). Technology forecast: 2001-2003—Mobile Internet: Unleashing the power of wireless

Siau, K., & Shen, Z. (2003). Building consumer trust in mobile commerce. *Communications of the ACM* (forthcoming).

Singh, G. (2000). Securing the mobile e-conomy. Retrieved from the World Wide Web: http://www.allnetdevices.com/wireless/opinions/2000/09/11/securing_the.html.

Thanh, D. V. (2000). Security issues in mobile commerce. *Proceedings of the 1st International Conference on Electronic Commerce and Web Technologies* (EC-Web 2000), London, 412-425.

Varshney, U., Vetter, R. J., & Kalakota, R. (2000, October). Mobile e-commerce: A new frontier. *IEEE Computer, 33* (10), 32-38.

Varshney, U., & Vetter, R. (2001). A framework for the emerging mobile commerce applications. *Proceedings of the 34th Hawaii International Conference on System Sciences.*

VoiceXML 1.0 Specification (2000, March). Retrieved from the World Wide Web: http://www.voicexml.org/spec.html.

Part II:

Technology Issues in Mobile Commerce

Chapter II

Mobile E-Commerce on Mobile Phones

Do van Thanh
Telenor – Norwegian University of Science and Technology, Norway

ABSTRACT

This chapter considers mobile e-commerce on mobile phones and in particular GSM mobile phones. It is aiming at clarifying the differences between e-commerce and mobile e-commerce. The powerful aspects of this new e-commerce form born with mobility are explained carefully. The chapter starts with a presentation of the generic business-to-consumer e-commerce followed by two typical examples of e-commerce on the World Wide Web. The mobile e-commerce is then described thoroughly. The fundamental differences with e-commerce are explained, and an ideal mobile commerce system is presented. The limitations in current wireless communication systems are clarified. The Wireless Application Protocol together with its impacts and limitations for mobile e-commerce are also considered. Different types of mobile phones will then be studied, and their limitations regarding security are shown. A solution providing adequate authentication and micro-payment, called Mobile ePay, is then presented with examples of operation. A mobile e-commerce receipt system enabling instantaneous delivery is described thoroughly. The chapter concludes with suggestions for future works.

INTRODUCTION

The rapid and constant growth of the Internet and especially the World Wide Web brought numerous valuable applications to private homes and, among them, business-to-consumer commerce. The private consumer suddenly obtained the possibility to browse and purchase goods across all borders. He is, however, still confined to his computer at home. Another dimension is, of course, to be able to do the same anytime and anywhere. Mobile e-commerce seems to be a natural step, but many issues need to be solved before it can be part of the private consumer's daily life. First, mobile e-commerce needs to be better understood in order to discover all its potential. Indeed, mobile e-commerce is more than the wireless extension of the Internet e-commerce. Second, the limitations of the wireless technologies need to be overcome in order to make mobile e-commerce sufficiently secure.

In this chapter, the mentioned issues will be considered thoroughly. Since mobile phone is undoubtedly the most popular and used mobile device, we will focus on mobile e-commerce on mobile phones, although the security issues presented apply to other mobile devices as well. The chapter starts with a presentation of the generic business-to-consumer e-commerce, followed by two typical examples of e-commerce on the World Wide Web. The mobile e-commerce is then described thoroughly. The fundamental differences with e-commerce are explained, and an ideal mobile commerce system is presented. The limitations in current wireless communication systems are explained. The Wireless Application Protocol together with its impacts and limitations for mobile e-commerce are also considered. Different types of mobile phones will then be studied, and their limitations regarding security are shown. A solution providing adequate authentication and micropayment, called Mobile ePay, is then presented with example scenarios. A mobile e-commerce receipt system enabling instantaneous delivery is described thoroughly. The chapter concludes with suggestions for future works.

BUSINESS-TO-CONSUMER E-COMMERCE

Generally, commerce refers to the exchange of goods for money. The merchant hands over certain products to the consumer in exchange for money. Actually, such an exchange consists of a set of actions that are carried out at a common place where both the merchant and the consumer are present. Figure 1 shows the actions performed in commerce.

1. The consumer examines the shop to make sure that it is trustworthy and worth visiting. The consumer is actually carrying out a visual **authentication of the merchant** and his shop. She enters the shop and starts browsing.
2. The consumer examines the product to make sure that it is what she wants. Both the functionality and the quality of the product are inspected by the consumer. The consumer performs the **verification of the goods**. She may

also consider different alternatives regarding price versus functionality and quality. She may also ask for other conditions related to guarantee or service.
3. If satisfied she decides to buy and tells the merchant.
4. Before handing the product over to consumer the merchant must make sure that he gets paid for the product. He asks for cash or payment schemes.
5. The consumer pays.
6. If and only if the payment is successful, the merchant delivers the purchased products to the consumer.
7. The consumer checks to make sure that she gets all the products she has paid for.
8. The consumer may ask for receipt for accounting or guarantee.
9. The merchant issues a receipt to the consumer.

Commerce is a transaction that can be considered as successful only when all the actions are successful. We observe that trust plays a crucial role for the success. The consumer must trust the merchant and believe that the products are as good as they look. The merchant must be sure that he gets paid by the consumer for the products he delivers.

In business-to-consumer e-commerce where the consumer and the merchant communicate indirectly via software entities and the Internet, trust must be somehow

Figure 1. Commerce as a set of actions

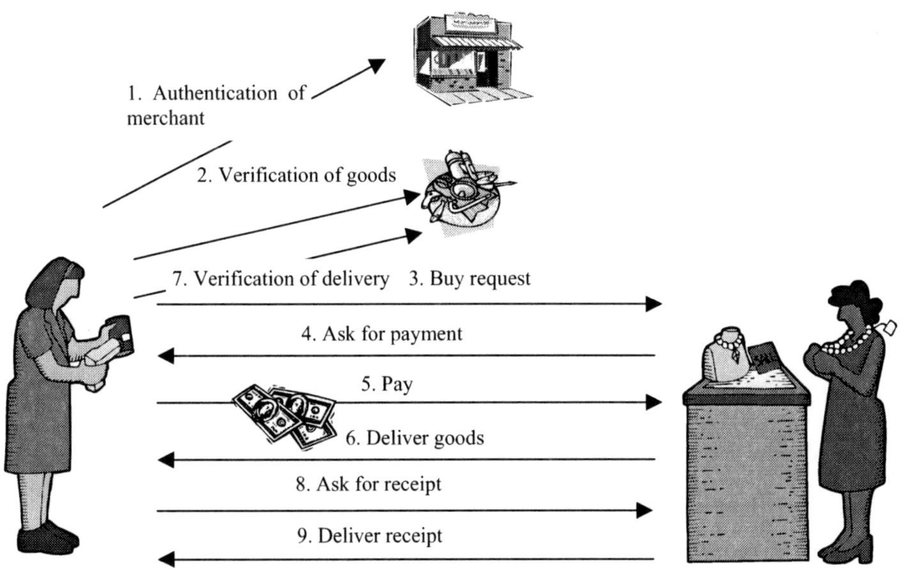

established between the two parties. In order to achieve trust the following security functions must be performed:

- **Authentication:** Each party needs to be able to authenticate its counterpart, i.e., to make sure that the counterpart is the one he claimed to be.
- **Integrity:** Each party needs to make sure that the received messages are not altered or fabricated by someone other than their counterpart.
- **Confidentiality:** Each party wants to keep the content of their communication secret.
- **Nonrepudiation:** Each party wants to prevent that the counterpart later on denies the agreements that he has approved earlier.

Usually, the two parties do not and do not have to know each other in order to do trading. In such a case, the *asymmetric cryptographic algorithm*, also called the *public key algorithm,* is more appropriate than the symmetric cryptographic algorithm.

Briefly, the public key algorithm uses a key pair, one private and one public, for encryption and decryption. What is encrypted by one key can only be decrypted by another. It should also be practically impossible to derive one key from the other. Confidentiality and integrity are preserved when the sending party encrypts the message with the recipient's public key since only the later has the corresponding private key to decrypt the message. Authentication and nonrepudiation are achieved when the sender encrypts the message or part of it with his private key. The receiver decrypts the message with the sender's public key and can be sure that it comes from the sender because only he has the private key. This later encryption scheme is known as *digital signature*, which usually also consists of a message digest (hash function) to reduce the size of the message to be encrypted and to optimize the signing process. There are currently several public key algorithms such as RSA's Public-Key Cryptography Standards (PKCS) #1-15 (RSA Laboratories, 1993), Elliptic Curve (IEEE, 1999).

It is worth emphasizing the difference between digital signature and electronic signature.

- *Digital signature* (ETSI 1998 Telecommunications Security) is a cryptographic transformation (using an asymmetric cryptographic technique) of the numerical representation of a data message, such that any person having the data message and the relevant public key can determine:
  - that the transformation was created using the private key corresponding to the relevant public key; and
  - that the data message has been altered since the cryptographic transformation.

  So based on a digital signature it is not possible to identify the signer of the message. An example of digital signature is PKCS #1.
- *Electronic signature* means data in electronic form in, affixed to, or logically

associated with a data message and used to identify the signer of the data message and indicate the signer's approval of the information contained in the data message. An example of electronic signature is PKCS #7.

The issue now is to be certain who owns what key pair. A certificate issued by a trusted authority, also called *Certificate Authority* (CA), attests that a public key belongs to an entity or individual with a certain name and attributes. Both certificates and keys need to be managed, i.e., generated, revoked, updated, recovered, etc., and a *Public Key Infrastructure* (PKI) is necessary for that. Unfortunately, no such global PKI exists yet, and, as we will explain in later sections, ad hoc solutions have been adopted in Web e-commerce.

Currently, the most popular business-to-consumer e-commerce is the Web e-commerce. Since our intention is not to give a deep presentation of Web e-commerce but only an elucidation necessary for the explanation of mobile e-commerce later on, only simplified views of Web shopping and Web banking are described.

Web Shopping

Web shopping is getting more and more popular, especially for books, music, films, etc. The procedure varies slightly depending on the visited website but can be summarized as follows:

1. A user visits a website of a merchant. He browses among the offers. Up to this point, no security measure is needed since everything is public.
2. He wants to order goods or services.
3. The Web server asserts its site identity by signing its server certificate, i.e., encrypt its server certificate using its private key. It will send both the encrypted certificate and the original unencrypted one to the browser. In this case the server must be a secure server, i.e., having a server certificate and being enabled for security. The browser uses the server's public key (from the server's certificate) to decrypt the signed certificate. It compares the result with the unencrypted certificate and in this way ensures that the owner of the certificate is the same who signed it.
4. The browser checks if the issuing CA is one that it accepts. The trusted CAs are specified in the list of so-called trusted root certificates. A root certificate is a certificate issued by a CA to itself, e.g., a certificate to VeriSign issued by VeriSign itself. Such a list is embedded in the browser. Some browsers like Microsoft's Internet Explorer allow the import of new trusted root certificates. If unknown, the browser informs the user that this server certificate was issued by an unknown CA.
5. The user manually (visually) authenticates that the site's certificate was issued by a trusted third party for the exact site the user is visiting.
6. The browser generates a session key, encrypts this key with the server public key, and sends it securely back to the server.

7. A secure channel is established, with the session key generated by the browser.
8. The user will be asked to enter his personal data, i.e., name, address, and email.
9. The user will be asked to enter his credit card number that will be charged for the purchase.
10. The server issues a receipt to the user or sends it back via email.
11. The merchant validates the credit card number and if valid ships the purchased goods to the user.
12. The transaction can be closed at this stage.

The procedure to establish the secure channel described above is in accordance with the Secure Socket Layer (SSL) protocol.

In Figure 2, Crypto Functions stand for cryptographic functions. Root Certificate **CA**$_x$ depicts the root certificate of Certificate Authority X. The requirements on the user's side are as follows:

1. His PC must have a browser.
2. The browser must be equipped with root certificates used in the authentication of the server.
3. It must have access to cryptographic functions that are capable of validating server certificates and capable of encrypting and decrypting for the secure channel.

The channel is secure in the sense that confidentiality and integrity prevail. However, it is not a trusted channel. Neither merchant nor user can be sure that they

Figure 2: Web shopping

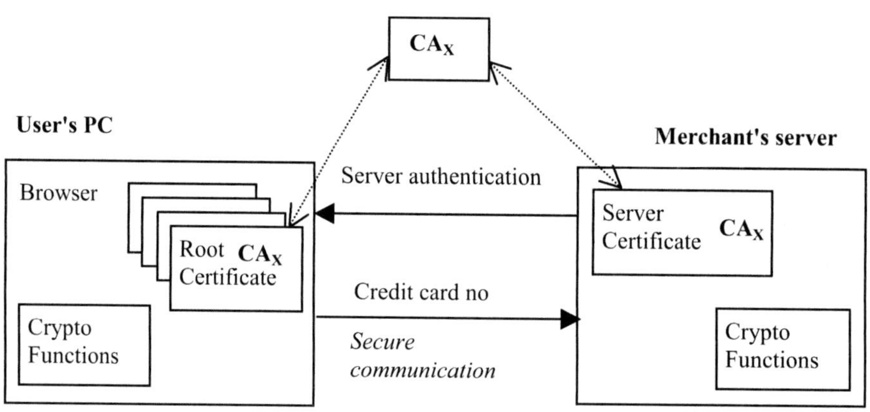

are dealing with the right counterpart. On the merchant side, only the Web server authentication is executed but not the merchant authentication. On the user side, no user authentication is done. It is worth noting that only the validation of the credit card number is done, i.e., the credit card number is valid and can be charged for the purchase. Nothing is said about whether the user is owner of the credit card and, hence, is entitled to use it.

The described Web-shopping scheme is used widely because it is simple and does not require much infrastructure and investment. However, it has the following limitations:

- The user has to trust the merchant's site. For well-known sites with good reputation, he can do that, but for unknown sites he faces a lot of risks. The site may be a fake shop that collects and abuses his credit card number.
- The merchant may deal with impostors that use credit card numbers from stolen cards or valid card numbers that are generated by an illegal process. In such cases, the validation of the card number is successful and the fraud can only be discovered long after the delivery of goods. The financial institutions refuse to cover losses for such cases, because the merchant has not verified that the user has a valid credit card and the signature is identical to the one on the credit card.
- The financial institutions are not satisfied, because the authentication of the user and the authentication of the merchant are skipped. The risk of frauds and the number of disputes are higher.

Visa and MasterCard have jointly developed the SET Secure Electronic Transaction protocol (Visa & Master Card, 1997) as a method to secure payment card transactions over open networks. SET, however, requires investments both on the merchant and on the consumer side and is not widely used.

Web Banking

Many banks in Europe have realized that by providing banking services such as paying bills, doing money transfers, balancing checks, etc. on the Web, they can reduce costs at the same time as offering better services to customers. However, they are very concerned about security and do not find the procedure used in Web shopping secure enough, since no client authentication is performed. In order to remedy the situation, the banks have adopted different authentication schemes.

Authentication using a set of numbered passcodes: The user receives from the bank by post a plastic card with a series of numbered passcodes printed on. The number of passcodes varies depending on the bank. The user is supposed to keep this card in a secure manner. When the user visits the bank's site, a secure communication is first established between the user's computer and the bank's server. Then, the user is asked to enter his username. The server will then ask him to enter for example passcode number *n*. The user consults his plastic card and enters the value of the passcode number *n*. If the passcode is correct, the user is authenticated.

Authentication using a passcode calculator: The user receives from the bank by post a calculator, which is capable of generating a one-time passcode. The calculator is secured by a PIN code chosen by the user at initialization. When the user visits the Bank's site, a secure communication is first established between the user's computer and the bank's server. Then, the user is asked to enter his username. The server will then ask him to enter the passcode. The user enters the passcode generated by the calculator. The server then compares this passcode with the passcode generated by another similar calculator at the server. If the two passcodes match, the user is authenticated. This method requires synchronization between the two calculators.

Authentication using software: Instead of a physical calculator the calculation function is delivered to the user as software on diskette or CD-ROM. The user installs it in his PC. Alternatively, the calculation function can be provided in a smart card, but in this case the user must have a card reader and associated software. When the user visits the Bank's site, a secure communication is first established between the user's computer and the bank's server. Then, the user is asked to enter his username. The authentication is then carried out by the user's client program (browser) and the merchant's server without intervention of the user. The client software generates the passcode and sends it to the server. The server compares with the code it has generated. If they match, the user is authenticated.

All the three schemes described above, although accepted by the banks because they provide sufficiently strong authentication, still have weaknesses as follows:

- The two first schemes are not very user friendly, since the user has to really concentrate in order to enter the numbers correctly.
- The user cannot be sure that the bank is performing the correct transaction that he wants.
- The bank, on its side, cannot prove that the user has requested a transaction, and the latter one can deny it later on.

What is Mobile E-Commerce?

Mobile e-commerce is e-commerce brought to mobile users via mobile devices such as palmtops, PDAs or, most dominantly, mobile phones. With an ever-increasing number of devices on the market, mobile phones will undoubtedly play a crucial role in promoting mobile e-commerce. Mobile e-commerce allows users to conduct e-commerce on their mobile devices: obtain marketing and sales information, receive ordering information, make a purchase decision, pay for it, obtain the service or product, and finally, receive customer support required.

Mobile e-commerce is more than a mobile and wireless extension of the Web-based e-commerce. It is an entirely new sales and promotion channel and is the enabler for a whole range of new services such as buying a Coke, paying for parking, buying train tickets, etc. via mobile phone. Most importantly, it is tailored to the users in many aspects. It follows the user and is available anytime and anywhere. Although mobility is a valuable characteristic to the user in general, it is especially precious for e-commerce because it enables a key factor, which is missing in other e-commerce forms, namely the ability to adapt to the user, his humor and his demands. In fact, the essence of commerce is to be able to satisfy the demands of the users. It is important not only to be able to offer whatever the user wants but also whenever he wants. Mobile e-commerce can also be customized such that it fits the preferences of the user in combination with time and location.

Another important aspect of mobile e-commerce is the ability to mix electronic media with other media such as newspaper, TV, radio, natural communication in any of the commerce phases, i.e., presentation, selection, ordering, payment, delivery, and customer care. For example, the mobile user can browse on his mobile phone and obtain the location of the closest shop. He goes there and buys a Coke. In this case, the presentation and selection are done electronically via the mobile phone, while the rest is done in a traditional way via natural communication. In another situation, the user buys groceries and pays via his mobile phone. The presentation, selection, ordering, delivery, and customer care phases are carried out in the traditional way and only the payment phase is done electronically.

At first glance, mobile e-commerce may appear to be identical to "wired" e-commerce extended with mobile wireless access, and the solutions used in Web commerce, e.g., Web shopping and Web banking, can be applied directly to mobile e-commerce. However, mobile e-commerce differs to "wired" e-commerce in the following aspects:

– **Instantaneous delivery:** The mobile user is of course interested in having services like Web shopping where the delivery of nonelectronic goods is carried out later. But, in addition, he may want to have the goods delivered to him immediately or in a short time. For example, after paying for a Coke via his mobile phone, he expects the can to roll out from the Coke vending machine. When paying for a cinema ticket he expects to be able to collect the ticket on the same day. It is therefore necessary to have mechanisms for quick user authentication and receipt delivery.
– **Micro payment:** For mobile users it is also being able to buy small things and to pay small amounts of money. The administrative charges for such payments must be small compared to the payments.
– **Mobile context:** The mobile user must in many situations be able to operate the services with only one hand. The user may be in environments that are distracting, e.g., crowded and noisy, and interactions with the e-commerce services must both simple and small in numbers. The payment scheme of Web

shopping described earlier where the user has to enter his personal data and his credit card number is hence not appropriate for the mobile user. A user-friendly payment scheme is required.

Do (2000) proposed an ideal mobile e-commerce system that is shown in Figure 3. Such a system shall support the following:

- User authentication
- Merchant authentication
- Secure channel, i.e., encrypted channel
- User-friendly payment scheme supporting micropayment
- Receipt delivery
- Simple user interface

Limitations in Wireless Communication Systems

The current wireless communications such as GSM (Global System for Mobile Communications) (Mouly and Pautet, 1992) is intended for voice communication only and is not optimized for data applications. In order to allow access to the Web from mobile devices, such as mobile phones, palmtops, PDAs, etc., the WAP forum

Figure 3: An ideal mobile e-commerce system

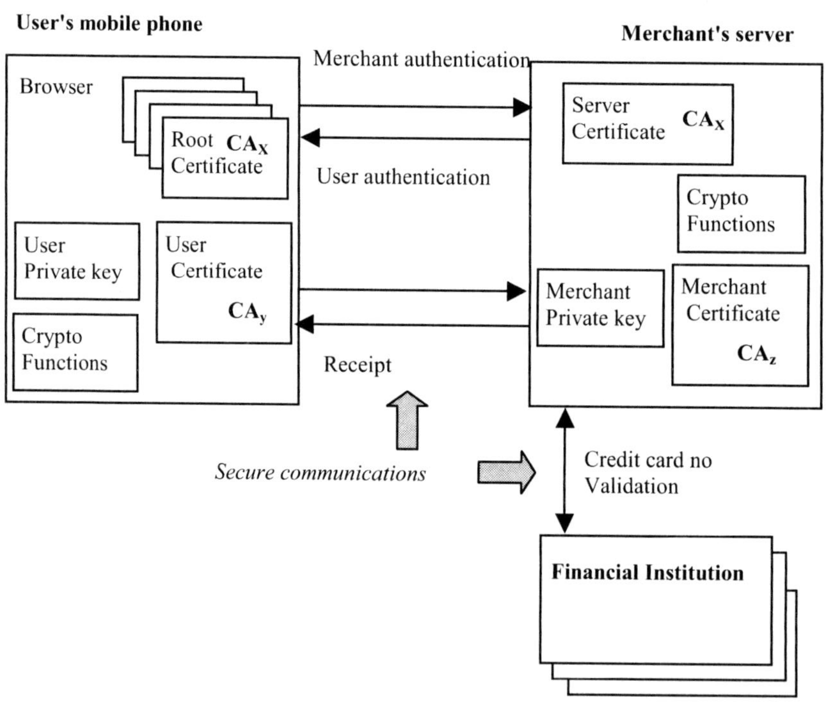

introduced the Wireless Application Protocol (WAP) (WAP forum, 1998). WAP enables a wide range of data applications such as information services, messaging services, location-dependent services, etc., but it is not sufficient in supporting mobile e-commerce since end-to-end security between the user's device and the merchant's server cannot be achieved. The physical limitations in the mobile phones in terms of storage and processing are the main obstacles. The WAP architecture and the limitations of the mobile phones will be presented in the following subsections.

WAP and Mobile E-Commerce

The *Wireless Application Protocol (WAP)* promoted by the WAP forum enables access to the Internet from mobile devices. Taken into account the limited bandwidth of the wireless link, the limitation of mobile devices concerning processing, storage, battery life, size and weight, WAP is optimized for the wireless environment. The architecture of WAP is shown in Figure 4.

Of course, WAP will contribute to the success of mobile e-commerce, but it is worth noting that mobile e-commerce exists also without WAP. For example, the first mobile e-commerce application in Norway, "the cinema ticket" that was jointly developed by Ericsson and Telenor Mobile, is not based on WAP. It is based on a SIM (Subscriber Identity Module) application toolkit where the commerce application is implemented on the SIM of the mobile phone. It is worth mentioning that WAP contains security specifications, but they are not sufficient because they do not provide end-to-end security. In the future, mobile e-commerce can be extended

Figure 4: The Wireless Application Protocol architecture

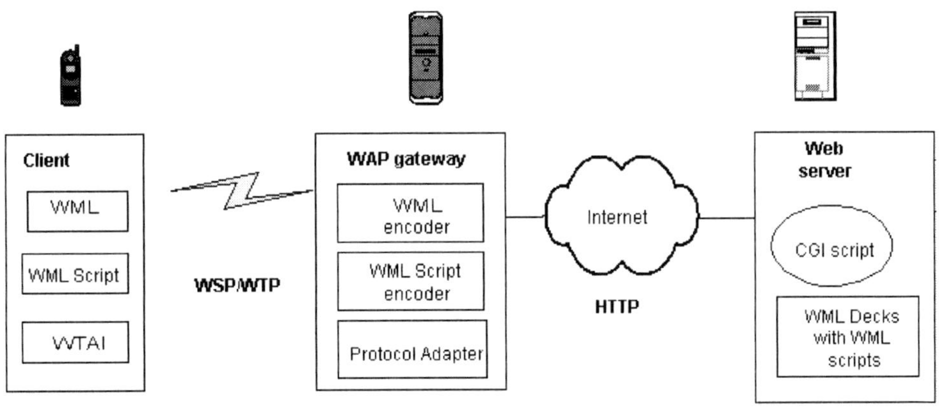

further through the adoption of newer technology such as Bluetooth, which allows local communications between devices without the need of an online connection with the network.

WAP has defined Wireless Transport Layer Security (WTLS) (WAP forum, 1998) aiming at providing privacy, data integrity and authentication between mobile phones and WAP gateways. However, WTLS does not provide end-to-end security between mobile phones and merchant servers unless the merchant server is equipped with the whole WAP protocol stack. Such a situation can happen for banks or large financial institutions, but it is not realistic to expect that every merchant server in the Web has such a WAP protocol stack.

Limitations of the Mobile Phones

An ideal e-commerce system places stringent requirements that are difficult to meet by the mobile phone itself. These are:

1. The mobile phone must be equipped with a browser that has interface to the cryptographic functions.
2. It must be capable of digitally signing a message using the user private key in order to participate in the user authentication. For that, it must have public key cryptographic functions such as RSA. It must have a tamper-proof storage for storing the user's private key. It must also have enough storage for the user's certificate.
3. It must be capable of authenticating the merchant. For that, it needs to have enough storage for root certificates and must be equipped with public key cryptographic functions.
4. It must have symmetric cryptographic functions for the establishment of the secure channel between the mobile phone and the merchant' server.

Let us consider successively a few different types of GSM mobile phones and see what capabilities they have and how to enable them to participate in mobile e-commerce. Although American mobile standards like TDMA (IS-54) and CDMA (IS-95) and Japanese digital mobile (PDC) system (Gibson 1999) are not considered in this chapter, the limitations identified for GSM mobile phones also apply to these mobile phones.

Standard GSM phones

A GSM (Global System for Mobile communication) phone (ETSI, 1998) consists of two components:

- An *ME (Mobile Equipment)* which is actually the "empty" phone with the display, keypad, microphone, and speaker;
- And a *SIM (Subscriber Identity Module),* which is a removable smart card. The SIM contains the *International Mobile Subscriber Identity (IMSI),* which unambiguously identifies the subscriber. Without a valid IMSI, the GSM

service is not accessible. The SIM also contains the security features for subscriber authentication such as authentication algorithm (A3), subscriber authentication key (Ki), cipher key generation algorithm (A8), and cipher key (Kc).

The ME is the master and initiates commands to the SIM, and there is no mechanism for the SIM to initiate a communication with the ME. A standard GSM phone does not meet any of the requirements mentioned (1-4) above and is not capable of engaging in mobile e-commerce.

GSM SAT-enabled Phones

The *SIM Application Toolkit (SAT)* provides mechanisms which allow applications, existing in the SIM, to interact and operate with any ME supporting the specific mechanisms required by the application. A browser, the public key cryptographic functions, and a user private key can be installed in the SIM. However, the SIM does not have enough storage capacity for all the certificates needed and is hence not capable of generating complete digital signatures. In addition, in order to communicate with merchant's Web server, the SAT phone needs assistance from an intermediary server that has similar functionality as the WAP gateway. We will not consider pure SAT phones, since more powerful WAP phones have emerged.

WAP Phones

The WAP phone is a mobile phone that has a WML browser and a WAP protocol stack on the ME. It is hence capable of communicating with any Web servers via the WAP gateway. The connection with the WAP gateway can be based on different bearers such as GSM circuit-switched connection, GPRS, SMS, USSD, etc.

The first version of WAP phones, called WAP 1.1 phones, do not have public key cryptographic functions for digital signature. However, a combined WAP-SAT phone will have a WML browser in the ME as well as public key functionality in the SIM. The only problem is the lack of interface between the browser and the cryptographic functions of the SIM. The browser is hence not able to invoke the cryptographic functions necessary for user authentication.

In the WAP 1.2 phone, there will be a Wireless Identity Module (WIM) which incorporates the SIM as well as local memory in the ME. Public key cryptographic functions and also the user private key can be stored in the WIM. There will also be an interface which allows the browser to communicate with the cryptographic functions. WAP 1.2 phones will be capable of generating digital signatures according to the PKCS#1 standard (RSA Laboratories, 1993), but they will not able to generate an electronic signature according to the PKCS#7 (RSA Laboratories, 1993) that is required in the validation process of the signature. It is possible to say that even WAP phones are unable to participate in mobile e-commerce by themselves—they need assistance from the network system.

In this chapter, in order to make the solutions simpler and easier to understand, only WAP 1.1 phones will be considered in later sections.

THE MOBILE EPAY SOLUTION

To allow mobile phones to perform digital signature, we present a solution which provides user authentication, merchant authentication and payment schemes while taking into account the limitations of the mobile phones. The solution proposes to use a proxy server, called *Mobile ePay*, which resides in the network. The Mobile ePay should be operated by a trusted party that could be an operator or a service provider. On behalf of the mobile phones the Mobile ePay is responsible of performing the tasks that the former is not capable of, such as:

- Storing the *user's* certificates.
- Generating *electronic* signature, e.g., PKCS#7 message format from digital signature, e.g., PKCS#1 format, generated by mobile phones.
- Validating *of* the merchant's servers.

In addition to the security functions the Mobile ePay also has payment functions such as:

- *Prepaid* account supporting micropayment.
- *Interfacing* with the payment systems of the financial institutions.

Tsalgatidou and Veijalainen (2000) proposed a similar solution with the mobile network operator as the owner of the proxy server. To illustrate the role of the Mobile ePay in our payment system two operations are described, namely user authentication for WAP 1.1 phones and payment from WAP 1.1 phones.

User Authentication

The user authentication as depicted in Figure 5 comprises of the following steps:

1. The user visits a merchant site.
2. The merchant server sends the content to the mobile phone via the WAP gateway.
3. The user wants to authenticate himself toward the merchant. The authentication request is sent to the WAP gateway, which sends it to the Mobile ePay. The Mobile ePay sends it to the merchant server.
4. The merchant server generates an authentication message, e.g., a random number, and sends it to the Mobile ePay, which sends it to the SMS-C (Short Message Center). The SMS-C delivers it to the SIM on the mobile phone.
5. The SIM asks for permission to sign.
6. If the user accepts, the SIM performs the signing, i.e., generating a digital signature in PKCI#1 format.
7. The SIM sends it back to the SMS-C, which sends it to the Mobile ePay.

8. The Mobile ePay generates an electronic signature in PKCS#7 format by using the received digital signature in PKCS#1 format.
9. The Mobile ePay sends the complete electronic signature to the merchant server.

Payment from WAP 1.1 Phones

The payment as depicted in Figure 6 comprises of the following steps:

1. The user visits a merchant site.
2. The merchant server sends the content to the mobile phone via the WAP gateway.
3. The user wants to buy. The request is sent to the WAP gateway, which forwards it to the Mobile ePay. The Mobile ePay delivers it to the merchant server.
4. The merchant server sends an offer to the Mobile ePay.
5. The Mobile ePay sends a request for payment type to the browser via the WAP gateway.
6. The user selects the payment type, e.g., prepaid account, credit cards, etc., and
7. The payment type is sent to the Mobile ePay via the WAP gateway.
8. The Mobile ePay sends the contract to the SIM via the SMS-C. The contract captures all the information about the transaction, e.g., user id, merchant id, merchandise, etc.

Figure 5: Mobile ePay role in user authentication

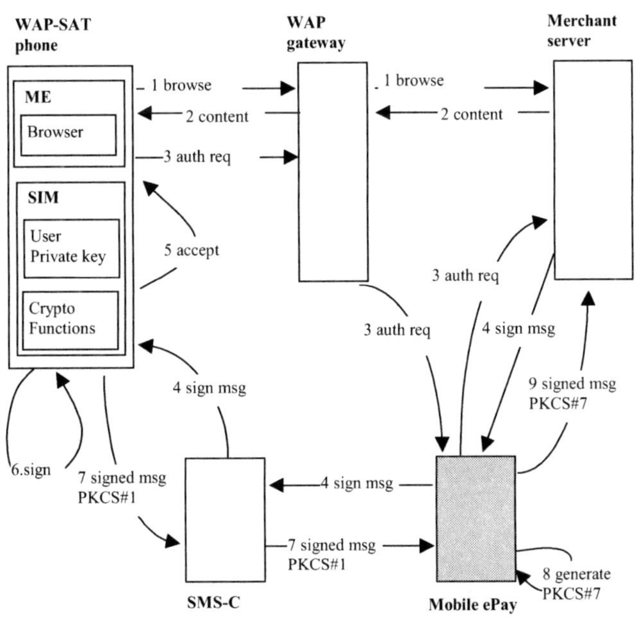

9. After asking for confirmation from the user, the SIM performs the signing.
10. The SIM sends the digital signature back to the Mobile ePay via the SMS-C.
11. The Mobile ePay executes the necessary transactions according to the payment type. This may include transactions towards financial institutions in case of payment by credit card.
12. The Mobile ePay sends a confirmation to the merchant server.
13. The merchant server returns a URL for the continuation of browsing.
14. The mobile ePay generates a receipt and sends it together with the URL for continuation to the browser via the WAP gateway.

The browser can then continue with the browsing from the received URL. The shopping is hence completed.

MOBILE E-COMMERCE RECEIPT SYSTEM

In Web commerce goods, except electronic services, are usually delivered later on. With mobile e-commerce, the user should be able to access the same commerce services with postponed delivery as the Web; but in addition, he must be able to

Figure 6: Payment from WAP 1.1 phones

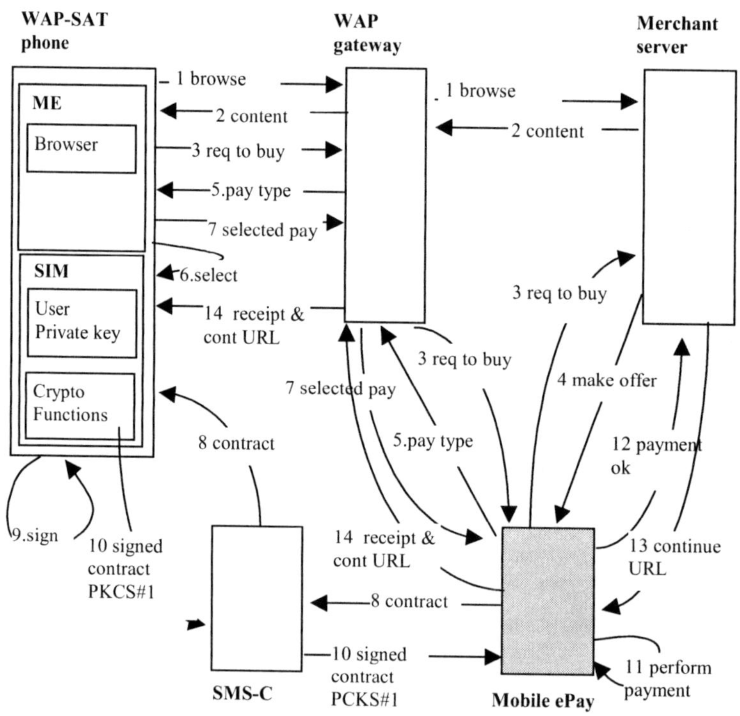

access commerce services with a short delivery time. For example, a user when on the move and thirsty wants to get the soft drink from the automat right after payment via his mobile phone. Another mobile user when visiting a city and wanting to see a movie expects to be able to collect the ticket at least before the beginning of the show.

In such situations, the delivery entity, that could be a human being or a machine, needs to receive the authorization for delivery quite rapidly. In addition, as in the case of the cinema ticket purchase, the user needs to receive some sort of electronic receipt that he shows to the delivery entity to get the cinema ticket. Such an electronic receipt must fulfill the requirements:

- It needs to be recognizable by the delivery entity.
- It can be used as proof showing that the holder of the receipt has made the purchase and goods can be delivered to him.
- It cannot be falsified.
- It cannot be duplicated or used twice.

Obviously, there is a need for a receipt system in mobile e-commerce.

Existing Receipt System

As we know, there is currently no such receipt system in the existing mobile e-commerce system. Systems for cinema ticket, such as the one offered by Telenor Mobile in Norway, have only a very primitive scheme for receipt. After the user confirms the acquisition of the tickets by entering his PIN code, he will receive a code, e.g., a four- or six-digit number. To collect his cinema tickets, the user tells the code at the ticket desk. The person in charge compares the code with the one he received earlier. If they match, he/she delivers the purchased tickets to the user.

An example of a current mobile e-commerce is depicted in Figure 7. The user uses a mobile phone equipped with a browser, e.g., WAP browser or a SIM Application toolkit browser, that allows him to browse on the World Wide Web via a gateway. The gateway can be a WAP gateway, an SMS gateway or any specific server capable of communicating with the browser on the mobile phone. The user visits a merchant's website. He contemplates the offers and selects the items that he wants. He pays for them through a payment scheme. The payment scheme may be, for example, based on a prepaid account, a credit or debit card, or a bank account. He receives a code from the merchant that he can use to collect the purchased items.

Such a system is simple but relies totally on the reliability of the merchant's system. It is only satisfactory if the delivery entity gets both the correct code and the correct information about the wanted tickets, e.g., theatre, movie, seats, etc. Otherwise, the user will not receive the tickets that he has paid for. In case of failure, the user has only a code that is insufficient to prove that he has bought the tickets. Of course he will not be charged for the tickets in such a situation, but this is not what he wants. It is quite frustrating not be able to watch the movie that one likes and has paid for.

Figure 7: Overview of current receipt systems

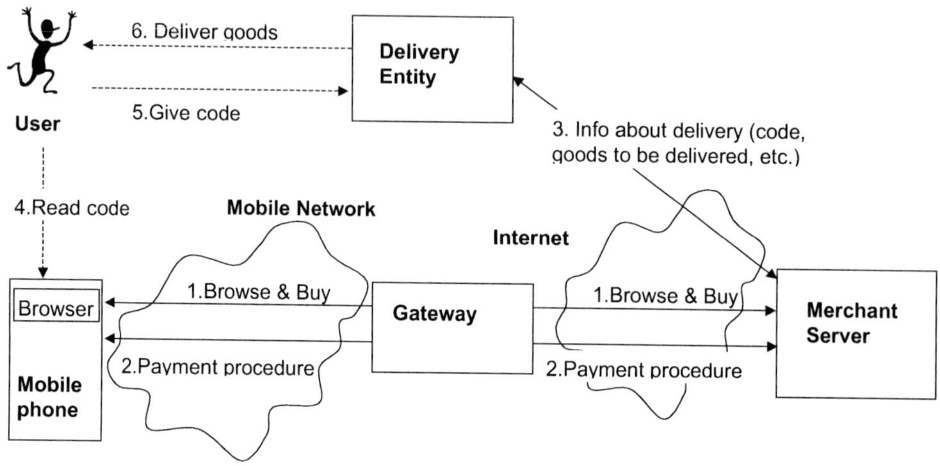

Ideal Solution for Mobile Receipt

As stated in the previous section, the current solution with a simple code is not sufficient since the user has to rely totally on the reliability of the merchant and his system. Although the merchant is honest and does not have the intention to play tricks on the user, if a fault occurs in his system the user will not have the goods delivered that he has actually bought. In addition, a mismatch can occur between the ordered goods and the goods that are delivered to the user.

Ideally, a contract stating all the details of the deal, i.e., the goods ordered, prices and quantity, etc., should be signed digitally by the merchant and then sent to the user mobile phone for local storing in the phone. At the delivery counter, the user can connect his phone via, for example, a cable, a socket, or wireless using Bluetooth or IEEE 802.11 to the delivery system and hand over the signed contract. The delivery entity verifies the signed contract and if valid delivers the goods to the user.

Such an ideal solution is however difficult to realize with existing technologies due to the following issues:

- A detailed digital contract is rather large, and the mobile phone may not have sufficient storage capacity for storing multiple contracts, which is necessary when the user buys several items.
- If the mobile phone is broken or stolen the user will lose all his contracts and hence may also lose all his purchases. Of course, the user can always claim to the merchant, but it is up to the merchant to honor the claim.

- The delivery entity must have sufficient capability to verify rapidly the digital contract, and this could be unacceptable from an economic point of view.
- In some situations, the merchant having the deal with the user may not be the same as the delivery entity, and a contract showing all the details about the user: address, prices, etc. may be inappropriate since the user's privacy can be a concern.

The Mobile E-Commerce Receipt System

In order to avoid the problems described in the previous section and in the same time to enable short time goods delivery that is usually required in mobile e-commerce we propose a system as shown in Figure 8. The system consists of the following entities:

- Mobile phone with a browser
- Gateway
- Delivery entity terminal
- Trusted Third Party
- Merchant server

A *Trusted Third Party* (TTP) is a new entity introduced between the user and the merchant. It acts like a neutral intermediary that gives equal protection to both parties, i.e., the user and merchant. It is worth noting that a TTP is a role that any actor, such as network operator, service provider, bank, financial institution, etc. may play. In addition, it will enable the short time delivery feature that is required in mobile e-commerce. Since the mobile phone does not have enough capacity for storing contracts, the TTP stores contracts on behalf of the mobile phone and the mobile user. Based on the contract, the TTP will issue and sign a simpler and smaller digital receipt that could be stored in the mobile phone. This digital receipt is then returned to the merchant's server, which sends it to the mobile phone. The digital receipt is stored in the mobile phone and will be used at the delivery of goods.

1. The mobile user browses on his mobile phone and visits a merchant's website. He selects the items that he wants and makes an order.
2. The payment procedure is carried out. Note that different payment schemes may be used according to the merchant's system and the user's subscription, e.g., prepaid account, credit card, debit card, bank account, etc.
3. The Merchant's server generates and digitally signs the contract using the merchant private key. The contract may contain the following attributes:
   - customer name
   - address
   - email
   - MISDN number of the mobile phone

Figure 8: The mobile e-commerce receipt system

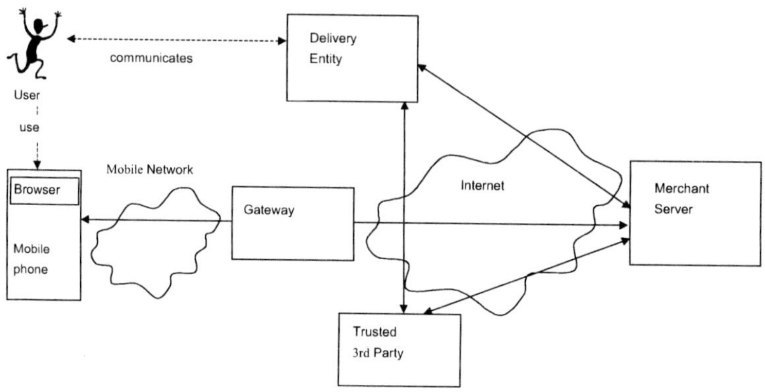

As shown in Figure 9, the system works as follows:

Figure 9: Mobile e-commerce with mobile receipt

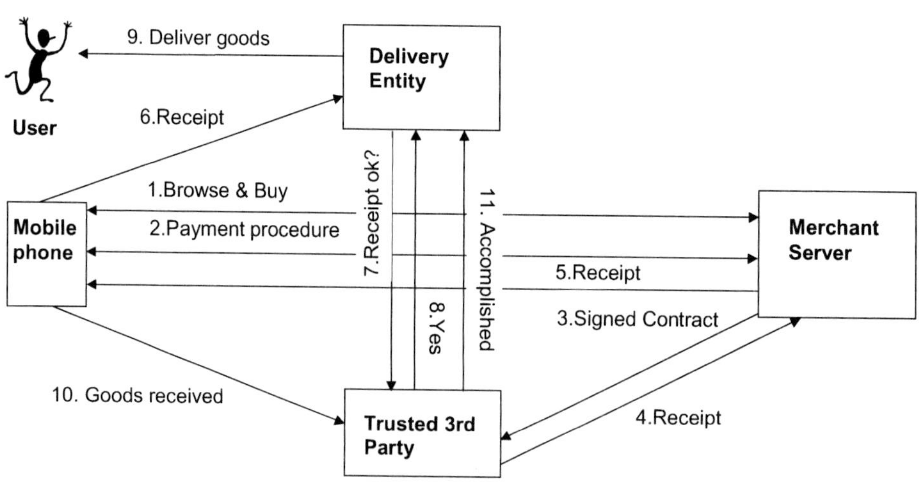

- credit card number and expiration date (in case of payment by credit card)
- merchant name and ID number
- address
- email
- date and time of contract
- contract ID
- delivery place (if necessary specify the delivery entities)
- earliest delivery date and time (if necessary)
- latest delivery date and time (if necessary)
- list of items with quantity for each item, unit price, part no
- total amount paid

The contract is then sent to the TTP.

4. The TTP validates the contract to make sure that it is valid and does originate from the corresponding merchant. The validation is done using public key cryptographic functions. If it is the case, the TTP will store it. Based on the digital contract the TTP will then generate and sign a receipt using its private key. This digital receipt may contain the following:
   - contract ID
   - TTP ID
   - TTP address

   The TTP will then send it to the merchant's server.
5. The merchant's server sends the digital receipt to the user's mobile phone that stored it.
6. At the delivery counter, the user connects his mobile phone to the delivery entity's terminal. This can be done via a wire, a direct contact, infrared or a wireless link such as Bluetooth, IEEE 802.11, etc. The mobile phone hands over the digital receipt to the delivery entity's terminal.
7. At this stage there are two alternatives depending on the capability of the delivery entity's terminal.
   a) It validates the digital receipt. If valid, it will fetch the corresponding contract either from the merchant's server or from a repository in order to find the list of purchased items. Move to step 9.
   b) It is not capable of performing the validation of the digital receipt by itself. It will then get in touch with the TTP by using the address specified in the digital receipt and send over the digital receipt for validation.
8. The TTP validates the digital receipt. If valid, the TTP will fetch the corresponding contract by using the contract ID specified in the receipt. It will extract the list of purchased items and send it together with an OK back to the delivery entity terminal.

9. The purchased items are delivered to the user. The delivery entity asks the user to acknowledge that he has received the goods. This can be done via verbal communication between the person in charge of the delivery or via the delivery entity terminal that sends an acknowledge request to the mobile phone via the link between the two devices.
10. The user acknowledges via his mobile phone that the goods have been delivered to him. The mobile phone sends an acknowledgement to the TTP. The acknowledgement can simply be the digital receipt digitally signed by the mobile phone using the user's private key.
11. The TTP validates the acknowledgement to make sure that it does originate from the right user and is not modified. If valid, the TTP will save it with the corresponding contract. The TTP will then send an OK to the Delivery Entity terminal.

The transaction is hence concluded.

The *Trusted Third Party* assumes the following responsibilities:

- Ensure that the interests of both parties are equally protected.
- Store the contract for the user such that it can be used in case of dispute.
- Issue a simpler receipt that can be used in the delivery phase.
- Certify that a trade is concluded successfully with a delivery of goods.

It has the following functions and capabilities:

- receive and validate contracts signed by merchant
- store and retrieve contracts
- issue and digitally sign receipt based on the received contracts
- validate digital receipts
- validate acknowledgements
- store and retrieve acknowledgements
- have access to necessary cryptographic function in order to perform signing, verification and validation of receipts and acknowledgement.

The *Delivery Entity's terminal* is located at every delivery counter. It assumes the following responsibilities:

- accept the digital contract and send it to the TTP for validation
- receive delivery information from the TTP
- ask for delivery acknowledgement

It has the following capabilities:

- communication with the mobile phones
- communications with the TTP and the merchant's server

The communications between:

- TTP and Merchant's server;
- TTP and Delivery Entity'sTerminal; and
- Delivery Entity's Terminal and Merchant's server

can go through secure channels on the Internet, i.e., encrypted or through dedicated networks. The communication between the mobile phone and the TTP goes through the mobile network, the gateway and the Internet. The communication between the mobile phone and the Delivery Entity's terminal can be via a cable, a socket, or wireless via infrared, Bluetooth, IEEE 802.11.

This solution has many advantages, such as:

- It enables short time delivery that is required in mobile e-commerce, while not requiring much capability either on the mobile phone or the delivery entity's terminal.
- It provides adequate protection to the user. In case of failure in the merchant's system, the contract digitally signed by the merchant that is stored by the TTP can be retrieved and used as proof. In the case where the mobile phone is broken or stolen the user does not lose the goods/services that he has paid for. The privacy of the user is achieved in the sense that information such as identity, personalia, credit card number, bank account, etc. is not revealed at the delivery entity.
- It provides adequate protection to the merchant. It ensures that purchased items cannot be delivered twice, since delivery acknowledgements are stored by the TTP.
- It is realizable without requiring much effort and resource.

CONCLUSION

In this chapter mobile e-commerce via mobile phone is carefully explained. The differences between e-commerce and mobile e-commerce are clarified. An ideal mobile e-commerce system is also presented. Unfortunately, the wireless world is characterized by limitations both on the wireless link and on the mobile phones, and such an ideal mobile e-commerce is difficult to realize. Two systems aiming at remedying the situation are presented. Taking into account the physical and functional limitations that prevent mobile phones from participating in mobile e-commerce, the first system introduces a proxy server that offers the necessary assistance to mobile phones. In addition to the security functions, the Mobile ePay also supports payment functions such as prepaid account, which interface towards financial systems. With Mobile ePay, the user can perform in a secure way any mobile e-commerce service, such as doing bank transactions and buying goods or services, from mobile phones. The second system offers a more reliable receipt, allowing the consumer to present and collect goods that he has paid for via his mobile phone.

The proposed solutions are far from perfect, and quite a lot of issues remain to be done such as time stamping for electronic signature, the relation between the private public key pair and the user (i.e., how many key pairs should the user have), and the relation between key pair and certificates (i.e., how many certificates can be associated to a key pair). The administration and storage of the user's key pairs are of utmost importance and need to be considered carefully. Another obvious issue is the one about the actor most suited to assuming the role of Trusted Third Party in the mobile receipt system.

REFERENCES

Do, V. T. (2000). Security issues in mobile ecommerce. Lecture Notes in Computer Science, Electronic Commerce and Web Technologies. 1st International Conference EC-Web 2000. Springer, 467-476.

ETSI (1998). GSM 02.17 V8.0.0 Digital cellular telecommunications system (Phase 2+); Subscriber Identity Modules (SIM); Functional characteristic.

ETSI (1998). GSM 11.14 Digital cellular telecommunications system (Phase 2+); Specification of the SIM Application Toolkit for the Subscriber Identity Module—Mobile Equipment (SIM - ME) Interface.

ETSI (1998). Telecommunications Security; Electronic signature standardization report Draft TR 101xxx V0.4.2 (1998-11).

Gibson, J. D. (1999). The mobile communications handbook. *CRN Press and IEEE Press.*

IEEE (1999). P1363 standard specifications for Public-Key Cryptography.

Mouly, M., & Pautet, M. B. (1992). The GSM system for mobile communications. Mouly & Pautet, 49 rue Louise Bruneau, F-91120 Palaiseau—France, 1992.

RSA Laboratories (1993). PKCS #1: RSA Encryption Standard. Version 1.5, Nov 1993.

RSA Laboratories (1993). PKCS #7: Cryptographic Message Syntax Standard. Version 1.5, Nov 1993.

Tsalgatidou, A., and Veijalainen, J. (2000). Mobile electronic commerce: Emerging issues. Lecture Notes in Computer Science, Electronic Commerce and Web Technologies. *1st International Conference* EC-Web 2000. Springer, 477-486.

Visa & Master Card (1997). SET Secure Electronic Transaction Specification—Book one: Business description. Version 1.0, May 31, 1997. Retrieved at: http://www.setco.org/download.html/#spec.

Visa & Master Card (1997). SET Secure Electronic Transaction Specification—Book two: Programmer's guide. Version 1.0, May 31, 1997. Retrieved at: http:/

/www.setco.org/download.html/#spec.
Visa & Master Card (1997). SET Secure Electronic Transaction Specification—Book three: Formal protocol definition. Version 1.0, May 31, 1997. Retrieved at: http://www.setco.org/download.html/#spec.
WAP forum (1998). Wireless Application Protocol architecture specification. Version April 10, 1998.
WAP forum (1998). Wireless Application Protocol, Wireless transport layer security specification. Version February 18, 2000.
WAP forum (1999). Wireless Application Protocol, Wireless markup language specification. Version November 4, 1999.

Chapter III

Transactional Database Accesses for M-Commerce Clients

Hong Va Leong
Hong Kong Polytechnic University, Hong Kong

ABSTRACT

Advances in wireless communication technologies in the past decade have led to the emergence of e-commerce applications that can be executed on mobile handheld devices. A major class of this new type of applications, commonly known as mobile e-commerce or m-commerce applications, relies on access to one or more database servers. Although conventional database technologies can still serve for applications in this evolving context, some of the techniques need to be adapted to the new environment to take advantage of the characteristics of the mobile environment or to combat the inherent limitations in such an environment. This chapter explores the appropriate mechanisms to be supported by database servers in the mobile environment and the generic architecture that can suit such a need. In particular, we focus our discussion on an important class of database applications, namely, transaction processing, which ensures the atomicity and other desirable correctness criteria of the database accessing activities. Furthermore, the concept of transaction processing is generalized to encompass accessing multiple databases while staying within the context of a mobile computing platform. A generic architecture that supports the necessary features is described. Relevant issues on the broadcast database and the disconnected processing of transactions are also considered.

DATABASE ACCESS IN A MOBILE ENVIRONMENT

Recent advances in wireless communication technologies, both the hardware and software, have resulted in the prosperity of the mobile computing paradigm. Mobile clients may be based on the resource-rich laptop computers communicating with the wireless network access points via a high bandwidth IEEE 802.11 wireless LAN, or resource-poor PDA or WAP (Wireless Application Protocol) devices communicating with a fixed network via low bandwidth wireless modems, using cellular phones. Classifying the mobile environment based on the computational power of the client device and the bandwidth of the communication network yields four different types of environments, as illustrated in Table 1.

In Table 1, the computational model exhibited by laptop computers with a high bandwidth wireless network is almost the same as the conventional distributed computing model, except for the added dimension of client mobility. Research, therefore, is focused on the management of the client location and providing location-dependent services (Imielinski and Badrinath, 1992). Moving away from the high bandwidth environment with relatively static mobile clients, quite a significant amount of research work in the past decade has been focused on the mobile environment where laptop computers interact and process data through low bandwidth networks using, for instance, wireless modems. The movement of clients is much more

Table 1: Different mobile computing environments

Computational Power / Bandwidth	High	Low
High	**Laptop under wireless LAN**	**PDA under wireless LAN or 3G infrastructure**
Low	**Laptop using wireless modem**	**PDA/WAP device using cellular phone**

frequent in such an environment. This is further complicated by the limited bandwidth of the unreliable communication channels, to the extent that mobile clients are often left disconnected from the network. Obviously, a disconnected client cannot access the file server or the database. Providing partial services upon disconnection is important, since a mobile client will need to access shared data. Research work addressing this issue has drawn the most share of attention in the past few years, as pioneered by the work on the Coda file system (Mummert, Ebling, and Satyanarayanan, 1995). In Coda, frequently accessed files are cached by the clients, and updates made during client disconnection are reintegrated with the file system upon reconnection.

The environment with resource-poor devices but a high bandwidth network will not be completely realized until the new 3G communication infrastructure becomes commonly established in several years time. We anticipate that by that time, the computational resources of those devices will have been improved to a level where they possess reasonable programming capability. We hope that by then there would be strong support for the JVM (Java Virtual Machine) on the Java Card or for well-developed platforms such as J2ME (Java 2 Platform, Micro Edition). In particular, the expected market penetration of wearable computers could form an ideal class of clients that access the Internet through the wireless network. Existing research in the mobile computing environment should therefore be dedicated to addressing the challenges in the use of relatively resource-poor devices accessing information via low bandwidth networks, since research has been conducted on this type of environment only quite recently.

Database systems constitute a very important component in commercial applications, ranging from the conventional to e-commerce applications, and then further to mobile e-commerce (m-commerce) applications; they serve as the information source and repository for these applications. Thus, database support for mobile clients forms a core component in m-commerce and it is application-independent. Those more application-specific issues, such as those arising from B2C exchanges, B2B exchanges or CRM, can be addressed separately. In practical m-commerce applications, we envision that most clients are moving around, and they are often far away from their home locations. Research and application focus should therefore be on the low bandwidth communication environment, as well as on the mobility of clients. In such an environment, a large client population communicates with various database servers over wireless communication channels (Alonso and Korth, 1993). Efficient access mechanisms to these mobile-client-enabled database servers (or simply mobile databases) must be devised. The major issues include the limitation in the bandwidth of the wireless channels, as well as their unreliability, and the frequent disconnection of clients from the wireless network (Imielinski and Badrinath, 1994). The movement of mobile clients and their location management should also be considered (Pitoura and Samaras, 2001). The new class of applications that arises from the mobility of clients and involves returning query results that are dependent on the location of a client are referred to as *location-dependent queries* (Madria, Bhargava, Pitoura, and Kumar, 2000). The querying need can be resolved by adding a geographical information system component to the database server (Choy, Kwan, and Leong, 2000) to enable geo-referencing. The relevant portion of the location-enriched database for a mobile client can be modeled as a view to that client (Lee, Leong, and Si, 2000a).

In this chapter, we are more focused on the database access issues and the provision of services for the PDA or WAP devices like WAP phones, though our design is intended to be generic enough to realize the potential of more powerful mobile clients. A typical mobile computing environment that is capable of supporting e-commerce applications is depicted in Figure 1, where different types of mobile clients are communicating with the Web server and connecting, possibly via intermediate computer systems, to the database servers that support data manipulation (through, for example, Java DataBase Connectivity or JDBC). In this example, there are two base stations that transmit data to and from the mobile clients, including laptops, PDAs and WAP phones. The base stations can communicate with a computer system, or even with a database system, through a wired network, thus enabling the mobile client to access the database system directly. Here, the mobile clients send and receive messages over the wireless channel as if they were regular TCP/IP connections. Alternatively, some base stations may be connected to a Web server through which other computer systems or database systems can be accessed. This is useful in providing Web access to clients.

In an environment such as that shown in Figure 1, the wired database servers are relatively fixed, and the movement of clients can be masked with the use of mobile IP or other mobile networking solutions. Alternatively, this can be handled with appropriate middleware.[1] In our design, we explore the notion of agents to allow

Figure 1: A typical mobile environment for m-commerce

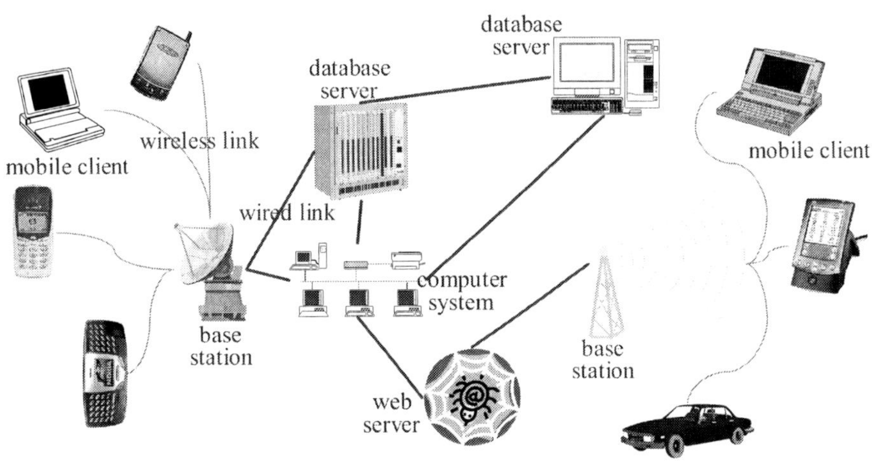

mobile clients to handle client movement with respect to wireless connectivity in a transparent manner. We would also like to simplify and standardize the architecture, which we will describe in the next section.

There are two major types of data access requirements for a mobile database: data dissemination and dedicated data access. Data dissemination is preferred in a mobile environment, since it can serve a large client population in utilizing the high bandwidth downstream channel to broadcast information of common interest, such as stock quotations, traffic conditions, special events, or the number of available seats at a theater performance. Dedicated data access has to be conveyed over dedicated communication channels with limited bandwidth. Mobile clients can query (dedicated data query) and update (dedicated data update) data items in the database. Owing to the bandwidth limitation, these dedicated data accesses are more expensive. Taking advantage of the efficient data dissemination mechanism, the proportion of dedicated accesses for data query would be small compared with data update. Although it is possible to handle a dedicated data query, such as querying for the availability of seats in a concert, through data dissemination, a dedicated data update is necessary in many m-commerce applications since clients often need to make a change in the database state for activities such as those to buy or sell a stock, or to reserve a seat in a show.

To disseminate data items in a database effectively, we can schedule the data items in the form of a broadcast disk (Acharya, Alonso, Franklin, and Zdonik, 1995; Tan and Yu, 1998). The scheduled broadcast can be viewed as the "air storage" (Leong and Si, 1997). The notion of a broadcast disk comes from the observation that a broadcast cycle resembles the rotation of a physical disk. A client accessing a certain part of the database must wait until those data items are broadcast, which is similar to the way a required data block rotates under the read/write head of a disk. The sequence of data items to be broadcast is referred to as a broadcast program (Acharya et al., 1995). The augmentation of the client cache and server database to the broadcast disk can make the structure analogous to the memory hierarchy as viewed by the mobile client, hence the name of "tertiary" air storage (Leong and Si, 1997). The concepts of broadcast disk and database cache hierarchy are depicted in Figure 2. A small broadcast disk containing information about five stocks is depicted. The broadcast program is (DELL, INTC, MFST, ORCL, SUNW). Proper indexes can be built to facilitate access to data items in the broadcast (Imielinski and Badrinath, 1994).

Observing that regular indexing is of an item-based nature, improvement can be achieved by ascribing semantic information to the indexing structure. This provision of semantics to the data organization in a broadcast disk results in semantic data chunks (Deshpande, Ramasamy, Shukla, and Naughton, 1998; Lee, Leong, and Si, 2000b). An example comparing conventional item-based data broadcast with semantic chunk-based broadcast is depicted in Figure 3. Using data chunks, a query for stocks priced between \$30 and \$70 can be resolved by accessing only the first

Figure 2: The broadcast disk and the cache hierarchy

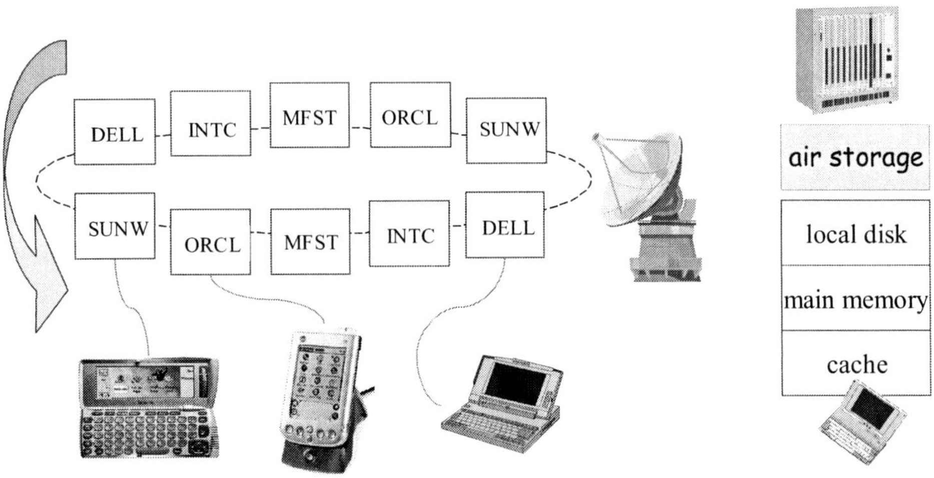

two data chunks, and the client is *certain* that the results are *complete* even if some parts of the broadcast have been missed. On the contrary, a client listening to the broadcast in an item-based broadcast cannot assert that it has obtained the appropriate data items unless all data items are contained in the broadcast disk, and it has not missed any part of the broadcast. For instance, if the client missed the records for Intel and Microsoft, it would not know that it had missed part of its answer. It has to send a complementary request to a server for the prices of Intel and Microsoft. However, a client missing the chunk [50 < price ≤ 75] would request information from the server for stocks with a price between $50 and $70, but a client missing only the chunk [75 < price ≤ 100] will know that it has obtained all of its answer. This desirable property, exhibited by chunk-based broadcast, is termed the *self-answerability* of client queries (Lee, Leong, and Si, 2000b). Such broadcast processing approaches are appropriate for resource-rich clients. To improve the data access efficiency, appropriate client caching mechanisms can be implemented (Chan, Leong, Si, and Wong, 1999). The cache of a client can be used to answer queries even in the event of a disconnection. The semantic data chunks can be used for answering queries, and the precise nature of missing data items from the cache can be asserted by the client. In the example above, a client knows that it has all the results if chunks for [25 < price ≤ 50] and [50 < price ≤ 75] are found in the cache, while a client under item-based broadcast will remain in doubt about the completeness of its cached data for the query.

To support data update, queries to databases are transmitted from the client to the server via a back channel. Since location-dependent queries form a significant portion of m-commerce services, mechanisms to improve the performance of

Figure 3: Conventional versus semantic data chunk organization

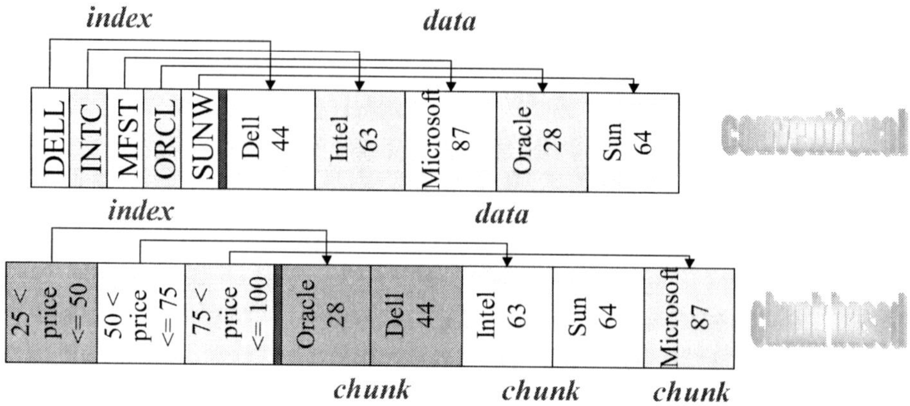

location-dependent queries should be implemented (Madria et al., 2000). For instance, typical information that is useful with respect to a location can be materialized in the form of a data warehouse, whose content can be updated dynamically with respect to changes in location (Lee, Leong, & Si, 2000a). In particular, the client view is defined based on a conjunction of location-independent and location-dependent predicates. It can be materialized through filtering based upon the location-dependent predicates.

Regular access to a database normally involves the querying and updating of database information. With multiple clients accessing the database, it is important to ensure the correctness of those concurrent accesses. This is realized and enforced through concurrency control protocols to support the notion of database *transactions*, which satisfy the ACID properties, namely, atomicity, consistency, isolation, and durability (Bernstein, Hadzilacos, and Goodman, 1987). The correctness criterion is the *serializability* of concurrent transactions. Intuitively, serializability means the possibility of rearranging concurrently executing transactions to a serial execution of those transactions. If this can be done, the concurrent transactions can

Figure 4: Problems in the absence of database transactions

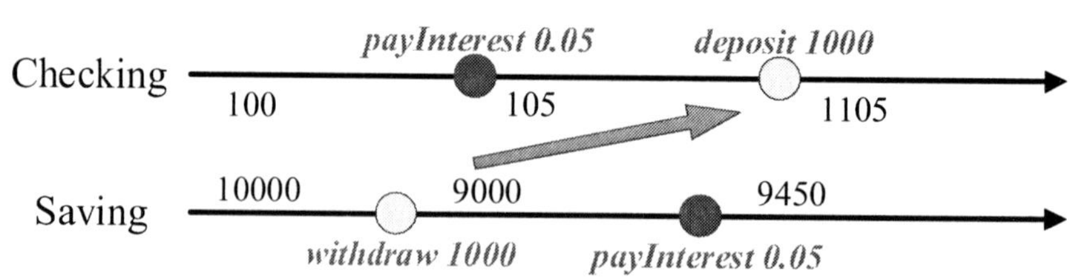

be executed in a sequential manner effectively, thereby eliminating interference and anomaly. Consider the space-time diagram in Figure 4 where Peter is transferring $1,000 from his saving account, Saving, to his checking account, Checking. While this transfer is in progress, the bank executes a program to calculate the interest payment of 5% for each account and perform the appropriate credit operation. It is unfortunate that Peter has lost the interest on his "phantom" cash of $1,000 while it was transferred. He received interest of $455 instead of $505, and the total balance became $10,555 instead of $10,605. If the transfer and the interest calculation were executed as transactions, it would not be possible for the interest calculation to occur in the middle because both transactions would access shared accounts in a "conflicting" mode. The concurrency control protocol, which ensures serializability, will delay the calculation of interest on the checking account until the deposit is made.

In a mobile environment, it is appropriate to relax the strong serializability requirement, which could be imposing overly restrictive constraints on the execution of transactions. It is often acceptable for a mobile client not to see the updates that were made by some concurrently executing mobile clients, in exchange for a faster execution and a higher probability of committing its transaction. In the above example, if the interest calculation transaction were replaced by a balance inquiry transaction, it might well be acceptable for Peter to see a total balance of $9,100 instead of $10,100 if he only needs to ensure that he has sufficient funds to cover his tuition fee, say, $5,000. Isolated-only transactions (Lu and Satyanarayanan, 1995) are proposed to reduce the impact of client disconnection. N-ignorance (Krishnakumar and Bernstein, 1994), bounded inconsistency (Wong, Agrawal, and Mak, 1997), and update consistency (Shanmugasundaram, Nithrakashyap, Sivasankaran, and Ramamritham, 1999) are some of the common weaker forms of correctness criteria for database accesses. They would try to ignore some of the operations in a transaction or allow them to be executed out-of-order in some controlled manner. Taking bounded inconsistency as an example, the concept of a resolution set is proposed for each operation, which may be serialized out-of-order. In particular, *commutative operations* such as deposit and withdraw (assuming sufficient fund) can be executed in any order. However, deposit or withdraw *conflicts* with balance. When a new operation balance is considered for admission to a set of admissible concurrent operations of deposit and withdraw, there can be a maximal difference in value equal to the sum of the amounts requested by the deposit and withdraw operations. With respect to Figure 4, let us replace the interest operation by the balance operation. Now, the resolution set for the balance operation to account Saving will be {9000, 10000}. The maximal deviation is $1,000. If this deviation is acceptable to the client at that moment, the operation is allowed to be executed and is included in the set. Performance studies on bounded inconsistency by Wong, Agrawal, and Mak (1997) indicate that there is a significant improvement to the degree of concurrency. N-ignorance is similar in nature to bounded inconsistency, except that the degree of ignorance is on a transaction basis rather than on the operation basis.

Regardless of whether serializability or its weaker form is adopted, a common avenue to improve transaction processing throughout in a mobile environment is to utilize the broadcast bandwidth effectively. The database can be broadcast and transactions can be processed against the broadcast database. This is very useful for read-only transactions (Pitoura and Chrysanthis, 1999), because it is sufficient to catch a consistent set of data items over the broadcast. To enable update transactions to be processed, the hybrid transaction processing protocol by Mok, Leong, and Si (1999) ensures serializability by performing validation for update transactions and utilizing the uplink channel to request for additional data items not readily available over the broadcast. Consistency across data items is ensured through the use of timestamps. We will discuss the consistent broadcast when we describe the broadcast manager later.

Shanmugasundaram et al. (1999) propose the notion of update consistency, a weaker form of serializability. In update consistency, a mobile client is only required to see updates made at server consistent with the values it reads, without having to follow the same serialization order as those observed by other mobile clients. For instance, assume that the server is broadcasting updated stock prices for SUNW and ORCL. Two mobile clients, *A* and *B*, inquire about the stock price. The global history H=read_A(SUNW,64) write_1(SUNW,65) commit_1 read_B(SUNW,65) read_B(ORCL,28) write_2(ORCL,29) commit_2 read_A(ORCL,29) may be resulted, where the server is executing update transactions T_1 and T_2 and the mobile clients are executing T_A and $T_{B,}$ respectively. Clearly, H is not serializable. However, for A, it observes a partial history, H_A = read_A(SUNW,64) write_1(SUNW,65) commit_1 write_2(ORCL,29) commit_2 read_A(ORCL,29), which is serializable with respect to updates made at the server. Similarly for B, the partial history H_B = write_1(SUNW,65) commit_1 read_B(SUNW,65) read_B(ORCL,28) write_2(ORCL,29) commit_2 is also serializable. Assuming that A and B are independent, both H_A and H_B will be acceptable; thus H should also be acceptable in order to improve system performance. H is said to be an update consistent history. The cycle-based algorithm by Shanmugasundaram et al. (1999) can guarantee update consistency.

In traditional database applications, it is often only necessary to control concurrent accesses to a single database system. With the rise of e-commerce and m-commerce, it is becoming more imminent that simultaneous access to multiple databases, which are maintained by different organizations, will become a norm rather than an exception. A collection of database systems organized in a coherent way, and functioning in a cohesive and collaborative manner, is often referred to as a federated database system (Fang, Ghandeharizadeh, McLeod, and Si, 1993; Lim, Hwang, Srivastava, Clements, and Ganesh, 1995). In certain applications, it is important to provide consistent accesses to those multiple databases with support for a transaction-like behavior across them. This stronger correctness criterion is called *global serializability* (Breitbart, Garcia-Molina, and Silberschatz, 1992). The distributed activity accessing the multiple databases (or multidatabase system) is called a *global transaction*. Consider the example in Figure 4 again. Now, assume that the two accounts are at two different banks and that the interest calculation becomes a request by the government to determine the balance of Peter's accounts for assessing taxation. There are four transactions: withdraw(bankB,1000), deposit(bankA,1000), balance(bankA), balance(bankB). A similar scenario to the one shown in Figure 4 would occur if each bank executes its transactions in a serializable manner but without coordination. The government would observe a balance of only $9,100. On the other hand, executing the transfer and balance enquiry as two global transactions would always produce the correct balance of $10,100.

In the past, creating a federated database system could involve a lot of coordination efforts, both at the system level and the enterprise managerial level. Furthermore, the execution cost of the global transactions, in terms of concurrency control and atomic commitment, can be very high. As a result of these factors, global transactions have not been widely adopted, despite their usefulness and convenience. Rather, multiple subtransactions were commonly executed on individual databases without achieving the global serializability standard. In other words, under existing systems, it is likely that the government in the above example would observe a total balance of $9,100 for Peter, and yet the accuracy of bank records is taken for granted.

Riding on the wave of the adoption of the Internet computing paradigm, more and more companies are willing to publicize their databases as part of their drive towards e-commerce. Under most cases, these databases can be accessed from outside the company via a Web interface, and they are gradually moving from an HTML representation to an XML representation of the data, thereby promoting data interchangeability and system interoperability. With these recent developments, the ability to access consistent information using global transactions in a federated database system becomes even more useful and more manageable. Although updates to databases are normally restricted in a nonfederated database environment, we can witness more and more databases becoming enabled for the execution of read-only transactions by external parties (limited to certain predefined database views). As a result, the use of global transactions for "loosely" federated databases could become more popular. Furthermore, the presence of a high proportion of read-only transactions can render the concurrency control for global transactions far more efficient. To further reduce the overheads of global transaction processing, one can relax the stringent global serializability requirement to allow a controlled degree of inconsistency, such as with N-ignorance (Krishnakumar and Bernstein, 1994) or update consistency (Shanmugasundaram et al., 1999) and exploit the abundance of read-only transactions. Relaxed requirements still provide a better correctness guarantee than most existing approaches which ignore global serializability completely.

To enforce global serializability, there are different well-defined concurrency control protocols. In the Optimistic Ticket Method by Georgakopoulos, Rusinkiewicz, and Sheth (1994), the concept of ticket updating is introduced, which causes direct conflict among different global subtransactions at local database systems. This can prevent the development of the indirect conflict that cannot be readily detected by the multidatabase system. In this approach, a global serialization graph is maintained to validate all global transactions before their commitment. In other ticket-based methods, tickets may be updated at different moments by a global transaction (Hwang and Son, 1996). The Implicit Ticket Method (Georgakopoulos, Rusinkiewicz, and Sheth, 1994) takes advantage of certain local database systems, which allow only serialization orders that are consistent with the transaction commitment order. It achieves global serializability by controlling the commitment order of global transactions, and hence their serialization order, without maintaining tickets.

Owing to the complexity of global transaction processing and the resource limitations of mobile clients, it is sensible to migrate the coordination effort to the proxy server and the Web server, thereby relieving the mobile clients from the complex processing. This is especially important in the context of WAP devices. As such, we propose a database access architecture that will *decouple* the transaction processing mechanism from the application logic. We make use of agents to combat the disconnection problem. The agent can act on behalf of the client for event-driven transactions, such as stock selling transactions or auctioning. Support for global transactions can occur primarily on the wired network through the agent. Proper exchange of data items can be conveyed to the client via the proxy architecture.

In the rest of this chapter, we first describe the architecture of our mobile global transaction processing system and then describe the major components as well as discuss some of the design issues in the next section. The support for global transactions with global serializability is described in the section, *Supporting Global Transactions for Mobile Clients*. More effective processing of read-only global

Figure 5: System architecture

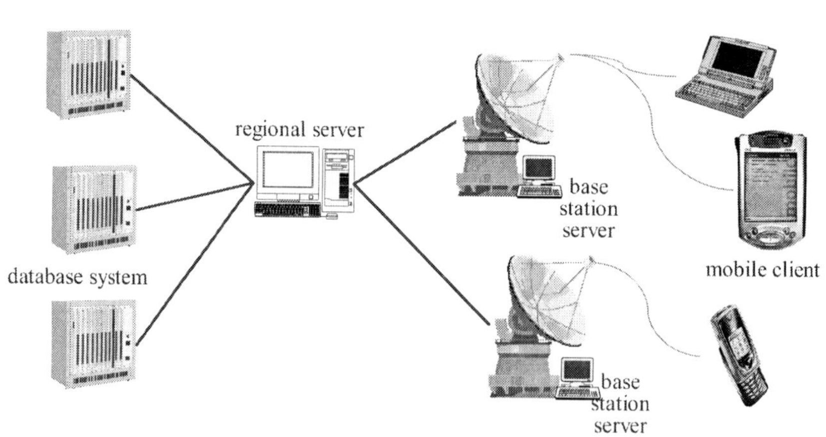

transactions that is achieved by exploiting the consistent broadcast cache is also examined. Next, the issues associated with providing services to disconnected mobile clients and handling the integration of data updates to the database systems are also investigated. Finally, we conclude with a brief discussion on some of the challenges for practical m-commerce systems.

SYSTEM ARCHITECTURE

The overall system architecture to support both conventional and global transactions generated by mobile clients in the wireless Web-based environment is illustrated in Figure 5. We adopt a geographic-based hierarchical structure (Choy, Kwan, and Leong, 2000) for our transaction processing architecture. There is one base station server (BSS) for each cellular structure and a regional server for a number of neighboring base station servers. The BSS is normally associated physically with the base station or mobile support station of the cellular wireless communication structure. It plays the role of a proxy server in the context of Web browsing, and this is similar to the role that the base station serves as the entry point for mobile clients in a cell or the role that the firewall/proxy serves as the first contact point to clients residing inside a department of an organization. To improve redundancy, fault-tolerance and load balancing capability of the BSSs, we could exploit multiple regional servers to organize these BSSs in a hierarchical manner (Choy, Kwan, and Leong, 2000), or we could even associate several regional servers with each BSS, with say, a primary server approach.

In our architecture, there are two types of mobile clients: resource-rich laptop computers and resource-poor WAP devices. They communicate with the BSS serving the cellular structure covering these mobile clients. The BSS deals with the mobile client interface issue and communicates with the regional server to carry out the necessary m-commerce activities. As with the architecture by Choy, Kwan, and Leong (2000), the regional server will possess better computational power, and it acts as a concentrator of requests from different BSSs. It houses the major components and computational processes for the dedicated functionality. Within the regional server, there is a transaction coordinator thread, which is normally realized as a Java servlet, for each active transaction to access the remote databases. This coordinator thread accesses the databases for either a conventional transaction, through mechanisms such as RPC (remote procedure call) or RMI (remote method invocation), or a global transaction by maintaining contact with multiple database servers; it executes a 2PC-like protocol (Two Phase Commit) (Bernstein, Hadzilacos, and Goodman, 1987) at the end of the transaction to install the updates. The presence of a high proportion of read-only subtransactions in the Web-enabled m-commerce environment is highly beneficial in yielding a higher commitment rate to the global transactions. This is due to the reduced degree of conflict among subtransactions at each local database system as well as the reduced cost incurred in the 2PC-like protocol for the database servers servicing read-only subtransactions. To support the execution of global transactions, individual local database systems should provide appropriate mechanisms that can serve as a bridge to the remaining system components. Architectural details and functionalities for the different system components will be discussed in the subsequent subsections.

Local Database System

A conventional transaction that accesses only one single database is referred to as a *local transaction*, in contrast with a *global transaction* that spans across several databases. Although a local database system inherently supports the execution of local transactions, additional components are incorporated to support global transactions. In our design, we exploit the *transaction server* to serve as the bridge between the local database system and the external world. As we will see in the next subsection, the transaction server interacts with the transaction coordinator at the regional server to support global transaction processing. To the database server, the transaction server acts as the database client. To the rest of the system, it serves as a participant site to the distributed transaction, which is under the coordination of the transaction coordinator. The local database system does not see any other system component beyond the transaction coordinator. The structure inside an individual local database system is depicted in Figure 6.

In Figure 6, the transaction server receives subtransactions containing database operations intended for it that were sent from the transaction coordinator; it then passes them to the transaction manager for execution. There is a division of labor in the design between the transaction server and transaction manager. Although the transaction manager is inherently a built-in component, the transaction server is a bridging component that a local database system should possess in order to process

Figure 6: Structure of a database system supporting global transactions

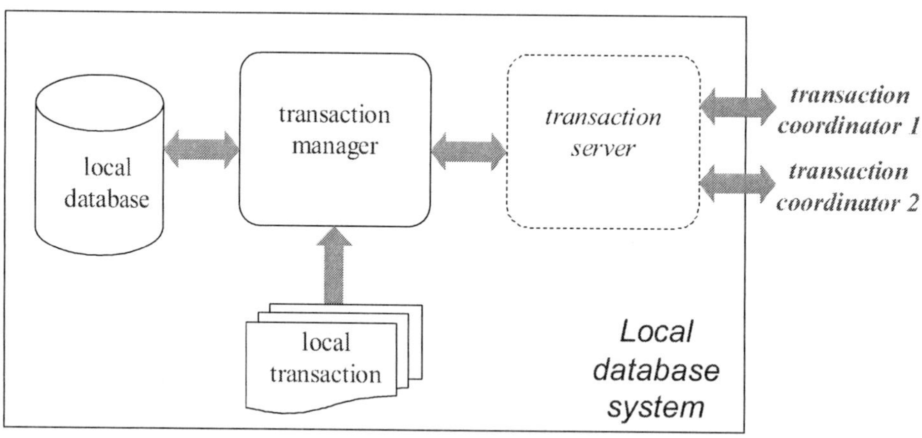

global transactions. The flexibility of such a design lies in the possibility of imparting minimal changes to the existing local database system. Additional processing need for information pertaining to the support of the global transactions is intended to be handled in the transaction server as much as possible.

The transaction server behaves like a "proxy" for the transaction coordinator at the local database system. It is considered as a "special" component (represented within a dotted box), because it may contain some foreign flavor. In particular, with an agent-based implementation, it may be realized as a *place* for foreign agents to visit (Yau, Leong, and Si, 2001), which pass operations on to the transaction manager and gather and process status information. The use of agents provides maximal flexibility to the local database system to support different types of global transaction processing mechanisms. In the Java context, it can be implemented by means of exported Java objects and interfaces, so that the transaction coordinator can send it the database operations and gather results by means of RMI. We have explored the possibility of using RMI for this purpose (Leung, 1998). In a more traditional context, it can be implemented as a process listening to a certain TCP port and accepting connections from the transaction coordinator for operation specifications through sockets.

The *transaction manager* in a local database system receives and processes operations for local transactions that originate from within its own organization. To support global transactions, the transaction manager must also handle operations from global transactions. The functionality and autonomy of the database system is governed by the behavior of the transaction manager, which coordinates and schedules the operations it receives from the two groups of transactions for execution, through its own concurrency control and recovery protocols. Quite often, no changes are needed at the transaction manager for global transaction processing. For instance, provided that the original local database system supports distributed transactions, the information pertaining to the implementation of the distributed 2PC protocol, such as the status of whether an active subtransaction can be committed or not (the ready-to-commit response), is automatically available from the system. With this return status from the transaction manager via the transaction server, the transaction coordinator at the regional server can ensure the atomicity or atomic commitment of the associated global transactions.

In certain global transaction concurrency control protocols, information pertaining to the local serialization orders of global subtransactions is useful to ensure global serializability of global transactions. One way to keep the serialization order of global subtransactions at the local database system totally under control is to have the transaction server force artificial conflicts on shared data. Nevertheless, this can lower the degree of concurrency of the global subtransactions. Therefore, although it is possible to maintain the serialization information at the transaction server without changing the transaction manager, it would often be far more efficient to make a minor modification to the local transaction manager, which can provide such information at minimal cost.

Regional Server

The regional server acts as a concentrator of global transactions from different mobile clients, which submit their requests via BSSs. Figure 7 depicts the architecture of the regional server. The core transaction processing mechanisms for global transactions are supported, together with the different mechanisms for data dissemination via broadcast scheduling, caching and refreshing, and transaction validation and disconnected transaction reconciliation. The *transaction controller* accepts formulated transactional operations from the applications at BSSs in a format that is independent from the user interface (regardless of HTML or WML). It also maintains relevant book-keeping information. To support transaction processing over a broadcast database, the *broadcast manager* maintains a set of mutually consistent data items that are ready to be broadcast to a collection of mobile clients via the BSS. It also maintains important data structures to keep track of the set of hot data items that are eligible to be scheduled for broadcast (data dissemination). The *data integrator* receives updates from mobile clients via BSSs that are made during the disconnection mode or made with respect to the broadcast database, and then confirms with their validity before passing them on to the remote database systems for installation.

The *transaction coordinator* behaves like the coordinator processing a distributed transaction. It interfaces directly with the transaction servers at various participating local database systems. To ordinary "local" transactions submitted by mobile clients for accessing a *single* database system, the coordinator merely becomes a transceiver, passing on the operations from the "local" transactions and returning the result sets. In general, the coordinator organizes the operations that compose a global transaction into subsequences of operations, creating different subtransactions, with one subtransaction per participating database. To improve

Figure 7: Structure of a regional server

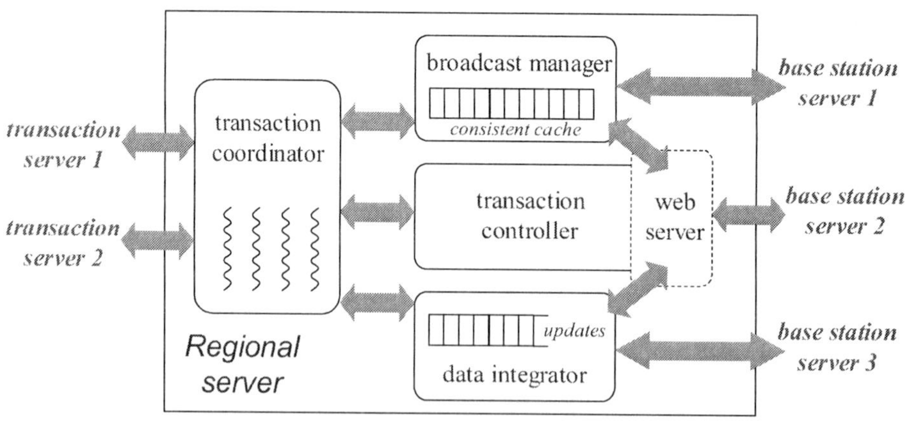

efficiency, we try to perform request batching. In particular, several operations composing the same subtransaction can be grouped into one single aggregate request to be transferred to the individual local database systems. This aggregate request can be in the form of a TCP message, an RPC, or an RMI. One or more result sets will be returned for an aggregate request that contains multiple read operations.

The *transaction controller* can be modeled in the form of a Web server, or it can be coupled with a Web server at the regional server to support the Web-enabled environment. Through the Web server, it can provide Web services, accepting client requests encoded in HTML or XML formats. In the past few years, XML has evolved very rapidly into one of the most important data interchange standards so that a proper support mechanism for XML has become essential. There are also an increasing number and different types of systems supporting XML-based data repositories, or even XML databases (Luk, Leong, Dillon, Chan, Croft, and Allan, 2002).

The transaction controller monitors, controls and schedules the execution of both "local" and global transactions, through cooperation with the transaction coordinator, broadcast manager and data integrator. It keeps track of the identities of mobile clients, their initiated transactions, and the set of operations from those transactions. This information is useful for various accounting and statistical purposes, such as determining the set of data items of high affinity to the mobile client population for potential broadcasting by the BSSs. Other useful information can include the readset and writeset of the transactions to be used later for validation purpose.

The *broadcast manager* maintains a consistent broadcast cache, ready to be broadcast via the BSSs. A consistent broadcast cache is a data structure that holds a collection of data items that are mutually consistent. In fact, a consistent broadcast corresponds to a *snapshot* of the database, or collection of databases, on a set of selected data items at a "quiescent" point (in the absence of executing transactions). Operationally, such a consistent broadcast corresponds to the result set returned by a read-only transaction reading the set of data items of interest. An example of a consistent broadcast of the balances for three bank accounts is illustrated in Figure 8. In this example, the respective balances for the three accounts, namely, (Checking,3100), (Saving,2000) and (TimeDeposit,13000), are consistent because they reflect a snapshot in which both transfer transactions of $2,000 from TimeDeposit to Checking (T1) and $1,000 from Saving to Checking (T2) have been performed, but the transfer transaction of $3,000 from TimeDeposit to Saving (T3) has not yet taken place. The sum of the balance remains the same as before, i.e., $3,100+$2,000+$13,000=$100+$3,000+$15,000. Note that a direct snapshot taken vertically (i.e., at the same time) may reflect balances of (Checking,2100), (Saving,2000), and (TimeDeposit,13000), and it is not consistent. In particular, the read-only transaction that returns such a consistent snapshot could be a global transaction that returns results from several database systems and potentially reads a large

Figure 8: An example of a consistent broadcast cache

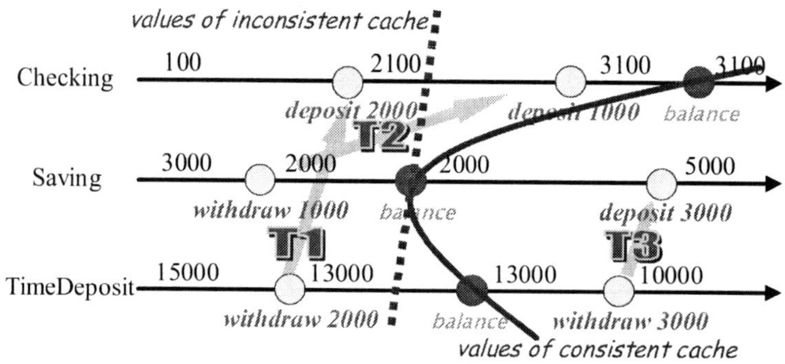

amount of data items. Techniques for returning the values for a large collection of data items by a read-only transaction can be found in Barghouti and Kaiser (1991).

The consistent broadcast cache is indexed by the identity of the individual database systems and perhaps also by the geographical location of the data items. Appropriate data items in the broadcast cache are selected by BSSs for broadcast to their mobile clients. Periodically, the broadcast manager determines the access patterns of the clients and the aggregate access patterns of data items from the statistical information maintained by the transaction controller. A possible way to determine the set of useful data items for broadcast is based on the approach by Leong and Si (1997). The data items can be organized in the form of a flattened B+tree for broadcast, as in some existing approaches (Imielinski, Vishwanathan, and Badrinath, 1994; Tan and Yu, 1998). Alternatively, one may consider structuring them in semantic chunks as proposed by Lee, Leong, and Si (2000b). Since the maintenance of item consistency and the selection of broadcast items are orthogonal to the way they are organized for broadcast, we concentrate on the discussion of the former and more important issues in this chapter. To ensure the recency and consistency of the set of data items broadcast, the broadcast manager can refresh the data items through a large read-only transaction. Mobile clients reading the set of broadcast items in the same broadcast cycle are guaranteed transactional correctness (Mok, Leong, and Si, 1999). Thus, they can choose to process read-only transactions based on the broadcast (Pitoura and Chrysanthis, 1999) or to read data items from the broadcast and to send update requests for data reintegration.

The *data integrator* manages an upstream update queue, which holds updates to data items made by update transactions that need additional processing, besides managing conventional global transaction processing. One kind of information received by the data integrator is the information pertaining to individual mobile clients. Another kind of information is updates made by a mobile client based on data values that it obtains from the broadcast. The updates must be installed back to the databases, after being validated for consistency with concurrently executing transactions. Finally, there are also updates made by a client in a disconnected mode. All these disconnected transactions must be validated and reconciled against the databases to ensure that they are consistent with transactions of which they are unaware.

The fundamental mechanism for data reconciliation, which is performed at the data integrator, has been discussed in depth by Davidson (1984). In particular, the concept is to determine a set of *undesirable transactions* that should be aborted. All the remaining transactions can then be committed. Intuitively, the set of transactions executed can be modeled as a serialization graph, which is a form of precedence graph. To reconcile updates made in several executions, the histories representing the executions are merged. If there is no cycle in the precedence graph for the merged history, then the executions are compatible and there is no undesirable transaction. When there is a cycle, we can select a certain number of transactions to be aborted, in a similar way to how we can select some victim processes to break a deadlock cycle. The victim transactions selected for abortion are called undesirable transactions. The difficult part is to determine the optimal set of undesirable transactions to be aborted. Unfortunately, finding the optimal set is known to be an NP-complete problem. Several suboptimal strategies have been proposed and evaluated by Davidson (1984), and the one used to break two-cycles appears to be close to optimal, in general.

One can draw an interesting observation to the three transaction-related components, which are communicating with the transaction coordinator. The transaction controller is generic for handling transactions from mobile clients. The broadcast manager appears to be executing large read-only transactions to prepare for consistent sets of broadcast data, maintaining appropriate indexing and timing information to the broadcast data for mobile clients. The data integrator handles update transactions, which are "pseudo"-committed at the mobile clients, ready to be confirmed for validity through the validation process. The actual updates by the transactions can be made within a relatively short period, since all the operations are known, together with the expected values or result sets. These components will exchange information about transactions with the transaction coordinator, which maintains the relevant information carefully to ensure that all transactions can be executed in a harmonious manner.

Base Station Server

Figure 9 depicts the structure of the base station server, which is responsible for communicating with the mobile clients through the Web-based interface. A laptop mobile client has a higher computational power and can also interact with the BSS through standard HTTP using HTML or even XML Web pages. However, a WAP device mobile client only possesses a low computational power and can only interact with the BSS through WAP using specially formatted WML pages. To enable the

BSS to cater for both types of mobile clients, the WAP gateway can be utilized to translate the standard HTML or XML Web pages into WML pages (Tai, 2001). Alternatively, two different sets of HTML and WML pages can be designed (Lo, 2001). To abstract out the commonality, we can associate the interfacing mechanism with each individual application by providing a simple annotation mechanism to guide the automatic translation. Application business logic is embedded in the applications, which are designed through the use of a set of well-defined application programming interfaces (API). We have developed an API suitable for the design and implementation of global transactions in the Internet environment (Leung, 1998) and in the WAP environment (Lo, 2001). Depending on the nature of the mobile client, the modules within the API interact with the appropriate proxy. For regular mobile clients, the modules interact with the server proxy, while for resource-poor WAP devices, they talk to the WAP gateway instead. In the former situation, there is a corresponding client proxy associated with the mobile client, as depicted in Figure 5. The pair comprising a server proxy and client proxy achieves a division of labor to improve the overall system performance. Technically, the pair of proxies can be referred to as a server intercept and a client intercept, and this is a variation of the traditional Web client/server model to alleviate the negative impact of the wireless links (Housel, Samaras, and Lindquist, 1998). In the latter case, the WAP gateway normally delivers the WML pages directly to the WAP devices, because we would not expect to get much processing power out of those devices. As such, the client proxy can be considered to be minimal or even null, with the transmission optimization capability defined precisely by the WAP and managed at the client side by the WAP browser.

Figure 9: Structure of a base station server

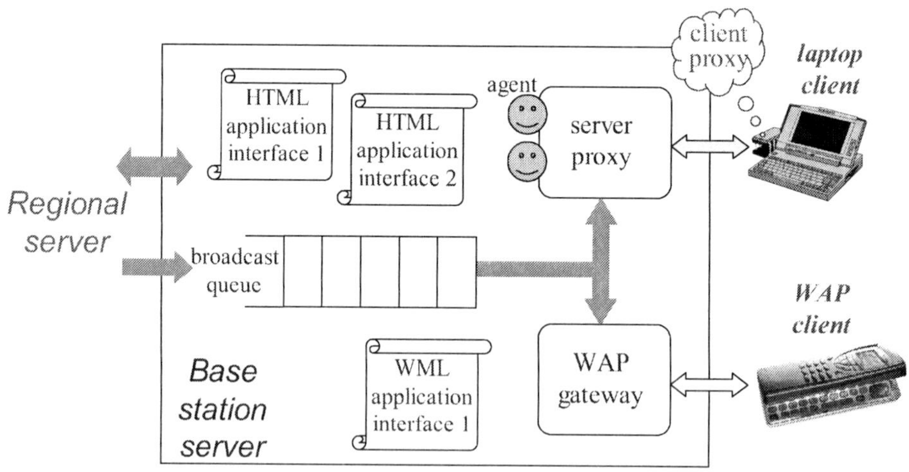

The server proxy is a standard proxy that interacts with the mobile client to support the necessary database services through the use of HTTP or other more efficient interaction protocols. An appropriate transcoding mechanism (Bharadvaj, Joshi, and Auephanwiriyakul, 1998) to reduce the size of the transmitted data can be implemented. The server proxy also helps to return updates, which are made to cached data items by mobile clients during disconnection for transactions, back to the regional server for reintegration. Upon reconnection of a mobile client, transactions executed in the disconnected mode must undergo the process of validation and reconciliation. This task is performed by the data integrator at the regional server. Finally, this server proxy can implement the data broadcasting mechanism for mobile clients. The data items to be broadcast are selected from the consistent broadcast cache maintained by the broadcast manager at the regional server. It can also select data items with respect to its geographical location and ignore those in other locations (location-dependent information). The data items selected to be broadcast are first stored in a downstream broadcast queue and then transmitted to the mobile clients using appropriate mechanism such as a UDP broadcast or a channel expressed as CDF (Channel Definition Format) documents in a subscription-based Web environment. To serve a mobile client that is a WAP device, the server proxy will be replaced by the WAP gateway. It translates the information into WML pages to be transmitted to the WAP device client.

The client proxy located at the mobile client will carry out the appropriate actions in correspondence with the server proxy, but it may not be necessary to support the full set of functionality. In particular, mobile clients can execute a number of read-only transactions using the broadcast data items by considering them to come from a broadcast disk (Acharya, Alonso, Franklin, and Zdonik, 1995), since the data items are guaranteed to be consistent by the broadcast manager (see section 3.2). Furthermore, the broadcast data items can be cached in client local storage with timestamp or cycle information, for future support of read-only or disconnected transactions (Chan, Leong, Si, and Wong, 1999). The client proxy for a WAP device can be considered a direct add-on to the WAP browser. Owing to the low computational power of the device, it may not be able to utilize the data broadcast well.

SUPPORTING GLOBAL TRANSACTIONS FOR MOBILE CLIENTS

There are two major types of transactions to be supported for mobile clients: local transactions accessing data in only one local database and global transactions accessing data in several databases. Although traditional concurrency control protocols can handle local transactions (assuming that all local database systems support local transaction processing), there must be some cooperation from the local database systems to support global transaction processing. We will first describe how global transactions can be processed in a regular distributed environment. We then proceed to consider making use of the broadcast data at the broadcast manager and performing data reconciliation at the data integrator to improve transaction processing efficiency.

Global Transaction Processing

The correctness criterion for global transaction processing is global serializability. There are many different mechanisms that can be used to ensure global serializability. A simple solution is achieved by means of timestamp ordering (thereby establishing a global ordering). Each global transaction is assigned a unique timestamp, which is carried to the local database systems. All transactions at a local database are serialized in their timestamp ordering. This ensures global serializability. Besides its simplicity, an added advantage of such an approach is its freedom from deadlock, which is expensive to detect and resolve. However, there is a drawback that timestamps generated by local transactions have to be kept roughly synchronized with those of global transactions. Otherwise, excessive abortion is experienced by the transactions with small timestamps. Furthermore, the degree of concurrency is not as high in timestamp ordering as in 2PL (Two Phase Locking), as evidenced from performance studies on various concurrency control protocols.

Another common solution is realized when the strict 2PL concurrency control protocol is adopted by all individual database systems. An important advantage of this approach is that it comes almost for free, since most practical databases adopt the strict 2PL as their primary concurrency control protocol because of its efficiency and higher throughput in practice. The major problem with 2PL is that it is deadlock-prone. Global deadlock detection involving a number of databases is much more difficult, and resource wait-for graph information has to be made available to the deadlock detection mechanism. Revealing resource wait-for information may not be desirable in practice, since it could expose too much internal information of the individual database systems.

A simple modification to the 2PL mechanism to avoid the necessity of global deadlock detection is to allow deadlocks to be detected and resolved locally, using an appropriate timeout mechanism to handle global deadlocks. Local deadlocks are detected and resolved periodically by individual database systems. Subtransactions for global transactions are associated with a deadlock timer. This timer is maintained by the transaction coordinator. When the timer goes off, the global transaction is assumed pessimistically to be involved in a deadlock, and it is aborted by the coordinator. The timer is reset when a new operation from the global transaction arrives or when a result set is returned. To cater for possibly belated transmission of the operations for a subtransaction from mobile clients, due to a poor wireless connection, this timer value can be adjusted adaptively. Note that this timer is not the deadlock timer that is set by the local deadlock detection mechanism at individual databases.

It is not always possible to assume that all participating database systems employ the 2PL concurrency control protocol. A conservative mechanism that can be easily implemented to operate on database systems with heterogeneous concurrency control protocols is the definition and use of the site lock, one for each database system. A global transaction must obtain a site lock before it can submit a subtransaction to the database system at a particular site. This simple mechanism prevents global transactions from executing on conflicting sites, thus dismissing the potential problem of being serialized in different ways on different database systems. A disadvantage is that global transactions can suffer from performance degradation for the loss of a certain degree of concurrency. Other concurrency control protocols can also be implemented (Georgakopoulos, Rusinkiewicz, and Sheth, 1994).

Utilizing the Consistent Data Broadcast

The broadcast paradigm is effective in a mobile environment for the dissemination of information to a large collection of mobile clients (Lee, Leong, and Si, 2000b). With a data broadcast comprising a consistent set of data items, read-only transactions can be processed at a low cost against the broadcast. The broadcast manager at the regional server maintains a consistent set of data items in its broadcast cache, whose contents could come from the result set for a large read-only transaction. We observe that mobile clients reading the set of broadcast items in the same broadcast cycle are guaranteed transactional correctness (Mok, Leong, and Si, 1999) because the broadcast represents a consistent snapshot of the databases. Taking Figure 8 as our example again, in which there is a consistent broadcast containing three data items: (Checking,3100), (Saving,2000), and (TimeDeposit,13000). A credit-checking read-only transaction, T_c, may be initiated by a credit card company that is only interested in the balance in Saving and TimeDeposit accounts. The read-only transaction can be executed directly using the broadcast values, and in this case it is a total balance of $15,000 at a particular point in time when both T_1

Figure 10: Utilizing the consistent broadcast cache

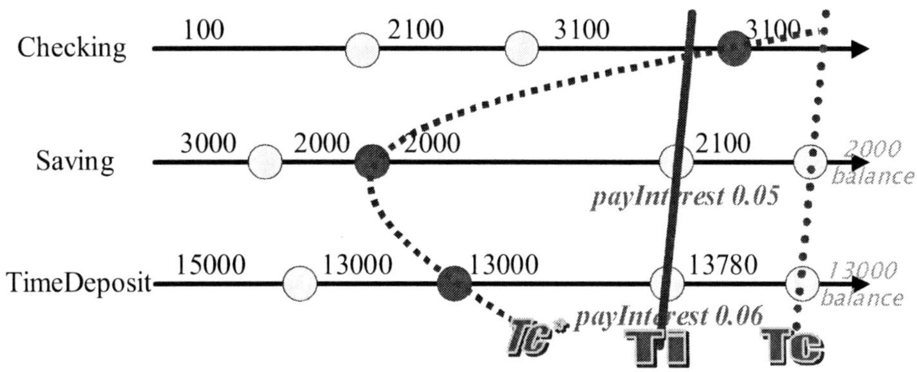

and T_2 have been committed. It is possible that by the time the credit-checking transaction T_c is issued, the transfer transaction T_3 has already been committed. The result is still correct because we can choose to serialize T_c before T_3. In fact, serialization of read-only transactions to an earlier point can result in a good improvement in performance (Pitoura and Chrysanthis, 1999). Note that even if the set of broadcast items were an inconsistent one, for instance, (Checking,2100), (Saving,2000), and (TimeDeposit,13000) blindly returning results for T_c would still reflect the correct sum of \$15,000 in this case, since T_c is not concerned with the balance of Checking. However, one is not always that lucky in avoiding the inconsistent items.

In the presence of updates to transactions processed through the broadcast data, the updates must be propagated back to the database systems for installation. In particular, one has to validate the updates against all concurrent accesses. In the research work by Mok, Leong, and Si (1999), an optimistic concurrency control protocol is adopted. We now modify the example from Figure 8, by considering an interest computation transaction T_i, as illustrated in Figure 10. We assume that the interest rate is higher if the total balance of all accounts exceeds a certain threshold, say, \$100,000. Furthermore, no interest is paid on the balance in a checking account. Thus, the readset of T_i includes all of the three accounts, and the writeset includes both Saving and TimeDeposit accounts. Though the read part is guaranteed to be correct, the updates need to be validated against concurrent updates occurring in the database systems. When validation is successful, the updates are installed through a short update transaction. Note that T_i will not interfere with a concurrent credit-checking transaction T_c even if both are processed against the same set of broadcast items. This is because, irrespective of the actual submission time of T_i and T_c, T_c can return results from the broadcast and can be serialized before T_i, as if it were transaction T_c^* in the figure. In fact, the issue of processing data updates and their installation to the database systems is quite similar to data reconciliation. As a result, we discuss this in further details in the next section.

To improve the flexibility of the data broadcast selection made by each BSS for its clients, the consistent broadcast cache may be maintained in a multiversioned manner. Several versions of the cache produced with the read-only transactions can be stored and indexed. This is particularly useful for time-stamp-based protocols, since a mobile client may have to read the correct data version in order to be able to commit. Using multiple data versions for time-stamp-based protocols can also be useful for data reconciliation upon the reconnection of a previously disconnected client, see the next subsection.

From another perspective, the set of data items in the consistent broadcast cache can be visualized in the form of a data warehouse, since it represents consistent information jointly available from a number of databases. In addition to using a simple read-only transaction, possibly reading a large amount of data items, the consistency of the data warehouse can be maintained by different view update mechanisms (Zhuge, Garcia-Molina, Hammer, and Widom, 1995). In a general model, the set of data items broadcast during two consecutive cycles may be slightly different. It is possible that a client reading data item *x* from cycle *c* can only read data item *y* from cycle *c*+1 (which is unavailable in cycle *c*). To determine whether the two items are consistent, the cycle number (or the identity of the read-only transaction) can be included for cross-cycle consistency checking (Shanmugasundaram et al., 1999). Mobile clients can choose to process read-only transactions based on the broadcast or to read data items from the broadcast and send update requests for data reintegration. Alternatively, those updates can be transmitted together for data reconciliation in a batch, as if the client had been disconnected to reduce the cost of using the uplink channel, at the expense of potentially more conflicts.

Mobile Client Disconnection and Data Reconciliation

An additional dimension in transaction processing with mobile clients is the possibility of mobile client disconnection. In other words, mobile clients might have held locks for the operations they have initiated and then been disconnected. Alternatively, mobile clients may have performed updates based on broadcast data items or even based on cached data items. Owing to disconnection, or otherwise, they may be unable to relay back those updates to the BSS and then the regional server for validation and installation. Furthermore, it is possible that multiple mobile clients have made conflicting updates to the same set of data items, thereby dictating the need for data reconciliation. These scenarios require validation of previously made updates, upon reconnection. This leads to the data reconciliation issue.

To combat the problem of mobile client disconnection, while holding locks for transactions, the locks held on behalf of mobile clients could be like a "lease." Data items whose leases have expired will be unlocked. Lease extension is required when the mobile client is still connected and accessing the data item. The lease mechanism can be implemented by the agents residing at the BSS, which act as surrogates for the mobile clients. This mechanism is, thus, transparent to the rest of the system. When the mobile client is still "alive" and connected, its agent can extend the lease for the client by continuing to hold onto the lock on the data item and requesting a reset of the deadlock timer. When the lease expires and the client has become disconnected, the agent releases the lock on the data item on behalf of the client. This arrangement can reduce deadlocks caused by mobile client disconnection. On the other hand, there are scenarios where mobile clients have delegated certain data access operations and decisions to the agents. With both delegation and lease renewal mechanisms, the deadlock timer can be set to a smaller value to terminate disconnected global transactions earlier. With the timeout mechanism at the transaction coordinator for global transactions, the autonomy of participating database systems can be preserved, as there is no modification required at the transaction manager.

The data integrator at the regional server is responsible for data reconciliation of updates made by mobile clients. As mentioned previously, there are three major kinds of information to be considered for data reconciliation, namely, updates to private or personalized information "owned" by a mobile client, updates made based on information received over the consistent broadcast, and updates made by clients in a disconnected mode that are based on cached values. The differences between the second and third kinds of updates are the timeliness of the updates and the volume of transactions to be reintegrated. Both kinds of updates must be validated against the databases before they can be installed. Updates of the second kind are of a more instantaneous nature, occurring on a per-transaction basis for each mobile client. Updates of the third kind are often of a batch nature, for a collection of interdependent transactions that are "pseudo"-committed by a reconnecting mobile client (Mok, Leong, and Si, 1999).

It is relatively easy to integrate private information or information pertaining to an individual mobile client. This kind of information logically belongs exclusively to a mobile client. A good example of this information includes personal records for the user of a mobile client stored in a database system for his/her own organization. Updates to such "private" information as the user profiling information are conveyed to the appropriate information repository. It is natural to assume that a mobile client is the one who knows best about itself. Thus, a collision in updates would be resolved for the "owner" in favor of other conflicting updates by other clients. One can signify a subset of data items to be of a personalized nature, through annotation at the databases. Those data items then become associated with a particular mobile client. The data integrator can always install updates made by the "owner" client to the databases, aborting other active transactions when necessary. In most cases, those data items are only updated by transactions from the owner and quite often by update-only transactions.

To reintegrate updates of the second kind, that is, those made by clients in a connected mode after reading data items from the broadcast, we can simply process the update transactions sequentially, since they normally arrive in a sequential manner. The simplest validation mechanism is to compare the readset and the writeset of the update transaction made by the mobile client with those transactions that have been committed, since the data items were read by the client, following the optimistic concurrency control protocol (Bernstein, Hadzilacos, and Goodman, 1987). In particular, the server will check backward for read/write conflicts and check forward for write/read conflicts. With a consistent broadcast cache, this process translates into checking the version number of the data items read from the broadcast cache with the current version number of those data items. Updates from validated transactions can then be installed to the databases by reexecuting the write operations of the validated transactions. To improve the commit rate of these update transactions, one can apply a reprocessing approach by attempting to reexecute an update transaction and comparing the values of data items returned with the values in the original execution. The update transaction can be committed if the values are consistent, even though there may be several update transactions changing the values of the data items in between (for instance, the system state remains the same when customer *A* releases a booked seat, followed by the booking of the seat by customer *B*). Such an improvement effectively extends the notion of a conflict in the readset and writeset in conflict serializability to the notion of a reads-from relation in view serializability. This is because transactions executing under the same views produce the same effects on the database.

Reintegration of the third kind of updates is similar, except that there is now a list of transactions to be reintegrated, and the transactions are only submitted to the server quite some time after their creation. This time interval increases the potential for conflicts to develop and for a cascading abort of the reintegrating transactions. The heuristics by Davidson (1984) could be employed to determine a near to optimal set of undesirable transactions for abort. As the consistent broadcast cache may contain multiple versions of the consistent set of data items, one can also employ the multiversion reconciliation mechanism to improve the performance (Phatak and Badrinath, 1999). Further research is required to bring a good solution to this complicated problem.

CONCLUDING REMARKS AND CHALLENGES

In this chapter, we have discussed the support of transactions, both local and global, to carry out consistent and atomic operations on multiple databases. This is achieved in the context of a Web-based mobile environment, intended to support primitive m-commerce applications. An architecture is described that is capable of delivering the necessary functionality. In particular, we have considered issues such as the support of global transaction processing and the separation of the interface design, the business logic and the application program development. Transactional access to databases has been isolated. As such, the same global transaction processing mechanism can be used to develop applications that support clients utilizing WAP devices and those utilizing regular Web browsers simultaneously. We have also taken into account the support for data dissemination in the form of a broadcast database, via the broadcast manager with a consistent broadcast cache. Processing of read-only transactions over the broadcast database can be very effective. On the other hand, updates made by transactions reading data items over the broadcast database and by disconnected transactions have to be validated and integrated back to the database systems through the data integrator. To reduce the reliance on the wireless network and the impact of client disconnection and migration, agents can be considered and utilized in the design of the architecture.

A number of challenges remain to be addressed before a seamless integration of the architecture realization into a practical m-commerce system can be achieved, and we name a few here. Security and authentication are the most important issues that have to be tackled, before an m-commerce application can be of any practical value. No one would feel comfortable conducting any activity involving a moderately large amount of money over an insecure network. Mobile clients such as WAP devices possess very limited computational capability, making the encryption overhead for good algorithms, such as RSA, too high to be practical.[2] Second, a good payment infrastructure must be set up before m-commerce can be useful. This infrastructure can basically be built upon the infrastructure available for e-commerce. This issue is currently being investigated in our research laboratory. Global transaction processing with payment via a bank (Lo, 2001) only constitutes one particular solution, but this solution requires preauthorization of the involved parties. Third, the use of agents can consume quite a significant amount of resources, especially the code and the state of the agents, and this becomes more serious when they migrate. Fourth, there are many potentials and yet problems to consider in the support of location-dependent queries. Some of them are concerned with not only local information in a time-dependent manner, but also information in locations exhibiting certain spatial relationships with the current location. Efficient spatial-temporal processing mechanisms should be provided. Finally, support for advanced applications other than database applications will be even more complicated, but this still needs to be addressed. Those applications may dictate different problem solving requirements, for instance, support for multimedia information retrieval and perhaps processing with real-time constraints. In conclusion, there is still a long way to go for m-commerce research and its practical deployment.

REFERENCES

Acharya, S., Alonso, R., Franklin, M., & Zdonik, S. (1995). Broadcast disks: Data management for asymmetric communication environments. In *Proceedings of the ACM SIGMOD International Conference on Management of Data* (pp. 199-210). ACM.

Alonso, R., & Korth, H. (1993). Database system issues in nomadic computing. In *Proceedings of the ACM SIGMOD International Conference on Management of Data* (pp. 388-392). ACM.

Barghouti, N. S., & Kaiser, G. E. (1991). Concurrency control in advanced database applications. *ACM Computing Surveys, 23* (3), 269-317.

Bernstein, P. A., Hadzilacos, V., & Goodman, N. (1987). *Concurrency Control and Recovery in Database Systems.* Reading, MA: Addison-Wesley.

Bharadvaj, H., Joshi, A., & Auephanwiriyakul, S. (1998). An active transcoding proxy to support mobile web access. In *Proceedings of the 17th IEEE Symposium on Reliable Distributed Systems* (pp. 118-123). IEEE.

Breitbart, Y., Garcia-Molina, H., & Silberschatz, A. (1992). Overview of multidatabase transaction management. *VLDB Journal, 1* (2), 181-239.

Chan, B. Y. L., Leong, H. V., Si, A., & Wong, K. F. (1999). MODEC: A multi-granularity mobile object-oriented database caching mechanism, prototype and performance. *Journal of Distributed and Parallel Databases, 7* (3), 343-372.

Choy, M., Kwan, M., & Leong, H. V. (2000). Distributed database design for mobile geographical applications. *Journal of Database Management, 11* (1), 3-15.

Davidson, S. B. (1984) Optimism and consistency in partitioned distributed database systems. *ACM Transactions on Database Systems, 9* (3), 456-481.

Deshpande, P. M., Ramasamy, K., Shukla, A., & Naughton, J. F. (1998). Caching multidimensional queries using chunks. In *Proceedings of the ACM SIGMOD International Conference on Management of Data* (pp. 259-270). ACM.

Fang, D., Ghandeharizadeh, S., McLeod, D., & Si, A. (1993). The design, implementation and evaluation of an object-based sharing mechanism for federated database systems. In *International Conference of IEEE Data Engineering* (pp. 467-475). IEEE.

Georgakopoulos, D., Rusinkiewicz, M., & Sheth, A. P. (1994). Using tickets to enforce the serializability of multidatabase transactions. *IEEE Transactions on Knowledge and Data Engineering, 6* (1), 166-180.

Housel, B. C., Samaras, G., & Lindquist, D. B. (1998). WebExpress: A client/intercept based system for optimizing web browsing in a wireless environment. *ACM Mobile Networks and Applications,* 3, 419-431.

Hwang, B., & Son, S. H. (1996). Decentralized transaction management in multidatabase systems. In *Proceedings of 20th International Computer Software and Application Conference* (pp. 192-198). IEEE.

Imielinski, T., & Badrinath, B. R. (1992). Querying in highly distributed environments. In *Proceedings of the 18th International Conference on Very Large Data Bases* (pp. 41-52).

Imielinski, T., & Badrinath, B. R. (1994). Mobile wireless computing: Challenges in data management. *Communications of the ACM, 37* (10), 18-28.

Imielinski, T., Vishwanathan, S., & Badrinath, B. R. (1994). Power efficient filtering of data on the air. In *Proceedings of the 4th International Conference on Extending Database Technology* (pp. 245-258). Springer-Verlag.

Krishnakumar, N., & Bernstein, A. J. (1994). Bounded ignorance: A technique for increasing concurrency in a replicated system. *ACM Transactions on Database Systems, 19* (4), 586-625.

Lee, K. C. K., Leong, H. V., & Si, A. (2000a). Incremental view maintenance for mobile databases. *Knowledge and Information Systems: An International Journal, 2* (4), 413-437.

Lee, K. C. K., Leong, H. V., & Si, A. (2000b). A semantic broadcast scheme for a mobile environment based on dynamic chunking. In P*roceedings of the 20th International Conference on Distributed Computing Systems* (pp. 522-529). IEEE.

Leong, H.V., & Si, A. (1997). Database caching over the air-storage. *The Computer Journal, 40* (7), 401-415.

Leung, H. C. (1998). Global transaction management in the Internet environment. Master's thesis, The Hong Kong Polytechnic University, Hong Kong.

Lim, E. P., Hwang, S. Y., Srivastava, J., Clements, D., & Ganesh, M. (1995). Myriad: Design and implementation of a federated database prototype. *Software—Practice and Experience, 25* (5), 533-562.

Lo, J. (2001). Mobile transactions based on WML pages. Technical report, Department of Computing, Hong Kong Polytechnic University, Hong Kong.

Lu, Q., & Satyanarayanan, M. (1995). Improving data consistency in mobile computing using isolation-only transactions. In *Proceedings of the 5th Workshop on Hot Topics in Operating Systems.*

Luk, R. W. P., Leong, H. V., Dillon, T. S., Chan, A. T. S., Croft, W. B., & Allan, J. (2002). A survey in indexing and searching XML documents. Journal of the *American Society for Information Science and Technology, 53* (6), 415-437.

Madria, S. K., Bhargava, B. K., Pitoura, E., & Kumar, V. (2000). Data organization issues for location-dependent queries in mobile computing. In *Proceedings of International Conference on Database Systems for Advanced Applications* (pp. 142-156).

Mok, E., Leong, H. V., & Si, A. (1999). Transaction processing in an asymmetric mobile environment. In *Proceedings of First International Conference on Mobile Data Access* (pp. 71-81). Springer-Verlag.

Mummert, L. B., Ebling, M., & Satyanarayanan, M. (1995). Exploiting weak connectivity for mobile file access. In *Proceedings of the 15th ACM Symposium on Operating System Principles* (pp. 143-155). ACM.

Phatak, S. H., & Badrinath, B. R. (1999). Multiversion reconciliation for mobile databases. In *Proceedings of International Conference on Data Engineering* (pp. 582-589). IEEE.

Pitoura, E., & Chrysanthis, P. K. (1999). Scalable processing of read-only transactions in broadcast push. In *Proceedings of the 19th International Conference on Distributed Computing Systems* (pp. 432-439). IEEE.

Pitoura, E., & Samaras, G. (2001). Locating objects in mobile computing. *IEEE Transactions on Knowledge and Data Engineering, 13* (4), 571-592.

Shanmugasundaram, J., Nithrakashyap, A., Sivasankaran, R., and Ramamritham, K. (1999). Efficiency concurrency control for broadcast environments. In *Proceedings of the ACM SIGMOD International Conference on Management of Data* (pp. 85-96). ACM.

Tai, M. (2001). Formulating WML Web pages for mobile clients. Technical report, Department of Computing, Hong Kong Polytechnic University, Hong Kong.

Tan, K. L., & Yu, J. X. (1998). Generating broadcast programs that support range queries. *IEEE Transactions on Knowledge and Data Engineering, 10* (4), 668-672.

Wong, M. H., Agrawal, D. & Mak, H. K. (1997). Bounded inconsistency for type-specific concurrency control. *Journal of Distributed and Parallel Databases, 5* (1), 31-75.

Yau, S. M. T., Leong, H. V., & Si, A. (2001). Multi-resolution Web document browsing in a distributed agent environment. In *Proceedings of International Conference on Mobile Data Management* (pp. 279-281). Springer-Verlag.

Zhuge, Y., Garcia-Molina, H., Hammer, J., & Widom, J. (1995). View maintenance in a warehousing environment. In *Proceedings of the ACM SIGMOD International Conference on Management of Data* (pp. 316-327). ACM.

ENDNOTES

1 Our research group is building a prototype for location-dependent querying based on Bluetooth and middleware, called BluePoint.

2 Computational optimization must be done very carefully, since it could draw on a timing attack on the mathematical property of the private key.

Chapter IV

Techniques to Facilitate Information Exchange in Mobile Commerce

Aslihan Celik
Santa Clara University, USA

Anindya Datta
Chutney Technologies, USA

ABSTRACT

Data management issues related to organizing and retrieving information from wireless channels have posed challenges for the database community. In this chapter, we discuss data dissemination to mobile clients and present solutions that address the bandwidth and energy limitations resulting from short battery life of the mobile units. We also add a subscription-based data access layer on top of this. Our solutions overall propose a secure and scalable wireless data dissemination architecture. Our broadcast organization and subscription-based access protocols are geared to work hand-in-hand to facilitate a complete content distribution solution via broadcasts.

INFORMATION AS A COMMODITY

As more and more e-commerce applications are brought to the mobile platform, the users have increasingly become reliant on mobile data. Mobile commerce applications cover a wide range from short and multimedia messaging and wireless mail, to downloadable multimedia. Financial services, business-to-business m-commerce, as well as consumer m-commerce are areas that will potentially use mobile data. With the extensive use of location-based services, more data will be available to information providers for relaying to the clients.

All these applications, when deployed, can quickly fill the airwaves and cause service disruption or service quality problems. Therefore, it is important to distinguish the types and priorities of data and design the information exchange protocols accordingly. In the rest of this chapter, we will concentrate on "rapidly changing data" that potentially has many users (e.g., stock market data) and present techniques to disseminate the data in an efficient manner.

The challenges of designing and implementing wireless networks are important ones for the telecommunications research community. At the same time, data management issues related to organizing and retrieving information from wireless channels have posed challenges for the database community as well. Some of the software problems, such as data management, transaction management, database recovery, etc., have their origins in distributed database systems. In mobile computing, however, these problems become more difficult to solve, mainly because of the narrow bandwidth of the wireless communication channels, the relatively short active life of the power supplies (battery) of mobile units, and the changing locations of required information due to client mobility. Solutions to such problems should also adequately deal with the requirement of securely accessing data from the wireless channels.

In this chapter, we will discuss data dissemination to mobile clients and will first present solutions that address the bandwidth and energy limitations resulting from short battery life of the mobile units. We will then turn our attention to add a subscription-based data access layer on top this. Our solutions overall propose a secure and scalable wireless data dissemination architecture.

This chapter is organized as follows. After a brief introduction to the mobile data concept, we first present a general architecture, requirements and trade-offs in designing a data dissemination application. We then present our solutions that provide 1) energy efficient and timely data delivery and access, and 2) subscription-based secure access to wireless data. We provide a brief literature review and outline the significance of our work and future research directions.

AN ARCHITECTURE FOR MOBILE INFORMATION EXCHANGE

In this section, we present the "broadcasting" paradigm in data dissemination. We first present a mobile architecture and describe the parameters. We then present the two problems that we are addressing in this chapter: the broadcasting problem and the subscription-based data access problem.

Mobile Architecture

A general architecture of a mobile platform was given by Dunham and Helal (1995) and is shown in Figure 1. It is a distributed architecture where a number of computers, *fixed hosts* and *base stations*, are interconnected through a high speed *wired* network. Fixed hosts are general purpose computers which are not equipped to manage mobile units but can be configured to do so. Base stations are equipped with wireless interfaces and communicate with mobile units to support data access.

The mobile units are battery-powered portable computers, which move around freely in a restricted area, referred to here as a *geographic mobility domain*. The size restriction on a unit's mobility is mainly due to the limited bandwidth of wireless communication channels. To manage the mobility of units, the entire geographic mobility domain is divided into smaller domains called *cells*. The mobile discipline requires that the movement of mobile units be unrestricted within the geographic mobility domain (inter-cell movement).

The mobile computing platform can be effectively described under the *client/ server* paradigm. Thus, sometimes we refer to a mobile unit as a *client* and sometimes as a *user*. The base stations are identified as *servers*. Each cell is managed by a base station, which contains transmitters and receivers for responding to the information processing needs of clients located in the cell. We assume that the size of a cell is such that the average query response time is much smaller than the time required by the client to traverse it. Therefore, it will seldom occur that a user submits a query and exits the cell before receiving the response.

Figure 1: A general architecture of a mobile platform

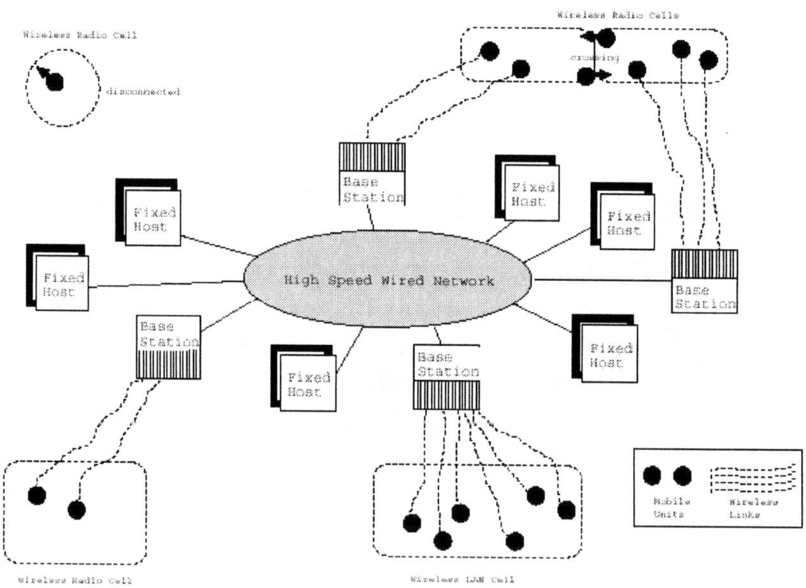

Clients and servers communicate through wireless channels. The communication link between a client and a server may be modeled as multiple data channels or a single channel (Imilienski, Viswanathan, and Badrinath, 1997). We assume a single channel since the objective is the efficient use of the overall bandwidth. We further assume that this channel consists of both an uplink for moving data from client to server and a downlink for moving data from server to client.

Database Architecture and its Characteristics

The data in this application is characterized as *rapidly changing* (Datta, 1994); users often query servers to remain up-to-date. More specifically, they will often want to query the server for their data item of interest. Typical examples of this type of data are stock, weather, and airline information. We assume the following for fully characterizing our mobile database.

1. The database is updated asynchronously, i.e., by an independent external process. Also, such updates arrive with high frequency, signifying that the database is rapidly changing. Examples of such information are stock, weather, etc.
2. Users are highly mobile and randomly enter and exit from cells. There is a parameter called *Residence Latency* (RL), which characterizes the average duration of a user's stay in the cell.
3. User reference behavior is localized; e.g., some stocks are more popular than others.
4. Servers are *stateless*, i.e., they maintain neither client arrival and departure patterns nor client-specific data request information (Imielinski et al., 1997). We assume a stateless server, because we believe that the cost of maintaining a *stateful* server in a mobile environment would be prohibitively expensive. We want to emphasize, however, that our scheme will work with stateful servers as well.

Broadcasting Problem

Wireless networks differ from wired networks in many ways. Database users over a wired network remain connected not only to the network, but also to a continuous power source. Thus, response time is the key performance metric. In a wireless network, however, both the response time and the active life of the user's power source (battery) are important. While a mobile unit is listening or transmitting on the line, it is considered to be in *active* mode. Assuming a power source of 10 AA batteries and a laptop equipped with CD-ROM and display, estimated battery life in active mode is approximately 2.7 hours.

In order to conserve energy and extend battery life, a clients slips into *doze* (standby) mode, in which it is not actively listening on the channel. Clients expend significantly less energy in doze mode than in active mode. Therefore, one of the major goals of our scheme is to minimize the amount of time a client must spend in active mode to retrieve the data items it requests.

The problem addressed may be captured by the following question: *given that users are highly mobile in their mobility domain, what are good strategies that the server can use to decide on what data to provide?* The assumption is that such strategies need to adapt to user demand patterns in the highly mobile environment. We are also interested in the question of retrieval strategies: *given that good strategies are found, what are good retrieval algorithms by which users can retrieve/download data from the server, with a minimum of energy expenditure?* The basic idea is one of "mixed broadcasting," i.e., automatic as well as on-demand broadcasting.

We define the following terms:

Access Time (AT): Access time refers to the time elapsed between query submission and receipt of the response.

Tuning Time (TT): Tuning time is the duration of time that the client spends actively listening on the channel.

The meaning of these terms is illustrated in Figure 2. Consider a client who submits a request at time T_0 and receives the response at time T_7. In this scenario, if the client listens continuously from the time a query is submitted until the response is received, then:

$AT = TT = T_7 - T_0$. On the other hand, if the client slips into doze mode intermittently, then TT is noticeably less than AT, significantly reducing battery usage. In this case, $AT = T_7 - T_0$, and $TT = (T_7 - T_6) + (T_5 - T_4) + (T_3 - T_2) + (T_1 - T_0)$.

This results in energy conservation, as the client is in active mode for only short periods of time. The question, of course, is how to determine the smallest possible tuning intervals. An ideal approach appears to be providing the client with *precise* knowledge of when its requested information will be broadcast.

Figure 2: Access and tuning times

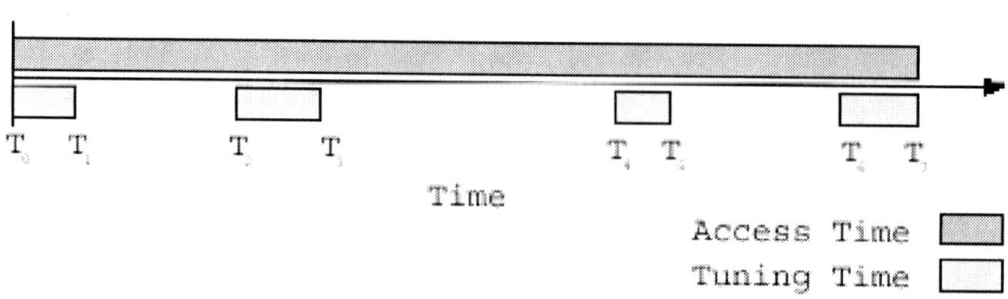

Our aim is to find optimal points in the two-dimensional space of *AT* and *TT*. This becomes difficult, because there appears to be a trade-off between *AT* and *TT*; attempts to reduce one tend to increase the other. For example, access time is directly related to the size of the broadcast, i.e., *AT* is smaller for a smaller broadcast size. On the other hand, providing information for *selective auto-tuning*, i.e., informing the user precisely where its required data is located in the broadcast, reduces tuning time. However, inclusion of such tuning information would increase the overall size of the broadcast by including overhead, which in turn could increase *AT*. Conversely, eliminating this overhead will reduce *AT* at the expense of an increased *TT*, because the user will not know precisely when to tune in.

Subscription-Based Data Access Problem

In this chapter, we address another critical problem: providing *secure access control* in broadcast schemes. To get a feel for this problem, consider the classical broadcast environment, where an *information server* broadcasts to a large number of clients using a shared channel. Each broadcast consists of a number of data objects that clients are interested in. Each client is interested in a certain number of these objects and *subscribes* to them. Subscription refers to a contract that each client enters into with an agent, which entitles the client to access a data object for a specific period of time. Once the contracted period for a subscription is over, the subscription is considered to have expired and the client cannot access the data object any longer without resubscribing. Therefore, broadcast protocols should provide adequate security and should scale well with the number of clients using the system.

Subscription-based access to broadcasts necessitates the use of encryption techniques in order to let only the legitimate subscribers to access the broadcasts. Therefore, the broadcast protocols should have the ability to distribute encryption keys to clients in an efficient manner. We present protocols that add a security layer on top of the basic broadcasting model discussed earlier. In this system, a server broadcasts data items over a shared communication channel, and clients tune in to the broadcasts to download their subscribed items. We add an access control layer that involves encrypting the data items and then adding smart controls on top of the encryption logic.

To enable the deployment of such applications, the following functionalities are necessary.

1. A client must only be able to access the data items that it is *subscribed to*. In other words, the access to all items that a client is not subscribed to must be blocked. An intuitively natural way to tackle this problem is by encrypting the data items in some way.
2. A client must only be able to access items as long as its *subscription to that item has not expired*. This is a non-trivial problem—clearly, in order to be given access to an item, assuming items are encrypted, a client needs to be provided with some sort of decrypting mechanism to retrieve this item. When its

subscription expires, however, the client is still left with the decrypting mechanism, which compromises security. One obvious way of course is to change the security mechanism of a data item every time a subscription expires to that item. This is, however, prohibitively expensive, given the large number of *(client, subscribed_item)* pairs present in a system of reasonable size.

3. *The protocol(s) must be scalable*, i.e., increasing the number of clients should not deteriorate the quality of service.
4. Finally, *the protocol that implements the above two functionalities must provide an adequate level of security*, i.e., it should not be easy to breach the security provided by the access control mechanism.

Essentially, the problem is one of secure data management in broadcasts. Given such an environment, in the rest of this section we stipulate distributed protocols to help the simultaneous achievement of the goals outlined above.

BROADCASTING SOLUTIONS

Two broadcasting solutions to the broadcasting problem specified above are the Variable Broadcast Size (VBS) and the Constant Broadcast Size (CBS) protocols. These techniques, first proposed in Datta, Celik, Kim, VanderMeer and Kumar (1997), seek to achieve a minimal tuning time while reducing the access time. However, it is important to understand the broadcast structure before explaining the protocols.

Broadcast Structure

Broadcast structure refers to the specific broadcast organization that we assume in developing our strategies. It is important to understand this structure in order to properly appreciate our protocols. We assume a (1, *m*) indexing strategy outlined in Imielinski *et al.* (1997). In this scheme, index information is provided at regular intervals in the broadcast. More specifically, a complete index is inserted *m* times in a broadcast at regular intervals.

Figure 3 illustrates our broadcast structure. A broadcast is a sequence of *data blocks* (containing data) and *index segments* (containing access information) as shown in Figure 3A. Using the (1, *m*) data organization methodology, an index segment appears every $1/m^{th}$ of the broadcast, i.e., there are *m* index segments. Clearly, each of the *m* data blocks is also of equal size. Each data block is composed of one or more *data clusters* as shown in Figure 3B, where each data cluster consists of a collection of tuples of a single data item of interest. For example, assume the broadcast consists of stock information, and each tuple is a triple <*stock_id, price, market*>. In such a scenario, the data items of interest would be represented by stock IDs. Consider a particular stock ID, e.g., IBM. All IBM records would comprise a data cluster. A data cluster may span more than one data block.

Data clusters are composed of data buckets (Figure 3C), which contain data records as well as some tuning information (denoted by the 5-tuple <X,Y,Z,N, E_B> in the figure) explained below.

We assume that each client has its own items of interest (e.g., clients are not interested in all stocks, but instead in specific ones). For the purposes of this study, we assume a client has a single data item of interest. As explained above, all records pertaining to this item appear in a specific data cluster which we refer to as the client's *Data Cluster of Interest* (DCI). Within the broadcast, the data clusters are organized in order of decreasing popularity, such that the most popular item will be broadcast first, and the least popular item will be broadcast last. This helps to reduce the access times for popular items.

Figure 3: Broadcast structure

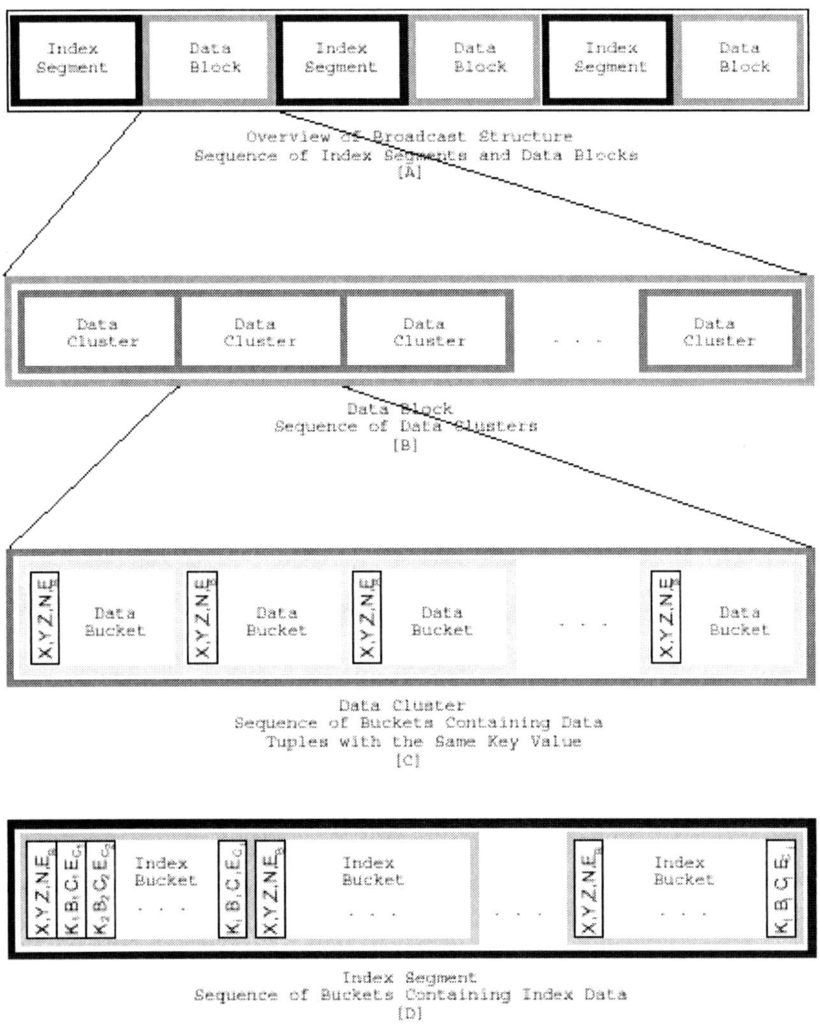

An index segment is a series of buckets containing index tuples and some other special tuning information. We first describe the index tuples. Each index tuple consists of a 4- tuple, $<K,B,C,E_C>$, that not only informs the client precisely where the DCI appears in the broadcast, but also provides information about updates to the cluster since the previous broadcast. The structure of the index segment is shown in Figure 3D. *K, B, C* and E_C are defined below.

K: The cluster's key value (e.g., for an IBM cluster, the key value is IBM).

B: The ID of the first bucket of the data cluster.

C: The offset from *B* to the first dirty bucket (bucket where changes have occurred since the last broadcast) of the data cluster. If all buckets in the data cluster are clean, **C** takes a default value of -1.

E_C: The time when the cluster is scheduled to be dropped from the broadcast.

The dirty/clean information (i.e., *B* and *C*) are included to handle the *rapidly changing data* scenario explained earlier in this section. We assume a tree-like structure for the index. Thus, clients must begin reading the index at the root in order to find the pointers to their DCIs.

As mentioned above and shown in Figure 3C and Figure 3D, all buckets, whether index or data, have a special tuple displayed as a 5-tuple $<X,Y,Z,N,E_B>$. This information is provided to orient clients as they initially tune in to a broadcast. The *X, Y, Z, N* and E_B terms are defined as follows.

X: An offset to the first bucket of the next nearest index segment.

Y: An offset to the end of the broadcast, i.e., the start of the next broadcast.

Z: Shows the bucket's type (data or index) and contains tuning information for items updated since the previous broadcast. It can hold one of four possible values: $Z = -2$ indicates an index bucket. $Z = 0$ indicates a data bucket and that the bucket is *clean*, i.e., unmodified since the previous broadcast. $Z = i$, where *i* is a positive integer, indicates a data bucket and that the bucket is *dirty*, i.e., modified since the previous broadcast. Moreover, the actual *i* value, i.e., the positive integer, is an offset to the next dirty bucket in the same data cluster. $Z = -1$ indicates a data bucket and that it is the last dirty bucket of the data cluster.

N = 0 Indicates a data bucket and that is the last data bucket of the data cluster.

N = i Indicates that this is not the last bucket of a DCI, and the offset to next data bucket of the same DCI is **i**.

E_B: The expected departure time (*EDT*) of the data item in the bucket. Obviously, the E_B value of every bucket in the same data cluster is going to be identical and equal to the *EDT* of the of the cluster key.

Protocols to Support Adaptive Broadcast Content and Efficient Retrieval

In the following, we describe two adaptive broadcast protocols which seek an optimal balance of access time (quality of service or average query response time) and tuning time (energy consumption) (Datta et al., 1997).

As mentioned earlier, *periodicity* is an important parameter in designing broadcast strategies. A periodic broadcast signifies that the broadcast "size" (i.e., number of buckets) is fixed. One can ascribe both advantages (e.g., predictability) as well as disadvantages (e.g., loss of flexibility) to periodicity. To study such effects, we describe two sets of protocols below for the periodic and the aperiodic cases. We refer to the periodic protocol as the *constant broadcast size* (CBS) strategy, whereas the aperiodic broadcast protocol is termed a *variable broadcast size* (VBS) strategy.

Finally, note that these protocols support a *mixed mode* retrieval policy, i.e., when a client arrives in a cell, it first tunes in to the broadcast to see if its DCI is already there. If not, the client explicitly sends a request to the server through the uplink for that item. Thus items may be found readily in the broadcast or may have to be placed "on demand." This policy has been deemed the most "general" policy in the literature.

Constant Broadcast Size Strategy

We first present the server protocol, i.e., the strategy used by the server in deciding upon the broadcast content. We then present the client protocol, i.e., how the client retrieves data from the broadcast.

CBS Server Protocol

In this strategy, broadcast size is limited, and the broadcast is periodic. Periodicity mandates an equal size for every broadcast (recall that we consider both size and time in terms of buckets). If there are too few requested items to fill the broadcast period, the broadcast will contain dead air. On the other hand, if there are more requested items than space in the broadcast, the server must prioritize requested items to decide which to include in the broadcast set. This prioritization mechanism should simultaneously satisfy two properties: *popularity consciousness* and *avoidance of chronic starvation.* Popularity consciousness means that items that are requested more often should have a greater chance of being included in the broadcast than less popular items. Avoidance of chronic starvation means that if a client requests a "less popular" item, it should not be chronically starved, i.e., the item should appear in the broadcast at some point during that client's residence in the cell. At a minimum, our protocol attempts (but does not guarantee) to provide access to a requested data item at least once during a client's probable stay in the cell; that is, within *RL* time of the request.

A system of priority ranking of items based on two factors, i.e., a *Popularity Factor* (PF) and an *Ignore Factor* (IF) is the following:

The popularity factor of item at time *T*, denoted by PF_X^T, identifies the number of clients in the cell at time *T* who are interested in *X*. When a client requests *X*, PF_X^T

is increased by 1. However, every time it is incremented, the system records the corresponding time. Let the timestamp of the ith increment be denoted by T^i_X. Then, a corresponding decrement of 1 is performed on the value of the popularity factor at time T^i_X+RL. This reflects the (anticipated) departure of the client whose request caused the ith increment.

The *Ignore Factor* (IF) is proposed to counterbalance the PF's effect. The IF ensures that less popular but long-neglected items get an opportunity to be included in the broadcast. The IF of a data item X at time *t* is simply the number of broadcasts that this item has not been included in the broadcast (i.e., ignored) since it was requested and is denoted by

$$IF_X^{T_i}$$

Let T_{Req} be the time the item was requested and P_B be the period of the broadcast preset under the constant size strategy. The ignore factor at a time T_i is defined as follows:

$$IF_X^{T_i} = \left\lfloor \frac{T_i - T_{Req}}{P_B} \right\rfloor + 1 \qquad \text{(EQ 1)}$$

Priority computation using IF and PF: An item's priority can be computed based on the following expression:

$$Priority = IF^{ASF} \times PF \qquad \text{(EQ 2)}$$

where *ASF* is an *Adaptive Scaling Factor* which is an exponential weighting factor based on a nearest neighbor approach. Its purpose is to increase the likelihood that items which have been ignored for a long time will appear in the broadcast. *PF* and *IF* differ largely in scale; if *ASF* is relatively low, *PF* dominates the priority expression (limited by the number of clients in a cell). If, however, an item has been ignored for a long time, we would like *IF* to dominate. A larger *ASF* value will achieve this. *ASF* is initialized to a base value for 1 for all data items and reset to this base value each time the item is included in the broadcast. It is incremented when the average time the clients have been waiting for a data item exceeds a preset value.

Having explained the underlying concepts, we are now prepared to describe the server protocol for constructing a broadcast. Prior to a broadcast epoch (the time at which a new broadcast is scheduled to begin), i.e., in its broadcast preparation stage, the server prioritizes all items which have been requested, i.e., items with a $PF > 0$, and sorts the items in order of descending priority. It then adds items to the broadcast set until the broadcast is full. For all requested but excluded items, their *IF*s are adjusted.

CBS Client Protocol

We now describe a client protocol designed to cleverly retrieve data from the broadcast in cooperation with the server protocol defined above. When a client senses the need for a particular data item, it begins the retrieval process by tuning in to the broadcast at an arbitrary time and reading a bucket. We remind the reader that the data cluster in the broadcast that holds the item of a client's interest is referred to as the *Data Cluster of Interest* (DCI).

The random initial probe in a continuous flow of broadcasts creates a large number of tuning possibilities. Note that because we assume a tree-like rather than a linear index structure, the client must start reading from the top of the index segment (i.e., the root). If it does not find a pointer to its DCI (i.e., DCI is not in the current broadcast set), then it requests the item and tunes to the initial index of every succeeding broadcast until either the DCI is found in the broadcast or it departs from the cell.

Variable Broadcast Size Strategy

Having discussed a periodic broadcasting strategy, we now turn our attention to an aperiodic broadcasting scenario. This strategy is called the *variable broadcast size* (VBS) strategy. Note that while the broadcast size varies across broadcasts, at the start of each individual broadcast the size is known; therefore, the start of the subsequent broadcast is known as well. However, the server has no knowledge beyond that, as it does not know what requests may arrive during the current broadcast.

VBS Server Protocol

The server protocol for VBS is much simpler than that for the constant size strategy. All requested items are included, i.e., all items with a *PF* greater than 0 are added to the broadcast set. The broadcast length changes as items are added and deleted. Items remain in the broadcast for *RL* units of time from their latest request and are subsequently dropped from the broadcast set. Within the broadcast, items (i.e., DCIs) are ordered based on descending popularity. Since no item is ignored, there is no notion of ignore factor in VBS.

VBS Client Protocol

The client protocol in this strategy is similar to that of the CBS strategy. The main difference is in the client's response to finding that its DCI is not in the broadcast. Here, if its DCI is not in the broadcast, or if it has missed its DCI and the item will be dropped when the next broadcast is composed, the client requests the item and exits from the protocol. Since its DCI is guaranteed to be in the succeeding broadcast, it begins the retrieval process at the beginning of the next broadcast and finds its DCI in that broadcast.

Performance of the CBS and the VBS Protocols

An empirical performance evaluation by means of an extensive simulation study reveals that the CBS and the VBS protocols perform differently under different system characteristics (Datta, VanderMeer, Celik, and Kumar, 1999). The *Access Time* (*AT*) metric is used to measure the quality of the broadcasting service. The smaller the *AT*, the faster the clients are receiving their data items from the broadcast. The *Tuning Time* metric, explained earlier, is originally proposed to approximate the

Figure 4: Experimental results: [A] AT curves for CBS and VBS hot clients; [B] AT curves for CBS and VBS cold clients

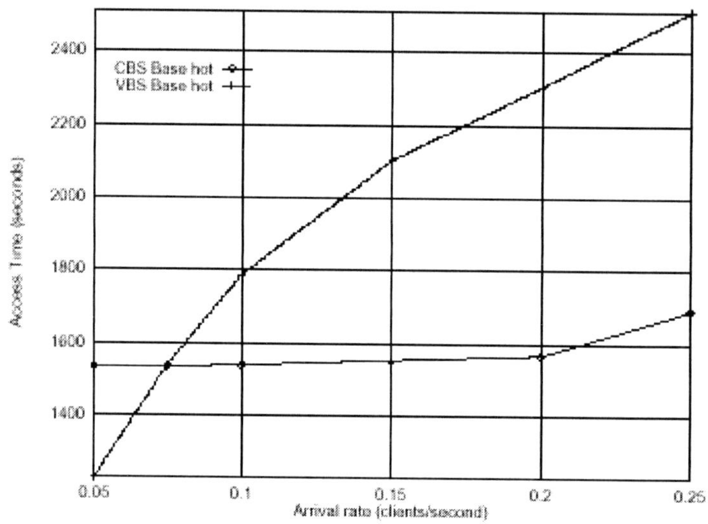

[A]

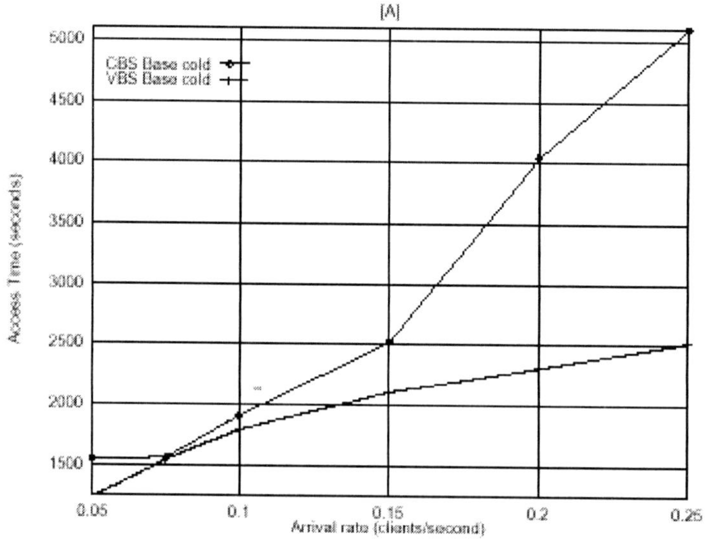

[B]

energy expenditure of the clients that download data from the broadcast. A more direct measure of the energy expenditure is the *Normalized Energy Expenditure* (NEE). NEE is simply the energy spent on average by a client to download a bucket of data. The simulation study allows us to keep track of the energy spent by each client's mobile unit (a combination of the CPU, the disk, the mobile data card, and the display). NEE is derived by dividing the total energy that the client has spent by the total number of data buckets that it has downloaded from the broadcast. In our simulations, a bucket is 128 Bytes.

Figure 5: Experimental results: [A] NEE curves for CBS and VBS hot clients; [B] NEE curves for CBS and VBS cold clients

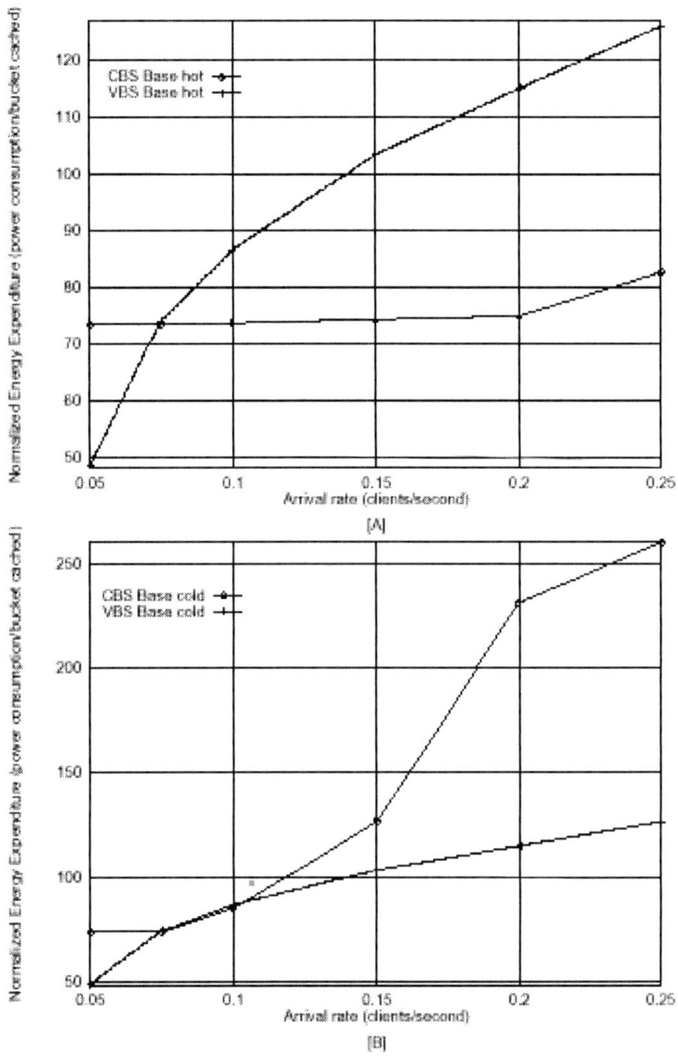

The simulation results distinguish between the *hot* and the *cold* items. Twenty percent of the data items are hot, meaning that they are more popular and requested more often then the cold data items.

The simulation is run for various client arrival rates that reflect the frequency of requests made to the broadcast server. The arrival rates are represented on the horizontal axis in the figures.

The access times for the CBS and VBS protocols corresponding to clients that requested hot and cold clients are shown in Figure 4A and Figure 4B, respectively. The NEEs for the CBS and VBS protocols are shown in Figure 5A and Figure 5B.

Both sets of curves show that for hot clients, CBS outperforms VBS at all but very low loads, i.e., where the CBS broadcast is not full. Here, the VBS dominates. For cold clients, for all load levels, the VBS protocol either outperforms or performs identically to the CBS protocol. This is reasonable, since the CBS protocol is optimized to provide better service to clients interested in more popular items at the expense of service for cold items.

SOLUTIONS FOR SECURE DATA ACCESS FROM BROADCASTS

In this section, we present two protocols that provide secure data access from the broadcasts by the clients: (1) SubScribe (Datta, Celik, Wright and Biliris, 1998), and (2) Drop Groups (Celik and Datta, 2000).[3] Both protocols rely on a security layer supported by the communication infrastructure. We explain the protocols and the broadcast structure that incorporates the security mechanism.

Protocols to Support Secure Data Access from Broadcasts

The protocols use encryption techniques to scramble the communication between the data server and the clients. Two types of encryption keys: (a) *client keys*, and (b) *data keys* are implemented. For the client keys, a public key cryptosystem such as RSA, proposed by Rivest, Shamir, and Adleman (1978), is used. For data keys, a symmetric system such as DES by Shepherd (1995) is appropriate. The public key of client c_j, which is assumed to be known by the server, is denoted by PK_j. c_j's corresponding private key, which is assumed to be secret to c_j, is denoted by SK_j. Messages encrypted with PK_j can only be decrypted by c_j using SK_j. Each data item D_i has a data key, denoted by DK_i, for use in the symmetric system. The data keys are initially known by the server only but will also be securely transmitted to subscribers of D_i. New data keys will be chosen as needed to ensure that only current subscribers can read a particular broadcast.

The SubScribe Protocol

In accordance with the basic cryptosystem described above, the SubScribe protocol operates as follows: When broadcasting D_i, the server encrypts it with DK_i, producing the ciphertext $DK_i(D_i)$ of D_i, denoted by T_i. T_i is included in the *data component* of the data block corresponding to DK_i. Subsequently, only clients knowing DK_i are able to access D_i. Thus, the data key DK_i is included (in an encrypted fashion) in the *key component* of the data block corresponding to D_i. In order to provide maximum efficiency, the data key is only changed when clients drop their subscriptions. The protocol has two components: a server side, or *information delivery* component, and a client side, or *information access* component.

SubScribe Server Side Protocol

The server side protocol is responsible for the delivery strategy for data items. It distinguishes between two types of data items: (a) data items whose subscriber set in the current broadcast includes every client who were subscribed to this item in the previous broadcast as well—these items are referred to as NODROP items, and (b) data items not satisfying the previous criterion, i.e., items which have lost some subscribers in the current broadcast. We will refer to these as DROP items. To include a DROP item, say D_i, in the broadcast, the server chooses a new data key, DK_i, and creates a data block for this item as follows:

1. **Data Component:** The server encrypts D_i with DK_i and includes the ciphertext in the data component.
2. **Key Component:** For each client in the subscriber set of D_i, the server encrypts DK_i with the client's public key and includes the ciphertext in the key component. In other words the key component of DROP items essentially becomes a concatenation of ciphertext chunks, where each ciphertext chunk represents an encryption of DK_i with a specific subscriber's public key.

To include a NODROP item into the broadcast, the server uses the same data key that was used in the previous broadcast for this data item. This is possible due to the fact that using this data key does not compromise security—all prior subscribers are still subscribed to this item. Also, in this case (potentially) substantial savings are realized as the key need not be sent to all the prior subscribers. The server composes the key component by encrypting the data key *only for the new subscribers*. Existing clients are notified of the fact that the data key remained the same by inserting a special offset value of -1 in an index in front of the data block for that data item. In both the DROP and NODROP cases, index records are created by the server for constructing the two types of indexes described before.

The Drop Groups Protocol

In the SubScribe protocol, the broadcast is organized in such a way that each data item is encrypted with its own key and is broadcast together with the key information intended for each subscriber. This key information for each client is obtained by encoding the data item key using the public key of the recipient. That is, the broadcast server sends off as many encodings as there are clients. This effectively renders the size of the broadcast unpredictable. Particularly when the number of subscribers is high, the key segment may become very large and significantly increase the size of the broadcast. Obviously, a longer broadcast means a reduced quality of service. Therefore, there is a scalability issue.

The Drop Groups (DG) Protocol is designed to bound the size of the key component in a broadcast regardless of the number of clients in the system.

DG achieves scalability by using a novel grouping criterion. DG assigns each client to predetermined groups and assigns each group a group key valid until the group changes. This is similar to the Group Key approach. In the Group Key protocol, subscribers of an item usually form a group and are given a group key valid until there are *drops* (i.e., subscription expiration) from the group. When a drop occurs, a new group key must be generated and distributed to remaining subscribers so that the dropped subscribers don't have access to new values of the data item. In DG, however, we propose to further divide the groups of a data item into subgroups using an additional criterion. The new criterion we use is the *time to drop*, which is simply the amount of time until a client's subscription for a data item expires. Therefore, two subscribers, A and B, of data item *i* are in the same group if and only if their subscription for *i* expires at the same time. Of course, in order to achieve this sort of grouping, we have to ensure that subscription expirations are bunched together at discrete epochs. This is done as explained later.

The choice of the time to drop as the grouping criteria is crucial. This is designed to remedy a major problem associated with the group key approach, namely, the key expiration problem in a dynamic environment. In this environment, the period of validity of a key is small, necessitating the generation of a new key frequently. Although key generation is rather fast and cheap, it is costly to distribute this new key to the clients.

In DG, since all subscribers in the same group will be dropped at the same time, it is never necessary to issue a new group key and distribute it to the group. Furthermore, when a new client contacts the subscription server, the client is given a group key for each data item that it is interested in. The client then listens to the broadcast prepared by the subscription server and downloads the data items.

The time continuum is divided into subscription epochs such that subscription expirations are only scheduled to happen at the end of an epoch. For example, if the epoch length is 1 hour, and client A wants to subscribe to a data item for 2.5 hours, it must choose to subscribe for either two or three hours. To limit the number of epochs, a limit is set on the *horizon* of subscription. For example, if a subscription epoch is one hour long, and the horizon is 24 hours, then there are 24 possible subscription epochs that clients may choose from. Note that real-world analogies exist for this scenario: readers may subscribe to journals between 1 and 24 months and receive issues monthly. The server can adjust the duration of an epoch depending on the popularity or subscription patterns of the clients of an item. Given such a framework, given a subscription horizon of H epochs, in the key component, there can be at most H group keys preceding the data item regardless of the number of subscribers in each group. Therefore, if there are d data items with H groups in each, then there will be dH group keys in the broadcast. At the end of each subscription epoch, d groups will be dropped, and d groups will be added, one group per each data item. Essentially, this bounds the number of groups. Clearly, this is a big step towards limiting the size of the broadcast, thus satisfying the scalability requirement.

Drop Groups Architecture

In this environment, there are two separate servers, the Subscription Server (SServ) and the Broadcast Server (BServ). When a client wishes to access the service, it first contacts the SServ that handles the subscription requests. After exchanging information with the SServ, the client listens to the broadcast prepared by the BServ until its subscription expires. Once the client contacts the SServ, it communicates the request to the BServ, which incorporates the requested data item in the broadcast.

The two servers need to maintain communication between each other mainly for exchanging information specific to data items. The client needs to contact the SServ but has no interaction with the BServ except for listening to the broadcast.

In DG, client keys are used for communicating with the Subscription Server. Authentication, subscription and the initial key exchange are performed using public and private keys. The group key of a data item D_i for epoch k is denoted by GK_{ki}. Thus, R_{ki}, the data key for D_i encrypted with GK_{ki}; i.e. $GK_{ki}(DK_i)$ is included in the *key component* of the data block corresponding to D_i.

Broadcast organization in DG

The organization of a broadcast implementing the DG protocol is shown in Figure 6. A broadcast starts off with a *broadcast index*, followed by a sequence of *data blocks*. The broadcast index segment and all the data blocks contain an *orientation header* (OH). The OH consists of a single element, namely an offset to the start of the next broadcast. This pointer is intended as a "tuning-aid" for clients.

The *Broadcast Index* (BI) precedes everything else in a broadcast. The BI consists of index records that hold pointers to each item's data block in the current broadcast. More specifically, an index record consists of a 2-tuple < *item id, offset to data block*>. A client obtains pointers to its desired data items from the BI and then sleeps, only to wake up at the desired points in time.

The BI is followed by a sequence of data blocks. A data block consists of four parts.

1. *OH* that contains the 2-tuple <*offset to item (OTI), flag*>. The flag bit is set if the item key has changed since the previous broadcast. When clients download the OH, if the flag bit is not set, they don't need to download the Key Segment if they have the key to the data item from the previous broadcast.

Figure 6: Detailed view of broadcast structure in DG

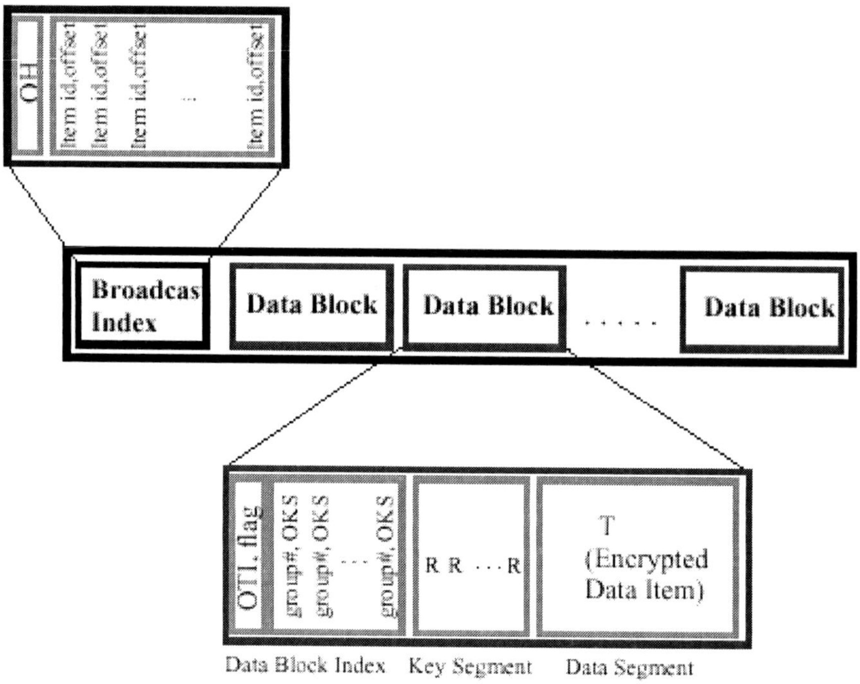

2. *Data block index* consists of the *<group number, offset to key segment (OKS)>*. The OKS element associated with client ID c_j indicates where c_j can find the key information for its group.
2. The second part in a data block is the *key component*. It contains key chunks for all the groups subscribed to D_i, i.e., it contains R_{ki} for all groups k that have at least one subscriber.
2. The last part of a data block is the *data component*. This contains T_i, a data item D_i encrypted with data key DK_i.

Therefore, a client, upon successfully downloading the key chunk and T_i, can obtain D_i.

RELATED LITERATURE

Our work is focused on disseminating data in wireless networks and on accessing the data securely. We use data broadcasting as the means for data dissemination. Alonso and Korth (1993) introduced database system issues in mobile computing. They also point out that the assumption of unlimited battery power for database queries is challenged due to short-lived batteries of the mobile units.

Dunham and Helal (1995) identified new database problems in a wireless environment. Imielinski and Badrinath (1994) were among the first to note that bandwidth limitations would make the data broadcasting a desirable alternative to interactive access. Vaidya and Hameed (1999) addressed, among other issues, the notion of broadcast scheduling, i.e., when to broadcast particular items. A very relevant work from our perspective is the work on *Broadcast Disks* (BD) performed by Acharya, Alonso, Franklin and Zdonik (1995). A BD involves the determination of a broadcast program containing all the data items to be disseminated and the subsequent transmission of this program in a periodic manner. BDs are constructed based on known access probabilities—this is very hard, if not impossible, to know in a mobile framework where clients move in and out of cells at their own will. Our strategies consider stateless servers, i.e., servers have no *a priori* knowledge of client movement or access patterns. For the abovementioned reason (i.e, clients moving in and out of cells), the broadcast content needs to be dynamic—BDs provide a static broadcast content. In the BD framework, client demands are completely satisfied, i.e., all the demanded items are included in the broadcast. In a mobile scenario however, the broadcast period may be too small for the inclusion of all items (particularly if some items are broadcast more often that others as in the BD framework)—this means that some method would need to be constructed that includes certain items in the broadcast in preference to others. The BD framework is unable to handle this. We have created a sophisticated prioritization mechanism to handle this issue. Finally, the BD framework exclusively considers latency as the primary performance metric—no attention is paid to energy conservation by the clients. We, on the other hand, simultaneously pay attention to both.

Kenyon and Schabanel (1999) proved that the broadcast scheduling problem that minimizes the average response time is NP-hard. Su, Tassiulas and Tsotras (1999) outline the properties of the optimal scheduling solution where the broadcast schedule is computed based on access probabilities of the users. Oh, Hua and Prabhakara (2000) proposed a mixed approach by broadcasting the popular data objects and providing other (unpopular) objects via the point to point method.

While mechanisms have been suggested to improve the security of wireless communications (Lo and Chen, 1999), there is not a widely adopted standard by the industry. As far as we are aware, no existing products support the subscription-based access control schemes studied in this work.

CONCLUSIONS AND FUTURE RESEARCH DIRECTIONS

In this chapter, we looked at (1) the problem of data organization and access in mobile networks using broadcasts, and (2) subscription-based data access from broadcasts. The protocols presented here provide efficient methods for designing a broadcasting application to counter the effects of infrastructural inadequacies such as low bandwidth and limited battery power. Section 3 concentrates on deciding the broadcasting strategy (indexing, broadcast organization, and broadcast period). Section 4 builds upon these results and prescribes a security layer on top of the basic broadcast structure. Any subscription-based access protocol may be implemented with either the constant broadcast strategy or with the variable broadcast strategy. Based on the characteristics of the data, network capacity, and customer needs, the right combination of protocols should be determined.

Our solutions can be implemented in a wide range of data and content delivery applications, ranging from financial data to wireless Internet services. The dynamic content in a Web page can be distributed to the subscribers via broadcasts. The users could cache the static elements of the page (frames, appearance, etc.) and obtain the freshest content (stock quotes, traffic, movie times, etc.) from the broadcasts. Broadcasting is also a good candidate for providing access to enterprise applications: mobile workers can subscribe to data items that they need to run the application, and any updates to the data items can be broadcast using our protocols. The security of such a system is enhanced if an encryption mechanism such as the one described in the Drop Groups protocol is used. The most general case for using our protocols is for content distribution to handheld devices, such as cellular phones and the PDAs (Personal Digital Assistants, such as Palm). Here, either the wireless carrier alone, or a content provider in alliance with the wireless carrier, could operate the broadcasting application.

The broadcasts should be prepared considering the current and near future demands of the clients. For example, assume that the broadcasts are prepared based on the requests of the clients within a geographic area. When new clients come into that broadcast area, their items of interest may not be included in the current broadcast. Then, these clients will have to resubscribe to the broadcast and wait until these items are broadcast. However, if the broadcasts are prepared preemptively, i.e., by estimating the incoming clients' requests, then resubscription can be prevented. To do this, the broadcast application needs to keep track of the movements of the clients in the wireless network. However, keeping track of client movements is both costly and difficult to manage. Therefore, trade-offs between preemptively including the data items in the broadcasts and managing the location information of the clients must be considered. We have started researching this area. We plan to derive analytical solutions that optimize bandwidth utilization and cost of operating the system.

In summary, the protocols and concepts presented in this chapter have a wide range of applicability in content distribution. Our broadcast organization and subscription-based access protocols are geared to work hand-in-hand to facilitate a complete content distribution solution via broadcasts.

REFERENCES

Acharya, S., Alonso, R., Franklin, M., & Zdonik, S. (1995). Broadcast disks: Data management for asymmetric communication environments. *Proceedings of ACM SIGMOD*, San Jose, California, pp. 199-210.

Alonso, R., & Korth, H. (1993). Database issues in nomadic computing. *Proceedings of the ACM*-SIGMOD.

Celik, A., & Datta, A. (2000). A scalable approach for subscription-based information commerce. *Workshop on Electronic Commerce and Web-Based Information Systems (WECWIS)*, Milpitas, CA, USA.

Celik, A., Datta, A., & Narasimhan, S. (2000). Secure data delivery protocols for information commerce in a push-based environment. *IEEE Transactions on Systems, Man and Cybernetics, 30* (4).

Datta, A. (1994). Research issues in databases for active rapidly changing data systems (ARCS). *ACM Sigmod Record, 23* (3).

Datta, A., Celik, A., Kim, J. K., VanderMeer, D., & Kumar, V. (1997). Adaptive broadcast protocols to support power conservant retrieval by mobile users. *Proceedings of the Thirteenth International Conference on Data Engineering* (ICDE), Birmingham, UK.

Datta, A., Celik, A., Wright, R., & Biliris, A. (1998). SubScribe: Secure and efficient data delivery/access services in a push-based environment. *Proceedings of the International Conference on Telecommunications and Electronic Commerce* (ICTEC), Dallas, TX, USA.

Datta, A., VanderMeer, D., Celik, A., & Kumar, V. (1999). Adaptive broadcast protocols to support efficient and energy conserving retrieval from databases in mobile computing environments. *ACM Transactions on Database Systems, 24* (1), 1-79.

Dunham, M. H., & Helal, A. (1995). Mobile computing and databases: Anything new? *ACM* SIGMOD Record, 24 (4), 5-9.

Imielinski, T., & Badrinath, B. R. (1994). Mobile wireless computing: Challenges in data management. *Communications of the ACM, 37* (10), 18-28.

Imielinski, T., Vishwanathan, S., & Badrinath, B. R. (1997). Data on air: Organization and access. *IEEE Transactions on Knowledge and Data Engineering, 9* (3), 353-372.

Kenyon, C., & Schabanel, N. (1999). The data broadcast problem with non-uniform transmission times. *Proceedings of the 10th Annual ACM-SIAM Symposium on Discrete Algorithms*. Baltimore, Maryland.

Lo, C.-C. & Chen, Y.-J. (1999). Secure communication mechanisms for GSM networks. *IEEE Transactions on Consumer Electronics, 45* (4).

Oh, J., Hua, A., & Prabhakara, K. (2000). New broadcasting techniques for an adaptive hybrid data delivery in wireless mobile network environment. *Proceedings of the IEEE International Performance, Computing and Communications Conference* (IPCCC 2000), Phoenix, Arizona.

Rivest, R., Shamir, A., & Adleman, L. (1978). A method for obtaining digital signatures and public key cryptosystems. *Communications of the ACM, 21* (2).

Shepherd, S. J. (1995). A high speed software implementation of the Data Encryption. *Computers and Security, 14* (4), 349-357.

Su, C.-J., Tassiulas, L., & Tsotras, V. J. (1999). Broadcast scheduling for information distribution. *Wireless Networks, 5* (2).

Vaidya, N. H., & Hameed, S. (1999). Scheduling data broadcast in asymmetric communication environments. *Wireless Networks, 5* (3).

ENDNOTES

1 Aslihan Celik is an assistant professor with the department of OMIS, Santa Clara University, California 95053, USA.

2 Anindya Datta is an associate professor with the DuPree College of Management, Georgia Institute of Technology, Georgia 30332, USA.

3 Another protocol, the SEMD, is omitted for brevity (Celik, Datta & Narasimhan, "Secure Data Delivery Protocols for Information Commerce in a Push-Based Environment," *IEEE Transactions on Systems, Man and Cybernetics, 30*(4), 2000).

Chapter V

Digital Rights Management for Mobile Multimedia

Sai Ho Kwok
Hong Kong University of Science and Technology, China

ABSTRACT

In the future, intellectual property protection will be a need for distributed media in mobile multimedia. With the constraints of mobile commerce and mobile technologies such as limited bandwidth and computing capability, new schemes of rights management emerge. Digital rights management (DRM) operations in these schemes differ from those in existing DRM solutions for electronic commerce. This chapter presents a general DRM framework for mobile multimedia based on current DRM, mobile network, mobile device, and payment technologies. The framework is partially referenced to the NTT DoCoMo i-mode model, which centralizes payment and maintains user information within the service center. This chapter also presents the basic operations of the general framework and illustrates how rights insertion, rights enforcement, and music sharing are realized under the framework.

INTRODUCTION

Mobile multimedia has been predicted to be a key service and application in mobile e-commerce (m-commerce) by many mobile operators and mobile service providers. At present, mobile multimedia is still in its infancy, accessed by relatively low-end mobile devices with limited bandwidth and resources. Several mobile service providers have already launched low-grade multimedia services with relatively low-end technologies to test the market, for example, a radio broadcast service from Orange in Hong Kong in 2000 (Orange). The owners, publishers and distributors of the distributed media pay very little attention to the issue of intellectual property protection since the market and business models are not actually there. With advances in mobile networks and technologies, the demand for mobile multimedia services will be much higher in terms of quality and convenience. Intellectual property protection, rights control and management of distributed media become concerns, similar to what happened in the market for distributed digital music.

Mobile multimedia services are different from multimedia services in electronic commerce due to the nature and properties of their client devices and communication networks. However, copyright protection and rights management are needed for the content distributed by these multimedia services over the wired Internet. Rights management technology makes various online payment schemes possible, such as pay-per-view, pay-per-download, pay-per-game, and so on. But until now, there has been no specific digital rights management (DRM) solution designed for mobile multimedia. This chapter will present a general DRM framework for mobile multimedia based on current DRM, mobile network, mobile device, and payment technologies.

BACKGROUND

What is Digital Rights Management?

The Australasian Performing Right Association Limited (APRA) [APRA] classifies different types of rights that music copyright owners possess and defines some common rights in music. These rights are listed below.

- The rights to reproduce their work, that is, record their music onto a CD, into a film soundtrack or onto a computer disk or reproduce their lyrics as sheet music or photocopy them. In the music industry the right to record a song onto record, cassette or CD is known as the ***mechanical right***; the right to include music with visuals (i.e., on the soundtrack of a film or video) is known as the ***synchronisation right***.
- The rights to publish their work; this means making their work available to the public for the first time.
- The rights to perform their work in public, for instance, performing their work live in pubs, clubs or at festivals or by playing a recording of their work or

showing a film containing their work. In the music industry the right to perform a work in public is part of the ***performing rights***.

- The rights to communicate their work to the public, for instance, by the Internet or by television or radio broadcasting. In the music industry the right to communicate a work to the public is part of the ***performing rights***.
- The rights to make an adaptation of their work, for example, by arranging or transcribing music or translating lyrics.

From the consumer perspective, end-user rights usually refer to usage against payment. Rights management addresses both owner and end-user rights. In general it refers to the problems associated with intellectual property rights, including copy protection. In particular it addresses the problems of assuring that, in a commercial setting, payment is made for a particular use of content and that the use made does not exceed the use authorized (Stewart, 1998). Online, rights management can be regarded as a process of honoring those copyright provisions, license terms and usage agreements established by the owners of the intellectual property in online media business (Anderson & Lotspiech, 1995). The explicit rights and obligations of the music owner are stated in a digital license (for example, how many times the licensee can play the licensed music), and DRM systems execute the rights granted in the license. However, note that DRM in this chapter addresses end-user rights only, as the owner's rights are not specific to the scope of mobile multimedia.

DRM standards have been developed for distributed digital content. For instance, the Secure Digital Music Initiative (SDMI, 1999), backed by the Recording Industry Association of America (RIAA) and 148 music and technology companies (as of October 2000), has been proposed to provide a secure environment for music distribution over the Internet. Another standard developed by the Moving Picture Experts Group (MPEG) is known as MPEG-21 (Bormans & Hill), dedicated to distributing digital multimedia content. In addition, there are commercial DRM systems especially for the wired Internet music business. They include *Windows Media Rights Manager* by Microsoft (Windows Media), *Electronic Music Management System* (EMMS) by IBM, *Intel Software Integrity System* (ISIS) by Intel, and *MetaTrust* by InterTrust Technologies (InterTrust, 2000).

The blooming of the online music business and the popularity of music sharing have attracted tremendous attention from the industry and for technology firms. A key to the success of any online music business model is effective digital rights management. The majority of music labels and related IT firms are positive about business models with DRM, and therefore early DRM solutions have been designed mainly for online music distribution.

For the online music business, DRM involves specifying and associating rights with the distributed music, placing controls on the music to enforce rights, enabling access checks, and tracking permissions usage and payment. For a general media service, the required capabilities include 1. rights specification and rights label management; 2. rights authorization; 3. content protection, rights enforcement, and trusted rendering; 4. rights tracking; and 5. a security and commerce infrastructure. Business transactions, such as payment, ordering, customer enquiry, etc. may occur between the concerned parties during content packaging, distribution, and usage (Ramanujapuram & Ram, 1998). Managing rights in all these transactions is necessary. Figure 1 summarizes the rights management capabilities involved in different transactions. To support all these capabilities, the DRM system needs rights insertion and rights enforcement operations. In addition, a license management mechanism is also needed in managing license documents.

Rights Insertion

Rights insertion is an operation to embed the identities of the concerned parties and assign business rules and conditions to the distributed content. The concerned parties in online media transactions include content creators, owners, distributors, and consumers. Their identities (IDs) can be represented through digital certification (usually for any individual ID) and digital watermarks (for the company's ID). Business rules and conditions are usually laid down in a license document, while the IDs are embedded in the digital media file containing the content. All the rights information, together with the file, is delivered to the consumer in a secured way. The rights insertion operation takes place at the production company or the distribution site, or sometimes both. A digital media file with an associated license document is known as a rights-protected media file.

To respond to new circumstances, opportunities and challenges in the electronic

Figure 1: Major transactions in digital rights management: content creation, content distribution, and content usage

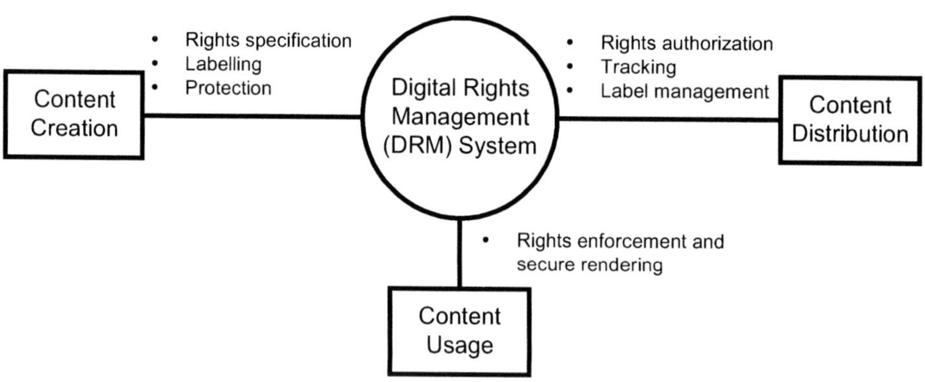

business market, additional, supplemental rules for the use of the media file may be needed for value chain partner, so that the distributor may enforce other rights to the content. A typical example is the consumer information, including consumer's certificate or consumer's keys, inserted into the licensed document during the transaction to certify that the consumer is the legitimate buyer. Another common situation is ownership transfer in media sharing.

Rights Enforcement

There are two types of rights enforcement, namely, active enforcement and passive enforcement. In the usual situation, DRM systems perform active rights enforcement before the rights-protected material is played or used. The active enforcement takes place within the media player as a built-in function. For example, the active enforcement of Windows Media Rights Manager by Microsoft [Windows Media] will fail to verify and follow specified business rules and conditions specified if the media file or the license document is corrupted by intentional or unintentional attacks. Some DRM systems may fail in detecting this, and the consumer may use other players to render the file. In case of the failure of active enforcement, passive enforcement is an offline ownership verification operation to check for the hidden owner identities represented by digital watermarks.

License Management

A license document can be a separate file or message embedded in a media file. The license document states all the terms and conditions concerning use of the licensed media file. These terms and conditions can be static or dynamic depending on the payment scheme. License management is a mechanism to execute the terms and conditions stated in the license. This requires coordination among the media player, the media file, and other supporting modules, e.g., the payment module. From the technical perspective, license management refers to issuing, hosting, and verifying the license.

There are two basic license management models—tethered and untethered models—used in commercial DRM systems. In the tethered model backed by Microsoft [Windows Media], Intel [ISIS], IBM [EMMS] and others, consumers must be online to purchase digital music. License distribution and management are handled by a license services center providing centralized license storage and centralized security. The advantage of centrally storing digital licenses is that it is easier to upgrade security as flaws are uncovered. In addition, the tethered model takes up less space on the consumer's computer and operates more discreetly than the untethered model. In the untethered model, specifically by InterTrust (InterTrust, 2000), consumers store licenses on their own computers and are able to make purchases offline. Payment is made at a later date. The untethered model is designed to promote music super-distribution models (Cox, 1996), where consumers can share files in a viral Napster-like fashion and where consumers can make micropayments on a song-by-song basis or on a subscription basis. The trade-off is that DRM becomes highly complex, takes up more memory, and is not user friendly.

The categorization of these models is based on where the DRM technology is hosted and how digital rights are distributed. The choice of a license management model depends heavily on the payment model in use. For instance, online payment favors the tethered model, while offline payment allows the untethered model. Existing commercial DRM systems support either the tethered model or the untethered model (Anonymous, 2000). An enhanced license management model, which combines tethered and untethered models in one DRM system, is being developed (Kwok, 2000). An appropriate license management model can facilitate content sharing. When a user shares purchased content with another user, the license creation, modification, and transfers that are required processes in sharing activity can be handled comfortably by the license management system.

Mobile E-Commerce

With the current 2.5G technologies of mobile devices and mobile communication networks, rights-protected digital content is usually delivered to mobile users through a low-quality audio channel. For example, Orange [Orange] has launched a radio broadcast service using the ordinary mobile/telephone network. The quality of the content is inevitably degraded subject to the limited bandwidth of the service network and the limited audio capability of the mobile devices. However, in the most popular service in mobile multimedia—music distribution—the media channel can be either the ordinary telephone network or an MP3 channel. The MP3 channel was invented and developed by Vitaminic [VITAMINIC], and it plays an important role of ensuring the quality of the music at an acceptable level. Higher music quality and higher quality of multimedia service can be achieved when a higher bandwidth channel or a dedicated media channel and a powerful mobile device are in use, perhaps when the 3G technologies become available.

DRM for M-Commerce

Today's mobile network and mobile device technologies constrain the sophistication of mobile multimedia services. These underlying technologies are fundamentally different from those used in Internet commerce, and they impose many limitations on the services. This explains why existing DRM systems cannot be applicable to DRM over the mobile environment in a straightforward way. Some of the most important technical and physical obstacles that inhibit DRM over the mobile environment are

1. *License management*: As explained above, there are three license management models available for DRM in e-commerce: tethered, un-tethered, and enhanced license management models. These models work fine in e-commerce applications, but not with m-commerce, in which the client mobile device usually has limited resources of both memory and processing power to handle

and process license document and rights-protected content.

2. *Limited storage and processing power*: Due to the limited resources of the mobile device, it is not possible to download the rights-protected content to the mobile device and play it there. DRM operations cannot be executed on the client side.
3. *Rights insertion*: A sophisticated consumer's ID, such as a private key or digital signature, cannot be kept on the consumer's device due to the storage limitation. Hence, the required consumer's ID must be provided by another party or uploaded for rights insertion, for example, through registration.
4. *Rights enforcement*: Active rights enforcement cannot take place on the mobile device because the device is not capable of intensive computation. However, passive rights enforcement is always possible when the rights-protected content requires rights verification offline using a normal PC.
5. *Payment*: Security has been a very important issue in handling payment through the Internet. Cryptography has been widely used in the protection of payment information. However, it is also proven that most of the cryptography techniques are breakable—it is just a matter of time. Besides, encryption and deencryption require elaborate computation, but mobile devices cannot support this. And, it is believed that the most secure payment method would be one using a private channel, such as a value-added network (VAN), and the payment method should involve minimal exposure of personal and credit card information over the public network—including the Internet and even the mobile network.

A GENERAL DRM FRAMEWORK FOR M-COMMERCE

The proposed framework is partially referenced to the NTT DoCoMo i-mode model [DoCoMo]. In this framework, the mobile devices can be ordinary WAP or i-mode phones with limited physical resolution in their display and limited storage memory, as these devices are dominant in the current mobile market. The DRM model in the framework is based on the license management scheme presented in Kwok (2000) and the design of commercial DRM systems in the e-commerce domain (EMMS, ISIS, InterTrust, 2000; Kwok, 2000; Kwok et al., 2000a; Kwok et al., 2000b; Windows Media, Bormans & Hill, SDMI, 1999), but all DRM operations, including rights insertion and rights enforcement, are executed at the service center. This DRM framework resolves many of the problems presented earlier in the previous section, and more importantly, DRM is realized and operated in the mobile environment. In addition, this framework can also enable other DRM-related activities, such as music sharing between mobile users using digital licenses.

Figure 2 presents a general framework for DRM for digital music distribution in a mobile environment. The center of the framework is a service center, which manages information to and from mobile users, information providers (IPs), and other concerned parties. The principal components include (1) a mobile network infrastructure, (2) a DRM system, (3) a payment system, and (4) a database. There are three types of parties involved in this framework: the information providers (both official and unofficial sites), the bank, and the mobile users. The communication channels between different parties and the service center are different from each other depending on the required security level. For example, a dedicated network is used between the bank and the service center, since highly confidential information is transferred through this channel, whilst the service center relies on the packet network for content delivery.

Mobile Network Infrastructure

The mobile network infrastructure is based on the NTT DoCoMo i-mode [DoCoMo]. It provides a network architecture that connects all involved parties to the service center, which is the mobile operator—NTT, in this case. The service center, being the only gateway for information delivery to mobile users, can provide value-added applications and services on top of the regular services offered by IPs. Value-added applications and services include billing service, a highly secure payment scheme, DRM service, and so on.

Figure 2: A general framework of DRM for m-commerce

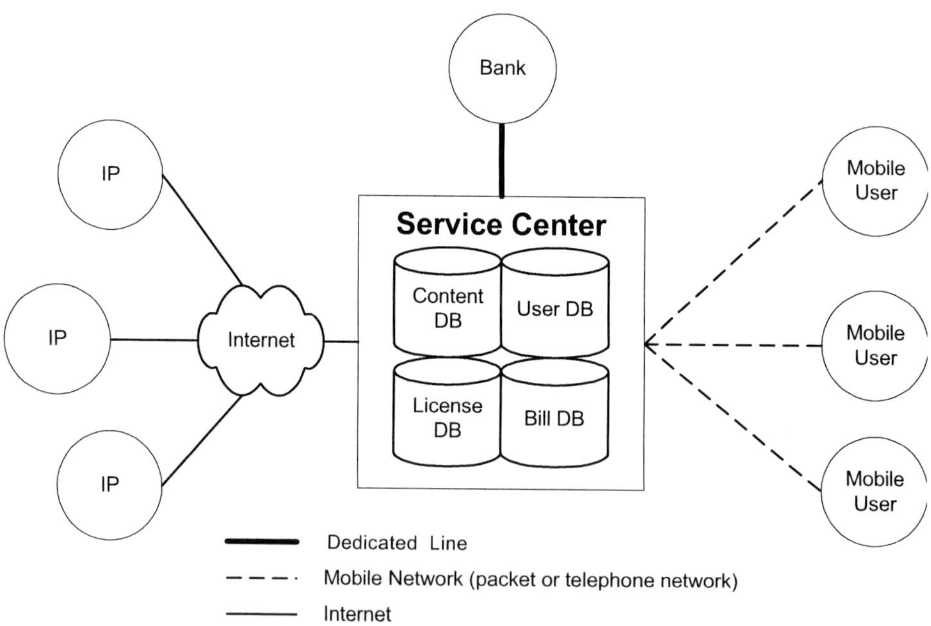

Network capacity, bandwidth, throughputs, and error tolerance vary with different telecommunication companies and communication networks. A 3G networking system could greatly improve many different aspects of the performance of the mobile network. Mobile multimedia, virtual reality and other high-bandwidth services could become possible. This general framework could also take advantage of these services and enhance the applications and services.

Payment

Payment is an undetachable component of m-commerce. The payment part of the general framework also adopts the DoCoMo i-mode [DoCoMo] and eCyberPay [eCyberPay] approaches. The concept of these approaches is to centralize the payment process within the service center and IPs and to require no confidential information from the consumer during transaction and payment. The concerned IPs receive payments from the service center, and the service center will bill the mobile consumers together with their monthly service charges at the end of the month. The major benefit of this payment method is that consumers do not need to provide any confidential personal information to the merchant through the mobile network. Instead, a highly secure payment channel—a dedicated network—is used in the payment process.

Database

Within the service center, there are a number of databases—content database, license database, bill database, and user database. The content and license databases are additional to the DoCoMo i-mode model, while the bill and user databases are not. These databases hold necessary information for various processing and operations, such as transaction, payment, and DRM.

The content database contains all the downloadable content provided by all official IPs (and perhaps unofficial IPs as well). The downloadable files are transferred to and held in the content database before purchasing and transactions. When a digital file is requested by a mobile user, the requested media will be retrieved from the content database and delivered to the user.

The license database holds license documents for all mobile users. Each license document states the owners of the content—creator, buyer, borrower, together with terms and conditions for use.

The billing database keeps records of all transactions, including information about the seller and buyer, together with the transaction date and charges.

The user database is a database about all registered mobile users—their personal and payment information. A mobile user must register with the mobile operator before accessing the mobile network and experiencing mobile services. Hence, each mobile user has a record in the user database.

DRM OPERATIONS

In this section, we first present the basic operations of the general framework, then illustrate how rights insertion, rights enforcement, and music sharing are realized under the framework. With the current 2.5G mobile technologies, media services may only refer to music distribution.

Rights Insertion

The rights insertion operation is outlined in Figure 3.

Step 1: A digital music distributor uploads digital music to the content database at the service center. The service center will insert the IP's identity (this may refer to a distributor's ID or a content provider's ID) and the service center's ID into the music using digital watermarking. The operation can be treated as the rights insertion operation.

Step 2: A mobile user browses a music catalog from the music distributor via his/her mobile device. When the user decides to subscribe to the service and purchase specific music, the distributor will then confirm and acknowledge the user.

Step 3: The distributor formally notifies the service center about the order by providing the IP's and the customer's (mobile user) identities—the caller's phone number. The service center will then verify the information. A way to verify the information is to cross-check the connection ID between these two parties.

Figure 3: Basic operations of the general DRM framework

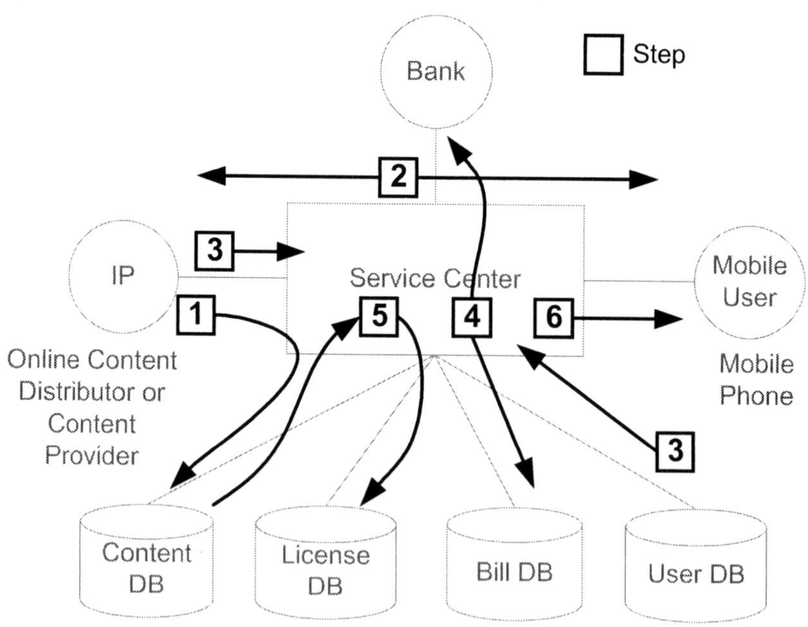

Step 4: The service center then updates the customer's monthly bill with the charge of the purchased music. A money-transfer process is activated to transfer money from the service center's bank account to the IP's bank account.

Step 5: The service center retrieves the selected digital music from the content database and inserts the customer's ID (as a digital watermark) to the digital music. This is also considered to be a rights insertion operation. The digital music becomes rights-protected music. A digital license is also generated and kept in the license database. The license contains the music usage agreement and other terms.

Step 6: The rights-protected music will be delivered to the customer via the packet network or the conventional voice channel (if a high-quality music channel, such as an MP3 channel, does not exist). This step can be executed immediately after the purchase and whenever the user requests it.

In the above steps, the rights insertion operation takes place at steps 1 and 5, where the distributor's ID or content provider's ID, service center's ID, and customer's ID are embedded in the media content.

Rights Enforcement

When a mobile user wants to listen to his previously purchased music, the user can make a request to the service center directly through his mobile device. Built-in software in the mobile phone can facilitate this. The service center will first verify the user's ID and the license terms. If the user has the right to listen to the music, the service center will execute step 6 and alter the license terms if a pay-per-view payment scheme is in use. This is regarded as an active rights enforcement operation. The active rights enforcement operation is transparent to the mobile user, as the operation takes place at the service center.

Passive rights enforcement is conducted by external parties and organizations. In Hong Kong, Customs and Excise officers administer the intellectual property law and are responsible for performing passive rights enforcement operations against any suspected copyright violation. The passive rights enforcement basically compares the embedded digital watermarks in the rights-protected music and the rights information kept in the digital license stored in the license database. Watermark extraction or watermark detection is used in the rights extraction or detection operation to obtain the embedded watermarks for verification. This is an offline operation, as depicted in Figure 4.

Music Sharing

Consider the case where User A wants to share his purchased digital music with his friend User B, with or without charge. Figure 5 shows the procedure to share rights-protected digital music from one user to another. Here are the required steps.

Figure 4: The passive rights enforcement operation in the general DRM framework

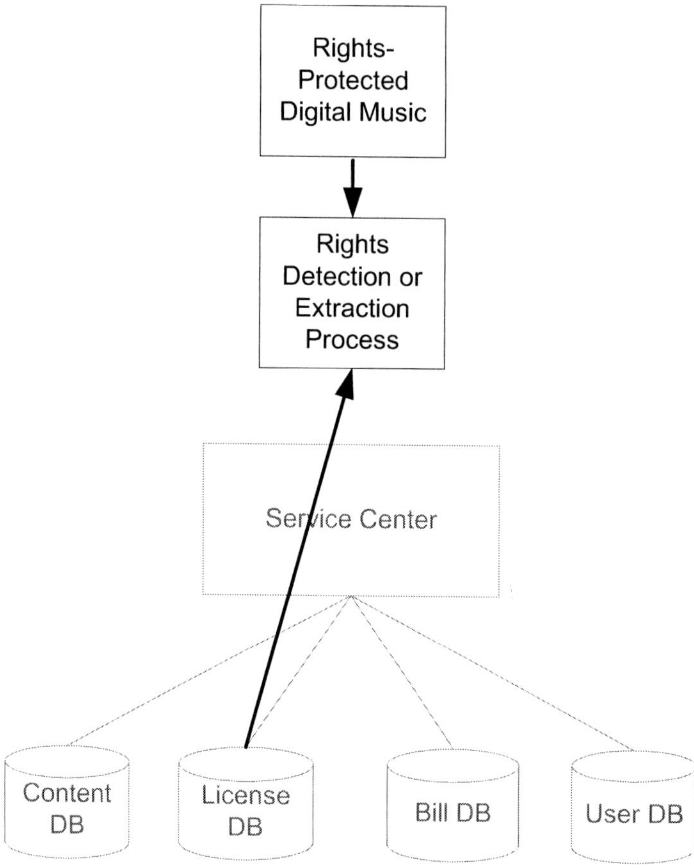

Step 1: User A informs the service center about his decision to loan his purchased digital music to another registered user—User B. This could be done in the portal site of the service center.

Step 2: The service center extracts the corresponding license from the license database and verifies its terms and agreements. The license terms must state that the purchased music is sharable before proceeding to the next step.

Step 3: If User A wants to charge User B for the usage, the service center may bill User B according to the instructions from User A and the agreement from User B. Otherwise, this will be skipped.

Step 4: The service center generates a "borrow" license for User B to enable User B to render the music. The "borrow" license enables User B to enjoy the music, but it may or may not be shareable with the third party, subject to the agreement specified by User A. The license for User A could be frozen if the original license prohibits concurrent use, while User B has the rights to render the music.

Step 5: User B can access and listen to the digital music, just like User A before.

It is noted that User B will not participate in the music sharing process if payment is not required. User A has the right to instruct the service center to transfer the ownership of his purchased digital music to another user as he wishes.

CONCLUSION

This chapter has outlined a general DRM framework for m-commerce. The framework is general in the sense that current 2.5G mobile technologies support this framework, but future 3G and even 4G mobile standards can also make use of it. Its applicability, usability and extensibility are justified, as the framework is based on existing DRM, mobile, and payment technologies. The concept of basic operations, such as DRM, payment and sharing, may remain unchanged, even as advanced mobile networks and mobile devices are adopted. The DRM framework will become more valuable when mobile multimedia really takes off in the future.

The proposed framework does not address some outstanding research issues. For example, user acceptance and user satisfaction with DRM have not yet been evaluated with real users.

Figure 5: Music sharing in the DRM framework

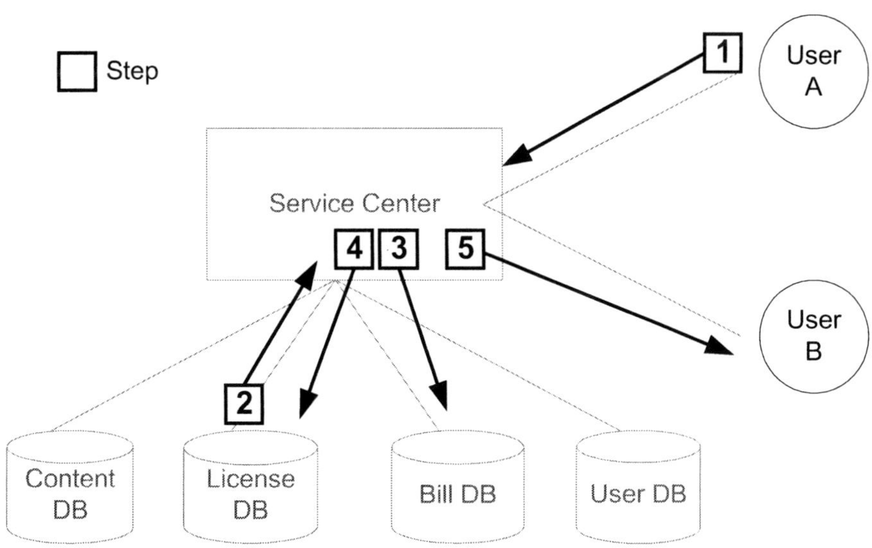

REFERENCES

Anderson, L. C., & Lotspiech, J. B. (1995). Rights management and security in the electronic library. *Bulletin of the American Society for Information Science, 22* (1), 21-23.

Anonymous (2000). The major players, partners in digital-rights management. Billboard, 112 (16), 103.

APRA. The Australasian performing right association limited. Available online at http://www.apra.com.au/index.htm.

Bormans, J., & Hill, K. MPEG-21 Overview, version 3. Available online at http://mpeg.telecomitalialab.com/standards/mpeg-21/mpeg-21.htm.

Cox, B. (1996). *Superdistribution: Objects as Property on the Electronic Frontier.* Addison-Wesley.

DoCoMo. NTT DoCoMo i-mode. Available online at http://www.nttdocomo.com/top.shtml.

ECyberPay. eCyberPay.com. Available online at http://www.ecyberpay.com.

EMMS. IBM's electronic music management system (EMMS). Available online at http://www.almaden.ibm.com/cs/madison.html.

InterTrust (2000). InterTrust, the MetaTrust utility, announces OpenRights Initiative. Intertrust press release.

ISIS. Intel® software integrity system enhances secure online distribution of music, documents, video and books. Available online at: http://www.intel.ca/ca/pressroom/releases/110999.htm.

Kwok, S. H. (2000). An enhanced license management model in digital rights management for online music business. *Proceedings of the International Conference on Information Society in the 21 Century: Emerging Technologies and New Challenges* (IS 2000).

Kwok, S. H., Wong, K. C., Tsang, K. F., Cheung, S. C., and Tam, K.Y. (2000a). Digital rights management in Internet open trading protocol (IOTP). *Proceedings of the International Conference on Electronic Commerce* (ICEC 2000), pp. 179-185.

Kwok, S. H., Yang, C. C., Tam, K. Y., and Wong, J. S. W. (2000b). An SDMI-based rights management system for electronic media using digital watermarking. *Proceedings of the International Conference on Electronic Commerce* (ICEC 2000), pp. 193-200.

Orange. Orangehk.com. Available online at http://www.orangehk.com/chi/index.jsp.

Ramanujapuram A., & Ram, P. (1998). Digital content & intellectual property rights. *Dr. Dobb's Journal*, pp. 20-27.

SDMI (1999). SDMI portable device specification, part 1, version 1.0. Available online at http://www.sdmi.org.

Stewart, T. (1998). *Designing Systems for Internet Commerce* (pp. 166-167). Addison Wesley.

VITAMINIC. News: VITAMINIC in WAP deal. Available online at: http://www.vitaminic.co.uk/news/0075.php3.

Windows Media. Microsoft Windows Media Rights Manager 7.1 SDK. Available online at http://msdn.microsoft.com/library/default.asp?url=/library/en-us/wmrm/htm/windowsmediarightsmanagersdk7.asp.

Chapter VI

Predicate Based Caching for Large Scale Mobile Distributed On-line Applications

Abhinav Vora, Zahir Tari, and Peter Bertok
RMIT University, Australia

ABSTRACT

Robust mobile middlewares are crucial for online applications as they provide solutions for the core issues of mobility, data interoperability, and security. This chapter describes our experience in designing such middlewares for one of the largest Australian transport companies (CMS Transport Systems). We focus on the design of a predicate-based caching technique for mobile object-based middlewares that optimises the performance of the mobile medium by better utilising the available bandwidth.

Several caching techniques have been proposed to improve system and application performance. Such techniques, along with consistency control mechanisms, are used to reduce the communication load between clients and servers, which is particularly important in wireless networks. Caching techniques are generally classified as either ID-based or predicate-based. In this chapter we propose a predicate-based caching scheme, in which the predicates are used in combination with updates and are broadcast by servers in a set of appropriate messages called cache invalidation reports. Each report/message contains information about the data items that have been updated in the server during a given period. A function mapping the predicate into binary representation is defined for each attribute. Because not all updates are relevant to a cache, there is a matching algorithm for detecting relevancy between the cache predicate and the predicates in the cache invalidation reports. The predicate-based cache invalidation reports inform the client cache manager concisely about items that need to be refreshed and about those that need to be discarded, and ensure efficient bandwidth usage.

MOTIVATION

With mobile computing devices becoming smaller and faster coupled with networks supporting higher bandwidth and reliability, mobile computing is playing an increasingly important part in how applications are designed today. Handheld devices enable users to actively participate in distributed computing while on the move. Communication in such distributed, in many cases wireless, environments is hampered by intermittent and weak connections; in particular, mobile devices encounter wide variations from high bandwidth, low latency through to low bandwidth, high latency and to possibly no connectivity at all (Forma and Zahorjan, 1994).

To avoid high cost of connection or overcome availability problems mobile clients often deliberately disconnect for some time and then reconnect to the network. It is as essential for mobile clients to keep operating while disconnected as it is to support operations when the connection is weak, since there may be clients that never get good connection and still need to operate. Caching is one technique that can be used to support disconnected operations as well as improve performance for mobile clients. Caching objects not only improves response time but also reduces the number of messages between clients and server during operation, which is particularly important due to the high costs and unreliability of connections in a wireless network.

Mobile and portable computing devices have relatively limited memory, processing, and power resources. This limits the ability of the mobile device to cache a large quantity of data. Caching and cache consistency algorithms need to take these factors into concern. Reducing the number and size of messages used for maintaining cache consistency can lead to performance improvement in the form of reduced battery and processing requirements in addition to the saved wireless bandwidth.

Existing caching techniques can be broadly classified into ID-based and predicate-based. Both use cache invalidation reports (server broadcast messages), which are used by the clients for maintaining the consistency of their cache. In addition to other information, a cache invalidation report contains a list of identifiers of data items that have been modified since broadcasting of the last message. ID-based approaches only tell clients which data items have been modified and do not contain any information about the new values. Clients need to query the server for new values before they can determine whether updates are relevant to them or not. If the size and the number of the cache invalidation reports are large, this approach can become impractical and impeding on the wireless environment. Predicates are used to represent aggregate states of updates in a server in the predicate-based approach. Mobile clients submit a query to a server in the form of a predicate (e.g., age < 20 and address = "Melbourne"). The server returns all the data that matches the predicate. Cache invalidation reports contain a timestamp and a list of predicate representations that reflect the current content of the objects that have been modified since the last report. These reports also contain a binary representation of the range of the new value (the client can use this to determine whether the update is relevant to it or not). The size of these messages is smaller than ID-based messages, thus making them an ideal choice in mobile environments.

Middlewares (Tari and Bukhres, 2001) provide the right software infrastructure for mobile environments, such as e-commerce/m-commerce, because they open the boundaries between intra-enterprise and inter-enterprise systems. Middlewares unite different applications, tools, networks, and technologies, giving users a common set of standard interfaces that hide the underlying implementation details. A mobile middleware is an enabling layer of software that connects m-commerce applications with different mobile devices, networks and operating systems while still preserving mobility transparency. However, current middleware platforms have profound limitations when used as a basis for m-commerce applications due to their inability to properly address the issues of mobility, performance and security. This chapter deals with the issue of performance of middleware platforms, i.e., how to make the processing of m-commerce applications efficient in terms of response time.

Object brokers, such as CORBA (OMG, 1995) and DCOM (Chung et al., 1998), are middlewares which offer the best integration of advanced software and data paradigms (e.g., object paradigm) in distributed environments by providing modularity, reusability and transparent handling of complex issues related to object binding and information processing. Object brokers are based on the integration of the distributed client-server computing (i.e., based on message-passing systems most commonly found in Unix-based environments) and object-oriented programming. An object broker plays the role of an object-oriented remote procedure call (RPC) application program interface. It provides common services, such basic messaging and an RPC-type communication between clients and servers, directory services, meta-description and location and host transparency. This integration has several advantages, such as transparency and extensibility. Different transparencies are supported (e.g., location, host) as clients do not need to be aware about where objects are located or how to interface their applications (with legacy code), or how to communicate with remote objects. Extensibility is the result of decoupling interfaces with implementations. Therefore existing legacy applications can be extended by inheriting from existing interfaces to support new requirements (whether it is new software requirements, such as a new language, to be supported by the new applications or changes in the requirements).

A detailed discussion and analysis of object brokers can be found in Tari and Bukhres (2001). This book provides a technical insider's view of all of the most technical issues related to object brokers, including architecture (e.g., portable adapter), performance (e.g., caching) and object services (e.g., trading, transaction and query).

Existing object brokers provide scalable solutions for distributed applications; however, they have severe limitations when dealing with m-commerce applications. The aim of the DOK project, which is currently under development at RMIT University, is to analyse and understand these limitations and to propose innovative solutions to deal with core issues of m-commerce applications. DOK intends to

- Extend the communication layer of object brokers, called IIOP (Internet Inter-Object Protocol), to deal with mobile objects, object consistency and disconnection.
- Extend object brokers to efficiently manage diverse data sources by designing advanced techniques for object prefetching and object pooling.
- Design sophisticated optimisation techniques, such as cooperative caching and dynamic load balancing, and integrate them with object brokers to provide better performance and scalability.

This chapter deals with improving the performance of mobile object brokers by caching data on the mobile clients' side. m-commerce applications that might want to use caching include transportation and logistics software managing and assigning deliveries and pick-up, clients having their schedule periodically updated from a control centre, emergency services, remote banking terminals (Mobile Funds Transfer System), etc. Caching data in the client environment offers substantial performance gain by reducing the need for remote access to data repositories and query processing.

The mobile object broker of the DOK system supports m-commerce applications by enabling communication across different software and database platforms, and by providing high performance and availability. This chapter describes a predicate-based caching technique with a server broadcasting approach, which uses predicates to evaluate updates in the cache invalidation report. A predicate mapping function is associated with each attribute, which produces a binary representation of the attribute. To detect the relevancy of an update, an algorithm matching the cache predicate and the predicates in the cache invalidation report is used. Predicate-based cache invalidation reports inform the client cache manager of items that need to be refreshed and ones that need to be discarded, resulting in efficient bandwidth usage.

In this chapter, we use the term *server* to refer to application nodes. A server could be a database server, application server or any type of service that maintains data and serves it to clients. The term *client* is used (interchangeably) to refer to mobile hosts, client application and client cache manager. A client cache manager is a part of the client software that manages the cached objects and is responsible for maintaining cache consistency.

This chapter is organised as follows. The next section gives an overview of caching in mobile-oriented systems. Section 3 reviews current caching techniques in wireless environments and puts our cache consistency approach in context. Section 4 explains the predicate representation and describes the structure of the proposed predicate-based invalidation report. Section 5 proposes a cache consistency protocol, and section 6 concludes the chapter.

BACKGROUND ON CACHING IN WIRELESS ENVIRONMENTS

A General Picture

Wireless environments are characterized by low bandwidth, high latency, being subject to network interference and frequent disconnection, which results in weak connections. Due to these limitations, it is important to reduce bandwidth requirements and to increase accessibility of data for mobile clients. Caching is one of the solutions enabling improved performance and reduced usage of bandwidth: by caching frequently accessed data objects at the mobile client side the communication between mobile clients and servers can be significantly reduced (Chan, SI, and Leong, 1998; Wu, Yu, and Chen, 1996). At the first request for a data item the server sends the data to the client. Subsequent requests for the same item can be satisfied locally, which reduces response time and conserves bandwidth. In addition, caching enables continuing operations at the mobile client while disconnected, since the mobile client can access data objects locally without communicating with the server.

A cache invalidation strategy ensures that updates to data on the server are consistent with the mobile clients' cache (Chan, SI, and Leong, 1998). In a wireless environment, the design of a cache invalidation strategy needs special attention, so that substantial reduction in bandwidth usage can be achieved. Several cache invalidation strategies have been proposed, which can be divided into two categories (Barbara and Imielinski, 1994): *stateful* and *stateless* methods. The stateful-server technique requires the server to know which clients are currently connected, and each client's cache state has to be known to and maintained by the server at all times. On the other hand, a stateless server does not need to know which clients are connected and what the state of the client's cache is. The responsibility of maintaining the cache relies with the client. Since mobile clients disconnect and reconnect frequently, the stateful server approach is impractical for wireless environments. The stateless server approach is more scalable and practical for wireless environments because the server does not have to keep track of which clients are connected or store the cache information of each connected client.

In both stateful and stateless cases, cache consistency needs to be addressed. Server broadcasting is an efficient method for maintaining cache consistency, since the cost of broadcasting a message can become independent of the number of clients receiving the message when using appropriate data transfer protocols. The stateless server approach broadcasts update information to clients in a message, called a *cache invalidation report.* Depending on the scheme used, invalidation reports use a specific data structure (to reduce the size of messages to be broadcasted) and contain specific information (needed to support the underlying consistency method). Figure 1 depicts the structure of a cache invalidation report used in one of important caching scheme, *Selective Dual-Report Cache Invalidation Scheme* (SDCI) (Kian-Lee et al., 2001). In this scheme, a pair of invalidation reports, an *object invalidation report* (OIR) and a *group invalidation report* (GIR) are sent every *L* time units. The GIR is a triplet of the form *(group-id, TS, ptr)* where *group-id* represents the group identifies, *TS* is the timestamp of the most recent update (excluding those in the OIR) of the group, and *ptr* is an offset to the starting position of the objects in OIR corresponding to the group identified by group-id. Each OIR contains a list of updated objects (in that specific group) and their timestamps.

There are different techniques that can be used to organise cache invalidation reports. A client uses the information contained in the cache invalidation report to keep its own cache consistent with the server's data. One way to do so is by comparing the signature or the timestamp of the item in the cache invalidation report with that of the locally cached data. The problem with this approach is that a report contains only information relating to which items have been updated on the server (and not their new values), and therefore this won't help a client in determining whether or not the updates are relevant to its cached data. Thus, the client would need to make remote invocations to the server to find out the new values. After the remote invocations are performed, if the client determines that the update is not relevant, then the remote invocations were unnecessary and therefore should not have been performed.

Figure 1: An example of structure of invalidation reports (SDCI model)

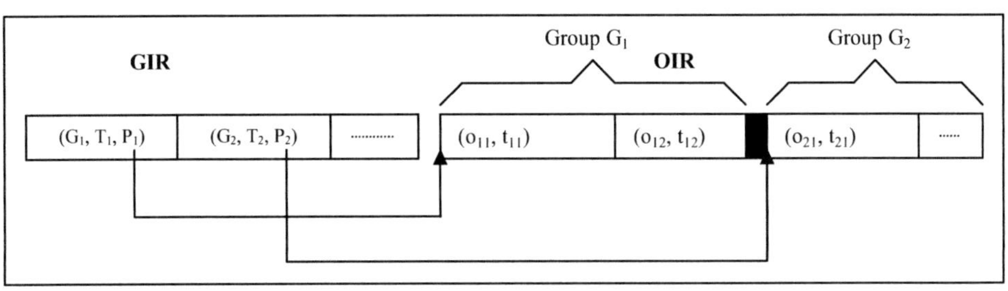

Figure 2: Mobile computing environment

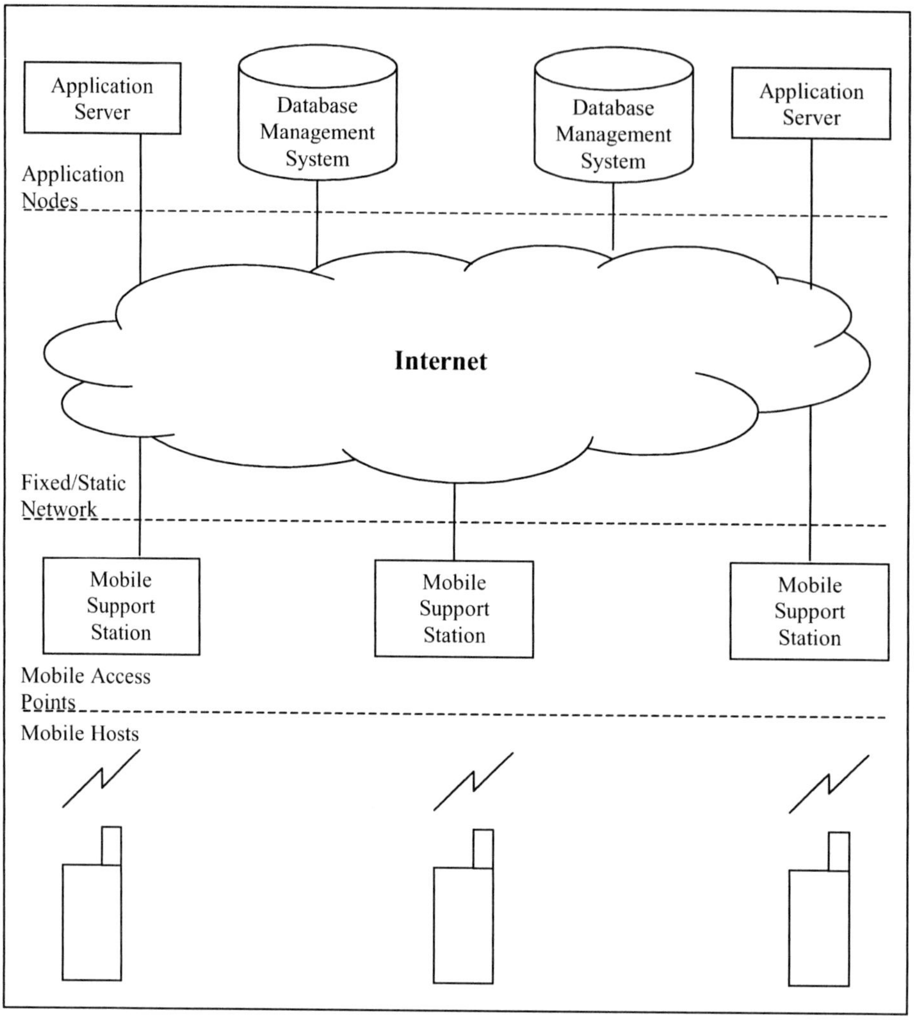

Example 1: *A mobile fund transfer system that caches a copy of customers' accounts might have a requirement that it only be notified of updates of customers' account balance, such as if the balance of an account is reduced. In existing caching solutions, even if the balance of an account is increased, the client using this account will still need to access remote data to check the validity of the updated information (before determining whether or not it is relevant). Our approach, however, does avoid requiring the client to access remote servers as the cache manager checks the predicate in the invalidation report and finds out that the update is not relevant (because the balance increased instead of decreased).*

A generic mobile computing environment is depicted in figure 2. In this environment, the Mobile Hosts (MHs) query and interact with the database servers and application servers that are connected to a static network. The mobile hosts communicate with the servers via a wireless cellular network consisting of Mobile Support Stations (MSS). A mobile host can be in either of two modes: awake or asleep. When the mobile host is awake (i.e., connected to the servers), it can send and receive messages. A MH can't be disconnected from the static network either voluntarily (to save network usage, battery, etc.) or involuntarily (unable to access the wireless network, etc.). We will not differentiate between the type of disconnection for any purpose.

Figure 3 illustrates the system components required to support caching in mobile environments. A preassigned static host (the mobile support station) of any mobile host maintains its Home Location Cache (HLC). If a mobile host is roaming, its HLC is duplicated at the MSS of its current cell. Thus, an MSS always maintains an HLC for MHs in its coverage area. HLC is a list of records (*x, T, invalid_flag*) for each data item locally cached at an MH, where *x* is is the identifier of the cached data item, *T* is the timestamp of the last invalidation of *x*, and *invalid_flag* is set to true for data items for which an invalidation has been sent to the MH but no acknowledgement has been received. The mobile host consists of at least a cache manager (client cache manager) that is responsible for maintaining the consistency of the cached data items. The cache manager also tries to provide mobility transparency to the client application (for example, by retransmitting a request if there are connection problems).

Figure 3: Caching system architecture

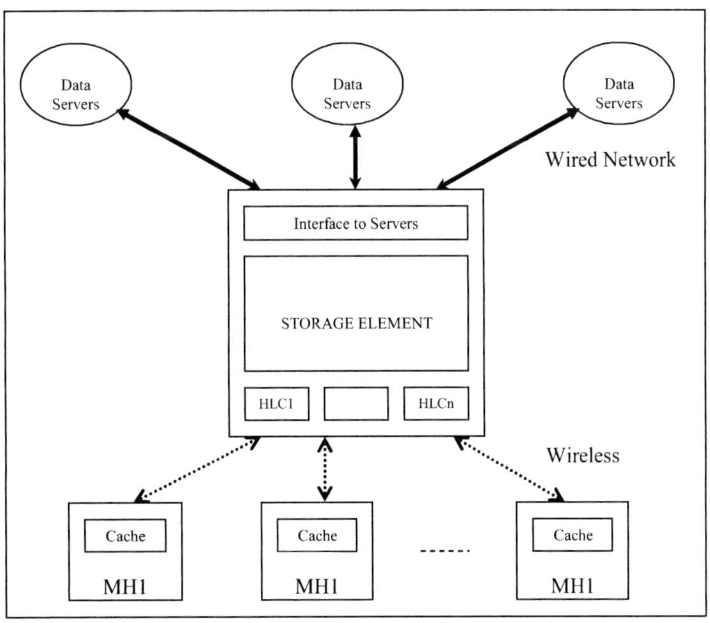

A mobile host can be disconnected (either voluntarily or involuntarily) from the static network. While disconnected (for whatever reason) the mobile host (client) satisfies application requests from the local cached data. Caching data on the client requires a consistency protocol that can ensure that cached data is consistent with the server data. For a consistency protocol to be suitable for wireless environments, it is important to minimize the number and size of messages exchanged between the mobile host and the application nodes for maintaining consistency of the cached data, to allow stateless servers, and to minimize the bandwidth requirements.

Updates are handled differently in different schemes. In general, when a client modifies a data item, this update needs to be propagated to the server at a certain time. All schemes do not guarantee that if a client updates a data item while disconnected, the update will be accepted by the server after reconnection. This is because the cached data item might have been modified by another client in the server side while the original client was disconnected. Therefore, the server needs to ascertain the validity of the client data before an update is accepted. Detection-based approaches (e.g., Franklin et al. & Adya et al.) defer the validity check of the data until commit time (which means that updates can be rejected by the server and the client would have to roll back the updates it applied), whereas avoidance-based approaches (e.g., Agarwal et al. & Goodman) check the validity of the data before it is accessed by the client (thus guaranteeing that the updates would not be rejected by the server).

In this chapter, we propose a cache invalidation strategy for wireless environments that substantially reduces both the amount of data transmitted by the server (as it includes information about the new values of updated items in the cache invalidation report) as well as the granularity of cached data from object to attribute. In the proposed approach, a server periodically broadcasts invalidation reports to clients so they can maintain consistency of their cached data. In order to reduce the information sent to (client) cache managers, cache invalidation reports contain only minimal information that enables these cache managers to make decisions (e.g., check whether the changes made at the server are relevant to the objects recorded in their caches). Relevancy in this context means that a cache manager will only consider those updates that are related to the cached information. For instance, when some attributes of an object are updated on the server side, the cache manager will discard the update if the update does not relate to the attributes cached by the client.

To allow relevant updates being detected by a cache manager, the new state of the updated data is included in the cache invalidation reports. In this chapter we propose an appropriate predicate-based data structure to model these reports. It not only informs each client about the attributes that have been modified in the server side but also includes the range of the new value. The latter information helps clients avoid requesting copies of attributes unnecessarily since a client is able to check the value of the updated data. We also incorporate object identifiers (OIDs) of objects whose attributes have been modified in the cache invalidation reports. Finally, relevancy is detected by matching the query predicate with the updated predicate (included in the invalidation report).

EXISTING CACHING TECHNIQUES

This section looks at the two main approaches for consistency control in wireless environments and puts our contribution in context. There are two basic approaches: *ID-based* and *predicate-based.* Both use the concept of cache invalidation report, which employs a message that a server periodically broadcasts to clients. Clients are responsible for maintaining the consistency of their caches. A cache invalidation report contains a list of identifiers of data items that have been modified in the server since the broadcasting of the last message. In addition to these identifiers, other information may also be added to a cache invalidation report to enable a (client) cache manager to check whether a client is aware of the new update (e.g., a report can append a timestamp to an identifier). The cache manager evaluates the combined information by comparing it with the cached information. The result indicates whether the version of the cached information is out-of-date or up-to-date.

ID-Based Server Broadcasting Approaches

Several techniques using server broadcasting have been proposed for wireless environments (Barbara and Imielinski, 1994; Hu and Lee, 1997; Jing et al., 1995; Lynch, 1999). These techniques differ in the way the content of a cache invalidation report is structured and how cache consistency is restored when clients reconnect to the server.

- In Barbara and Imielinski (1994), Timestamps (TS), Amnesic Terminals (AT), and Signature (SIG) techniques are presented. These techniques are designed for connections of different durations; hence, depending on client connection, relevant technique can be used.

  The TS approach uses timestamps to allow a client to check its awareness of the most recent updates on the data items. The server timestamps each report with a timestamp as the initialisation of the broadcast. A client verifies the freshness of its cached item by comparing its timestamp with that of the item in the cache invalidation report. For instance, if the last report is broadcast with a timestamp T_i and the client determines that the cached item is still valid after receiving the report, then the cache gets timestamped with the value T_i (marked as valid up to this time). In case the cache is invalid, the obtained copy will have a timestamp of this last request. In this approach, the server agrees to notify clients (or mobile units) about data items that have been modified within a given (fixed) period.

Figure 4 shows the broadcast of invalidation messages in the TS and AT schemes. *W* is the window size (time) between invalidation messages, and L represents the message latency. AT broadcasts only the identifiers of items updated since the last report. The server has the obligation of informing clients about the identifiers of items that are changed since the last invalidation report. As in TS, the server builds a list of items to be broadcast. Upon receiving the invalidation report, a mobile unit compares the items in its cache with those of the report. The client (or mobile unit) removes items whose identifiers appear in the report from its cache. The client cache manager will have to request a fresh copy of the data item (from the server) when the client application program wishes to reference the item. The client considers cache items not mentioned in the report to be valid; i.e., they require no further communication with the server.

SIG is based on data compression techniques for file comparison. It computes checksums of the values of items, called *signatures*. This technique computes one signature per item (or per page) and a set of combined signatures calculated as exclusive OR of the individual checksums. As for TS and AT, the server periodically broadcasts invalidation reports, i.e., these combined signatures. By comparing the signatures of the items in the invalidation report and in the cache, the client (or mobile unit) determines whether the cached copy is up-to-date.

With these techniques every client (or mobile unit) maintains a variable T_{last}, the timestamp of the last report received. If the difference between the current report timestamp (broadcast from the server) and T_{last} is greater than the fixed broadcast period, then the entire cache is dropped as the items in it are not valid and the entire cache is refreshed by fetching data from the server.

- Bit-Sequence scheme (BS) is proposed by Jing *et al.* In this technique, a set of timestamps is associated with a sequence of bits, which is included in the cache invalidation report. The association between the timestamps and the bit-sequences gives information about the freshness of the client cache. While BS is more bandwidth efficient in handling disconnection than the techniques in Barbara and Imielinski (1994), it is more complex and its cache invalidation report is much larger.
- Other techniques attempt to provide a more efficient caching support based on the techniques described above for mobile clients after reconnection. The improved techniques either modify the TS and AT algorithms (Hu and Lee, 1997) or use a combination of different cache invalidation reports, such as TS and BS, adaptively (Lynch, 1999). The former technique includes an update history window to reduce the likelihood of invalidating the entire cache upon reconnection. The adaptive techniques change the type of cache invalidation report according to system status, e.g., according to query/update pattern and disconnection frequency.

*Figure 4: Barbara and Imilienski's TS/AT scheme (*L*: broadcast latency;* W*: window size (time between invalidation reports);* k*<1 for the TS scheme and* k*=1 for the AT scheme)*

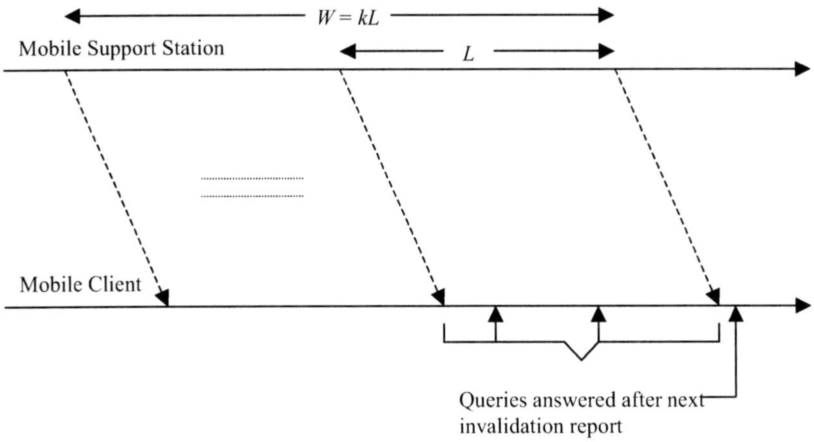

In general, ID-based server broadcast techniques can work reasonably when the size and the number of messages is small. They are, in general, suited for ID-based operations, such as ReadObject (ID) and WriteObject (ID). However, they have serious limitations when the number and/or size of messages exchanged between clients and servers is large. These techniques use cache invalidation reports in a limited way, as these reports only tell clients (mobile units) which data items are changed and do not contain any information about the new values. As a consequence, clients will not be able to check whether the new update is relevant, as the new updated value is not supplied. The client has to contact the server to find out the update information. Because the size of the cache invalidation report (as well as the number of these messages) is large, the whole ID-based approach can become impractical. However, if a cache invalidation report contains extra appropriate information about the update of the data item, this will help clients avoid refreshing the cached data item just because it has been modified. The client can judge from the extra information whether the update is relevant to it or not, and based on that it can refresh the cache.

Predicate-Based Approaches

Predicates have been suggested to represent aggregate states of updates in a server and could be used for cache consistency control (Barbara and Imielinski, 1994; Forma and Zahorjan, 1994). Using predicates to query items in the cache and also including predicates in the cache invalidation report can be combined with adding a timestamp to the predicate similar to the ID-based TS techniques. A predicate with a timestamp in the cache invalidation report informs mobile clients that at least one object that satisfies the predicate has been updated on the server at the specified timestamp. No objects are updated after that time.

A mobile client initially submits a query in the form of a predicate. The server processes the query using an appropriate access method (e.g., index) and returns all data items that satisfy the query predicate. The server periodically broadcasts cache invalidation reports to all mobile clients in its coverage area. When a client receives a report, it retrieves the predicate that matches that of the cache and compares the timestamps. If the timestamp in the report is more recent, the client has to send the predicate as a request to the server and retrieve all data objects for that predicate. The client (or mobile unit) then caches information with the timestamp of the last update of the cache content. The cache invalidation report contains a timestamp and a list of predicate representations that reflects the current content of the attributes of objects that have been updated since the last cache invalidation report.

If there are many objects satisfying a predicate and only few items are updated, retrieving all items of the predicate results in unnecessary information requests, wasting scarce bandwidth in the wireless channel. Furthermore, it is not practical to include all possible predicates in a single cache invalidation report when the number of bits in the predicates is large. An improvement of such a strategy is to select a subset of predicates to be included in the report. This set has to cover all data items in the server for completeness. The tradeoff is that exact matches of the predicate in the report and the predicate representing the client's query are no longer guaranteed. Without exact matches of predicates, there is a greater probability of a false alarm, when the client thinks a data item is out-of-date but it is not. When a false alarm occurs, the client would present a predicate to the server and then realise that the items are still up-to-date and the retrieval was unnecessary.

Example 2: *If a client caches 2 objects, say x and y, of a certain type that satisfy a certain predicate, though the same predicate actually matches 3 objects at the server (objects x, y and z). Now if object z is modified at the server, an invalidation report matching the predicate would be sent. The client would have to update its cache by sending the predicate as a request to the server and retrieving updates. This would result in a wasted retrieval since the client's cache was already consistent (since object x and y were not modified).*

Our Approach: The Predicate Match (P-Match) Approach

Predicate-based approaches detailed earlier have serious limitations, as they do not address the issue of exact match guarantee in a practical way. Using predicates to represent aggregate information about a set of items is not appropriate, as it is not possible to know which items have been updated on the server. A client will therefore need to request the content of these items from the server and then discard the information because the updates are not relevant.

It is common that a (mobile) client requests an item and keeps it in its cache while only a few parts of the item are of interest to the client's application program. This occurs when an object is chosen as the unit of data request. In such a case, time and bandwidth are wasted on transmitting the unused part of the object. In addition, the unused part of the object occupies some of the cache space that could be used for more relevant information. In terms of bandwidth and cache space usage, it is more efficient to perform caching at the attribute level, so the client requests and caches only a minimal amount of data, such as the new updated value of an attribute. To cache at attribute level, the update information in the server should also be reported at the attribute granularity.

Figure 5 shows the architecture of the proposed predicate-based caching scheme. Mobile clients submit a query in the form of a predicate (step 1 in the figure). The server returns the results for the query to the client (step 2). The contents of the client cache are represented by a predicate, called the cache predicate. The cache

Figure 5: The architecture of our caching scheme (where circled numbers represent the different steps in the caching process)

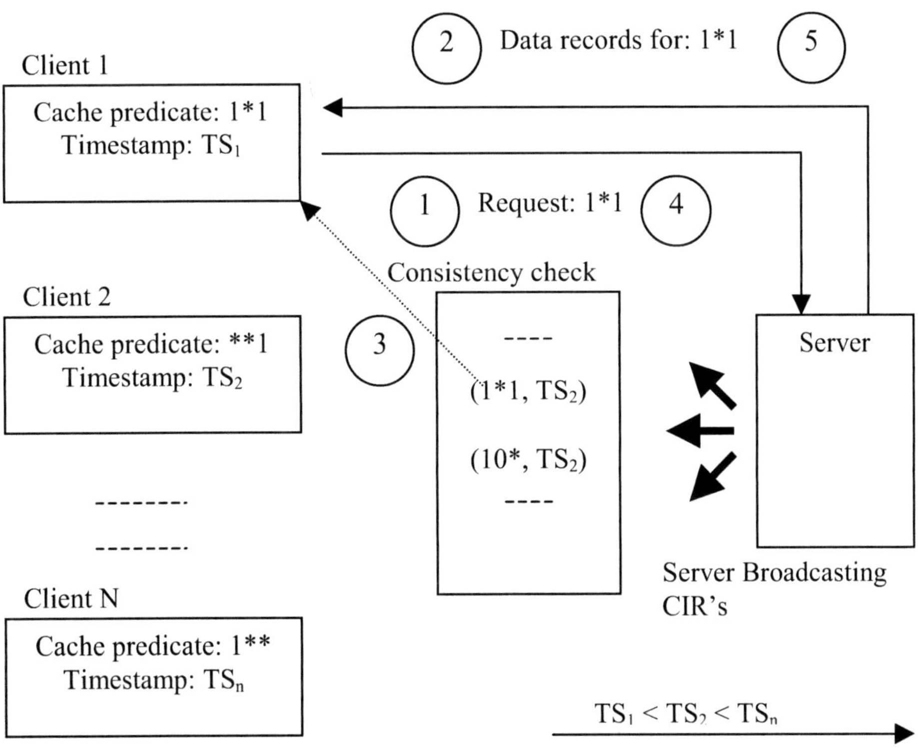

predicate corresponds to the submitted query. For example, if the cache predicate of a client is *'1*1'*, then the client caches data records satisfying the predicate *'1*1'*. The timestamp of a cache predicate is the last update time of the cache. The server broadcasts Cache Invalidation Reports (CIR), which are organised with predicates and their timestamps.

A mobile client receives the CIR (Cache Invalidation Report) and retrieves the corresponding predicate, which is matched with the client's cache predicate. Then, the client compares the timestamp of the cache predicate and that of the matched predicate (step 3). As depicted in Figure 5, since the timestamp of the cache predicate is less than the timestamp of the matched predicate ($TS_1 < TS_2$), then the client cache needs to be updated. Consequently, the client requests new data records for *'1*1'* (step 4), and the server returns the new data records to the client in step 5. The timestamp of the cache predicate is updated to reflect the new TS, i.e., it is updated to TS_2. Steps 3, 4 and 5 are repeated for all clients.

All updates performed on the server side need to be reflected in cached objects; however, several of these updates do not create any consistency conflicts. For example, an application performing calculations about profits made does not need to know where the stock of a product is located, and any change in the location of the stock is not relevant for this client. However, this update is important for an application that keeps track of stocks of products in the company.

We have designed a predicate-based cache invalidation report that reflects the updates made to attributes. This approach introduces a multiattribute mapping table for every object type. This data structure provides a binary representation for a range of attribute values, which is included in cache invalidation reports broadcast to clients. One of the advantages of this binary representation is that it is smaller in size than the actual value, which makes it more suitable for transmission over wireless networks. At the same time, it provides information relating to the new values of updated attributes for relevancy judgements by the cache manager. The proposed predicate-based report structure allows the detection of broadcast update relevance to cache information by a simple predicate matching technique. The predicate matching technique can inform the cache manager which cached attributes need to be refreshed.

CACHE INVALIDATION REPORT (CIR)

Predicate-Based Representation

This section describes the data structure we have designed to model the content of attributes so that (i) relevant updates can be "picked" up by clients and (ii) less bandwidth is used to broadcast cache invalidation reports. Each attribute is associated with a mapping function. By applying the mapping function to the value of the attribute, the resultant bit streams become the binary representations of the value of the attribute. The concatenation of the binary representation of all the attributes of an object can therefore represent the content of the object. Such representation gives a general picture of object content instead of providing its exact value, but it is smaller in size than the exact values.

We illustrate the use of the attribute mapping function on a "Product" object, which consists of four attributes: A_1: sales amount, A_2: price, A_3: stock quantity, A_4: location. Each of the attributes has a mapping function associated with it. Let F_1 be the mapping function for attribute A_1 "sales amount", F_2 for attribute A_2 "price", F_3 for attribute A_3 "stock quantity", and F_4 for attribute A_4 "location."

$$F_1(x) = \begin{cases} 00 & \text{if } x < 100 \\ 01 & \text{if } 100 \le x < 200 \\ 10 & \text{if } 200 \le x < 500 \\ 11 & \text{if } 500 \le x \end{cases} \quad (1)$$

$$F_2(x) = \begin{cases} 000 & \text{if } x < 10 \\ 001 & \text{if } 10 \le x < 50 \\ 010 & \text{if } 50 \le x < 100 \\ 011 & \text{if } 100 \le x < 200 \\ 100 & \text{if } 200 \le x < 500 \\ 101 & \text{if } 500 \le x < 1000 \\ 110 & \text{if } 1000 \le x < 1500 \\ 111 & \text{if } 1500 \le x \end{cases} \quad (2)$$

$$F_3(x) = \begin{cases} 0 & \text{if } x < 10 \\ 1 & \text{otherwise} \end{cases} \quad (3)$$

$$F_4(x) = \begin{cases} 00 & \text{if } x = \text{"Sydney-City"} \\ 01 & \text{if } x = \text{"Sydney-Airport"} \\ 10 & \text{if } x = \text{"Melbourne-City"} \\ 11 & \text{otherwise} \end{cases} \quad (4)$$

An example of an object of type Product is given with the following attribute values:

Attributes' values	Binary representation	
$A_1 = 199$	$F_1(A_1) = 01$	
$A_2 = 49.95$	$F_2(A_2) = 001$	**(5)**
$A_3 = 10$	$F_3(A_3) = 1$	
$A_4 =$ "Melbourne-City"	$F_4(A_4) = 10$	

The predicate-based representation of this Product object becomes "01 001 1 10." Although the bit stream does not give the exact value of the object attributes, we can infer from the bit-stream that *between 100 and 200 pieces have been sold, the current price is less than 50, there are more than 9 pieces in stock, and the product is located at the Melbourne City Store.*

This information is useful for the client cache manager to determine if the value of some attributes has fallen into the value range that is relevant to the client's application, e.g., stock level is too low.

The proposed predicate-based representation can be improved by using only the part of an object that has been updated, i.e., the cache invalidation report includes the predicates of modified attributes together with additional information, such as (i) the attributes to which predicates are applied and (ii) the order in which the attributes are represented. Details of the proposed data structure for cache invalidation report are presented in the next section (section 5).

The length of the binary representation is proportional to exactness: the more exactly a predicate reflects an attribute, the more bits are required to represent the value. (The number of bits, k, required for n possible ranges is $2^k = n$, where $k = \log_2 n$.) The proposed data structure is scalable in terms of increasing the number of value ranges for more precise update information. In general, a predicate mapping function takes an attribute value as parameter and returns a binary value representing the range in which the value falls. Let F be the mapping function, x be the value of a data item, and P be the set of binary representations for all the possible values of x. The general definition of the attribute-mapping table is as follows:

Definition 1: *Attribute Mapping*
Let x be in the mth range of all the possible value ranges for the attribute and m □ (0, 1, 2, …, n-1). The value of the yth bit of the binary representation for x, where y □ (k-1, k-2, …, 0), is as follows: $F(x)[y] = (m - \square_{j \square y \square 1}^{j \square k} (F(x)[j] * 2^j)) / 2^y$ and $F(x)[k] = 0$. **(6)**

A concatenation of predicate representation (see above product example for the predicate representation) for objects creates predicate streams that can be broadcast. Client and server each has to keep a copy of the mapping table so that predicate streams can be constructed and interpreted.

Structure of the Cache Invalidation Report

This section describes the components of the proposed cache invalidation report. A typical data item is an object, which consists of a set of attributes and has a unique object identifier OID. An OID represents an object instance and identifies the object being updated. To represent all N attributes of an object being updated we need N bits: each bit corresponds to an attribute and the value is 1 if the attribute has been modified, 0 otherwise. The predicate representing the current value of each modified attribute is then appended to the end of the N-bit flags in the same order as follows:

$$[\text{OID}, N\text{-bit-flags}, \sum P(i)] \quad \textbf{(7)}$$

where N-bit-flags(i) = 1.

Using again the previous example, after a sale the attribute "amount of sale" of product OID-i is changed to 200 and the number of products left becomes 9. The N-bit flag is therefore 1010, indicating that the 1st and the 3rd attributes have been updated. The cache invalidation report for this object becomes:

OID-i 1010 10 0 **(8)**

The entire report is then the concatenation of such each individual report plus a timestamp. The OID in the cache invalidation report serves two purposes. First, it is used for identifying the object that has been modified. Second, the type information conveyed by the OID allows the client to find the corresponding functions and interpret the changes in attributes.

When the client cache manager receives a cache invalidation report, it checks the OID-i to find out basic type information about the object, such as set of attributes and their order. This information is used to interpret the N-bit flag that corresponds to the set of attributes and their order. The 1st part of the N-bit-flag tells the cache manager which attributes of the object have been updated in the server. The cache manager checks if updates of those attributes are relevant. If the cache invalidation report contains relevant updates, the cache manager uses the corresponding predicates to check whether the predicates match the client's cache query. Details of the predicate matching algorithm are provided in the next section.

CACHE CONSISTENCY

This section describes the cache consistency protocol. We consider client data to be consistent with the server data as long as the server data satisfies the client predicate condition. If the data cached by the client is updated and if the updated value is within the client specified threshold (client predicate), we consider that client cache to be consistent with the server data, *for example, if the client caches a copy of an object with attribute age = 10 and the client predicate is (age < 18). Now if this attribute were to be modified to 11, we would consider the client data to be consistent with the server data (even though in the classical sense the data is inconsistent) as long as the predicate (age < 18) is not violated.*

Typically, the cache manager performs the following steps to enforce consistency:

- If a client receives a cache invalidation report, it checks each item in the report.
- If the timestamp of a predicate included in the cache invalidation report is older than the cached predicate, the client cache manager ignores the report.
- Otherwise, if a predicate matches the predicate of a query, the attributes in the matched set are added to the set of attributes to be requested from the server later. These matched attributes may or may not be in the cache. If they are in the cache, the attributes have become outdated. If they are not in the cache, the attributes have become relevant to the client after having been updated at the server. The cache manager, therefore, is able to detect attributes that have recently become relevant.
- If a predicate does not match the predicate of the cache query, any attributes specified in the predicate must be purged from the cache if they exist there. Attributes included in the report but not in the cache are ignored.
- Finally, the timestamps of the remainder of the client's cached attributes are updated to the current timestamp.

For two predicates to match each other, two criteria have to be satisfied. First the predicates must be relevant, i.e., the set of attributes on which the predicates are defined must overlap. Second, each attribute in the intersection of the two relevant predicates must have the same binary value, i.e., they share the common value ranges of the attribute.

The query predicate also has type information and *N*-bit flags at the front of the predicate. This type information is the same as the one contained in an OID. The *N*-bit-flags, where *N* equals the number of attributes of an object, indicate which attributes are relevant to the query. The relevant ones are marked as "1," and "don't care" attributes have the value of "0." For example, a query predicate to a type of object could be as follows:

$$[\text{Type-o}, N\text{-bit-flags}, \sum P(i)] \qquad \textbf{(9)}$$

where *N*-bit-flags(i) = 1.

Suppose an object has 4 attributes, and the type is identified by type-o. The query predicate represented as

$$[\text{type-o}, 0110, P(2)P(3)] \qquad \textbf{(10)}$$

which means that the cache manager requires the 2nd and the 3rd attributes of objects of type-o if the 2nd attribute satisfies predicate P(2) and the 3rd attribute satisfies predicate P(3).

To check if the predicate in the cache invalidation report matches the query predicate, the following algorithm is used.

As shown in algorithm 1, upon reconnecting to the network, the mobile client first waits for the server to broadcast a cache invalidation report. After receiving the report, the client compares the timestamp of its cache to that of the report. If the timestamp of its cache is older than the timestamp of the report minus the server's broadcast period, the current report is not useful to the client.

The mobile client then submits its query to the server with the timestamp of the last cache update. Upon receiving the query, the server divides the attributes that have been updated since the last cache update time of the client into 2 sets. The first set contains the attributes satisfying the query. The second set has the attributes that

Algorithm 1: Predicate matching algorithm (clients disconnection time is shorter than the servers broadcast period)

```
Notation
        T_q:                object type information in the cache query.
        T_cir:              object type information inside the OID in the predicate in the
CIR.
        N-bit-flags-_q:     N-bit flag of the predicate of the cache query.
        N-bit-flags-_cir:   N-bit-flag of a predicate in the cache invalidation report.
        P_q:                predicate of the query.
        P_cir:              predicate in the cache invalidation report.
        MatchSet:           set of modified attributes in a predicate of the cache invalidation
                            report that matches the query.
  Input:                    P_cir, P_q
  Output:                   Boolean (match or no-match)

BASIC-MATCH Algorithm
        if T_q ≠ T_cir
        then return No-MATCH;
        else
                Relevant-bits = N-bit-flags-_q bit wise AND N-bit flags-_cir
                if Relevant-bits == 0
                then  // P_cir is NOT relevant
                      return No-MATCH
                else
                        for each i in Relevant-bits
                        do if P_q(i) == P_cir(i))
                           then MatchSet = MatchSet + {i}

        if MatchSet == {}
        then return No-Match;
        else return MATCH
```

do not satisfy the query. The server then sends the actual values for the first set and only the identifiers for the second set. The client keeps the first set in its cache and uses the 2nd set to remove items no longer valid from the cache. Formally, the mobile client uses the following algorithm.

Algorithm 2: Predicate matching algorithm (clients disconnection time is longer than the server's broadcast period)

```
Notation
        T_i:             timestamp of the current broadcast.
        T_l:             timestamp of the broadcast the client last received.
        w:               interval between two consecutive broadcasts.
        P_j:             predicate of the item j in the cache invalidation report.
        P_q:             predicate of the cache query.
        oid-j:           OID if the item j.
        RequestSet:      set of items requested from the server to satisfy the cache query.
        RemainderSet:    set of items in the cache not reported in the cache invalidation
                                 report.
        CIR:             Cache Invalidation Report.
        Cache:           All the objects of the client cache.
  Input: Cache
  Output: nil

 P-MATCH Algorithm
        RemainderSet := Cache;
        if T_i – T_l <= w
        then
        for each j ∈ CIR
                do
                        if BASIC-MATCH(P_j, P_q)

                           then RequestSet := RequestSet ∪ MatchSet(j)

for each attribute a, a ∈ Pj and a ∉ MatchSet(j)

do
                        if a ∈ Cache
then Cache = Cache – {a}
                           send RequestSet to the server
                           RemainderSet = Cache – RequestSet;
                for each attribute k, k ∈ RemainderSet
                do
                        T_k := T_i;
T_l := T_i;
else send (P_q + T_l) to the server
```

Algorithm 1 is used only when the client's disconnection time is shorter than the broadcast period of the server. This is usually the case with intermittent connections. If the disconnection is longer, the server is responsible for the update information relevant to the client based on the client's cache query and its last updated time.

CONCLUSION

This chapter proposed a caching technique for mobile-oriented middlewares for m-commerce applications. The method not only maintains the consistency of cached items but also intelligently determines if data items are relevant to the client applications. By identifying required data items and their updates, the cache manager can reduce the number of data item retrievals, resulting in efficient utilisation of bandwidth and cache space.

In the proposed caching scheme the server periodically broadcasts cache invalidation reports, similarly to other solutions in ID-based type retrieval systems in wireless environments. The proposed algorithm improves efficiency by employing a predicate-based match for content-based data retrieval. Information about the content of updated data items helps the cache manager in selecting and requesting useful data items that have been modified and thereby maintaining their freshness, while discarding items no longer useful. The usefulness of an updated data item is determined by matching predicates of the item in the cache invalidation report with the cache predicate.

An attribute-mapping table is used to reduce data length in cache invalidation reports. A good set of functions in the mapping table can carry considerable amount of information about content of updated data items, which underlines the usefulness of predicates in cache invalidation reports. In general, to provide more precise information about the new values of updated data items the number of attribute value ranges needs to be increased. While more attribute value ranges require more bits in the predicate representation, the complexity increases only by log_2.

One limitation of the proposed approach is that the size of a cache invalidation report will increase linearly with the number of updates during the relevant broadcasting period. If the number of updates is large, additional policies will be required to provide further improvement in efficiency. Future work will focus on combining cache consistency with the selection of "what to cache" in order to keep net communication costs, mainly in terms of used bandwidth, between server and mobile clients as low as possible. The approach proposed here, in particular with the outlined improvements, promises efficient practical support for mobile clients retrieving data from a server via slow or weak links.

REFERENCES

Adya, A., Gruber, R., Liskov, U., & Maheshwari, U. (1995). Efficient optimistic concurrency control using loosely synchronized clocks. In *Proceedings of ACM SIGMOD Conference on Management of Data,* pp. 23-28.

Agrawal, A., Simoni, R., Hennessy, J., & Horowitz, M. (1988). An evaluation of directory schemas for cache coherence. In *Proceedings 15th International Symposium on Computer Architecture,* pp. 208-289.

Barbara, D., & Imielinski, T. (1994). Sleepers and workaholics: Caching in mobile distributed environments. *Proceedings of ACM* SIGMOD, pp. 1-12.

Chan, B., SI, A. & Leong, H. (1998). Cache management for mobile databases: Design and evaluation. *Proceedings of IEEE 14th International Conference on Data Engineering,* pp. 54-63.

Chung, P. E., Huang, Y., Yajnik, S. et al. (1998). DCOM and CORBA side by side, step by step and layer by layer. *C++ Report, 10* (1), 19-30.

Forma, G. H., & Zahorjan, J. (1994). The challenges of mobile computing. *IEEE Computer, 27* (6), 38-47.

Franklin, M. J., Carey, M. J., & Livny, M. (1997). Transactional client-server cache consistency: Alternatives and performance. *ACM Transactions on Database Systems, 22* (3), 315-363.

Goodman, J. R. (1983). Using cache memory to reduce processor-memory traffic. In *Proceedings 10th ACM Symposium on Computer Architecture.*

Hu, Q., & Lee, D. L. (1997). Adaptive cache invalidation methods in mobile environments. *Proceedings of 6th IEEE International Symposium on High Performance Distributed Computing Environments,* pp. 264-273.

Jing, J., Bukhres, O., Elmagarmid, A., & Alonso, R. (1995). Bit-sequences: A new cache invalidation method in mobile environments. Technical Report CSD-T-94-074, Department of Computer Sciences, Purdue University.

Kian-Lee T., Jun C., & Beng C. O. (2001). An evaluation of cache invalidation strategies in wireless environments. *IEEE Transactions on Parallel and Distributed Systems, 12* (8), 789-807.

Lynch, N. (1999). Supporting disconnected operation in mobile CORBA. [Online]. Msc Dissertation, ftp://ftp.cs.tcd.ie/pub/tech-reports/reports.99/TCD-CS-1999-66.pdf.

OMG (1995). Common object request broker: Architecture and specification. Revision 2.0.

Tari, Z., & and Bukhres, O. (2001). *Fundamentals of Distributed Object Systems—The CORBA Perspective.* John Wiley & Sons.

Tari, Z., Tari, K., & Setiawan, S. (2002). *CODAR—A POA-Based CORBA Database Adaptor for Web Service.* In D. Taniar (eds,) Forthcoming. Hershey, PA: Idea Group.

Tari, Z., & Stokes, J. (1997). Designing the reengineering service for the DOK federated database system. *Proceedings of the IEEE International Confer-*

ence on Data Engineering (ICDE), Birmingham, pp. 465-475.

Tari, Z., Hamidjaja, H., & Lin, Q. T. (2000). Cache management in CORBA distributed object systems. *IEEE Concurrency, 8* (3), 48-55.

Tari, Z., Tari, K., & Dupin, V. (2002). An object and query cache management system for CODAR database adapter: Concepts, architecture and implementation. *Proceedings of International Conference on Enterprise Information Systems* (ICEIS), Spain. To appear.

Tari, Z., & Craske, G. (2000). A query propagation approach to improve CORBA trading service scalability. *Proceedings of the International Conference on Distributed and Computer Systems* (ICDCS), Taipei.

Tari, Z., & Fry, A. (2001). Controlling aggregation in distributed object systems: A graph-based approach for aggregation. *IEEE Transactions on Parallel and Distributed Systems, 12* (12), 1236-1255.

Wu, K. L., Yu, P. S., & CHEN, M. S. (1996). Energy-efficient caching for wireless mobile computing. *Proceedings of the 12th International Conference on Data Engineering,* pp. 336-343.

ENDNOTE

1 This project is fully supported by the ARC Linkage-Project grant no. LP0218853 awarded by the Australian Research Council (ARC), 2002-2004.

Part III:

Information System and Application Issues in Mobile Commerce

Chapter VII

Modeling Static Aspects of Mobile Electronic Commerce Environments

Jari Veijalainen
University of Jyvaskyla, Finland

Mathias Weske
HPI University of Potsdam, Germany

ABSTRACT

Mobile phones and other small and powerful portable devices have revolutionized personal communication and affected the lifestyles of the people in the industrialized world. Following credible estimates, in a few years there will be one billion of such portable devices. An emerging trend is the electronic commerce performed using mobile terminals, often called mobile commerce. Mobile commerce environments are characterized by high complexity, including myriads of technical and organizational aspects. This property makes it difficult to distinguish the more fundamental issues, structures, and concepts in mobile commerce from the hype. To capture the fundamental aspects of mobile commerce environments, we have developed a model. It covers fundamental static aspects of the m-commerce environment and their relationships. We distinguish four spheres of concern: Regulatory Frameworks, Business Models, Enabling Technologies, and the Global Infrastructure. Rather than providing technical details of m-commerce environments, our aim is to model invariant properties that will evidently persist for years to come. Making use of the abstraction capabilities provided by the object-oriented approach, the model is represented by OO structure diagrams.

INTRODUCTION

Mobile phones and other small and powerful portable gadgets have revolutionized personal communication and affected, considerably, the lifestyles of the people in the industrialized world. In a recent development, voice capabilities of mobile phones are augmented with data capabilities of increasing speed, and stand-alone Personal Digital Assistants (PDA) are equipped with additional communication capabilities. The Mobile Electronic Transaction Forum (2001) indicates that small-size mobile terminals are currently converging and evolving into Personal Trusted Devices (PTD) that allow users to access mobile Internet services and run applications at any time and at any place. The telecom industry estimates that there will be 500 million Internet-enabled mobile phones in 2003. The number of these mobile Internet-enabled PTDs is expected to exceed the number of fixed-line Internet users around 2003. Mobile commerce (m-commerce) is an important emerging application class in the wireless Internet environment.

M-commerce involves numerous domains, including network technology, business, government regulation, and standards. The main contribution of this chapter is organizing these aspects of mobile commerce environments and describing them using an object-oriented approach. This chapter focuses on the static aspects of mobile commerce environments; dynamic aspects are discussed briefly, mainly in settings where they have implications on the static structures. However a complete modeling of the dynamic aspects of mobile commerce environments—such as functioning of a particular protocol—is outside the scope of this contribution.

The overall setting we have in mind is shown in Figure 1. At the center is the global network infrastructure (called Wireless Backbone) that carries all kinds of high-volume data traffic. Currently, it is mostly the Internet. At the edges there are different wireline and wireless access technologies, such as wireline telecom networks, wireline local area networks (IEEE 802.3), wireless local area networks (IEEE 802.11), standardized by the LAN/MAN Standards Committee 802 of the Institute for Electrical and Electronics Engineers or wireless telecom networks (GSM, 3G) specified by the European Telecommunications Standards Institute (ETSI) (2000), and Bluetooth, specified by Bluetooth Consortium (Bluetooth, 2002). In computer network technology (as e.g., Tanenbaum, 1996 points out), these access networks represent OSI-layers 1-3. The electronic commerce (e-commerce) services are offered by the servers and are accessed by the terminals or other servers through the wireless or wireline access networks. Most electronic commerce services require end-to-end connections between the terminal and server at OSI-layers 4-7. This is necessary especially due to authentication and authorization.

Differences in the technologies at layers 1-3 are mostly uninteresting for the e-commerce services, including m-commerce services. This is true for data transfer and general connectivity, but the access network types have some important differences that suggest that the division in Figure 1 into the wireless and wired worlds will persist. First, wireless terminals are inherently mobile. This makes roaming between diverse technically compatible wireless access networks possible. Interoperability of roaming terminals and local e-commerce services obtainable "at spot" are a new emerging issue that must be resolved again and again. Second, wireless networks, especially telecom networks, tend to have much smaller transfer capacity than wireline access networks. This causes performance problems, should e-commerce services initially designed for much faster networks be used through mobile terminals. Third, the mobile terminal should be as portable as possible which means at the same time that they have much smaller physical dimensions, less memory, slower processors, smaller displays and keyboards (if any), and smaller batteries. For these reasons their usability for e-commerce applications designed for wired desktop terminals are far from good. Last but least, the possibility to dynamically position a mobile terminal with increasing accuracy opens up possibilities for new e-commerce services, often called Location Based Services or LBS. These make much sense for mobile terminals, but not much sense for fixed terminals, and are thus the crucial difference between m-commerce and internet e-commerce.

Rather than going into the details of the technical specification of wireless networks, we refer to the various 2G and 3G standards that are widely accessible. The interested reader is referred to ETSI (2000) and two recent contributions by Helal, Haskell, Carter, Brice, Woelk, Rusinkiewicz (1999) and Siau, Lim (2001), discussing the specific properties of these wireless network generations concisely.

In addition to nationwide wireless networks, a new type of short-range network, called personal area network (PAN), was developed recently. These networks aim at providing fast and convenient access within short ranges, rendering unnecessary the cumbersome temporary installation of wires between devices. For example, synchronizing the PDA agenda can conveniently be supported by a personal area network without fiddling with cables or managing docking stations. Wireless access to printers, and wireless link between an earplug and a telecom terminal, are other typical applications of this new technology. As mentioned in a Durlacher Report (2000), Bluetooth is an important product in this context.

For our purposes we do not make a distinction between a PAN or a more conventional terminal as a way to access the global infrastructure. A car as a mobile small-scale surrounding environment of a person that also has wireless access to the global infrastructure belongs to the same category as a PAN. Such devices or networks are modeled below as a terminal or a PTD.

As sketched in Figure 1, the new possibilities of mobile applications come from the ability to offer Location-Based Services that can be regarded as real-time and spatiotemporal. In particular, location-aware services and location-dependent services are available and described as follows: Location-aware services are able to answer queries, where locations of objects in a coordinate reference system and possibly metrics are used ("where is the X closest to Y"), but the objects do not move on earth. Location-dependent services use the actual, real-time ("now") location of the terminal or the object to be tracked to answer location-related queries ("Where is X nearest to me?"). The results can then be used to offer more complicated

Figure 1: Wireless and wireline access networks and the global network infrastructure

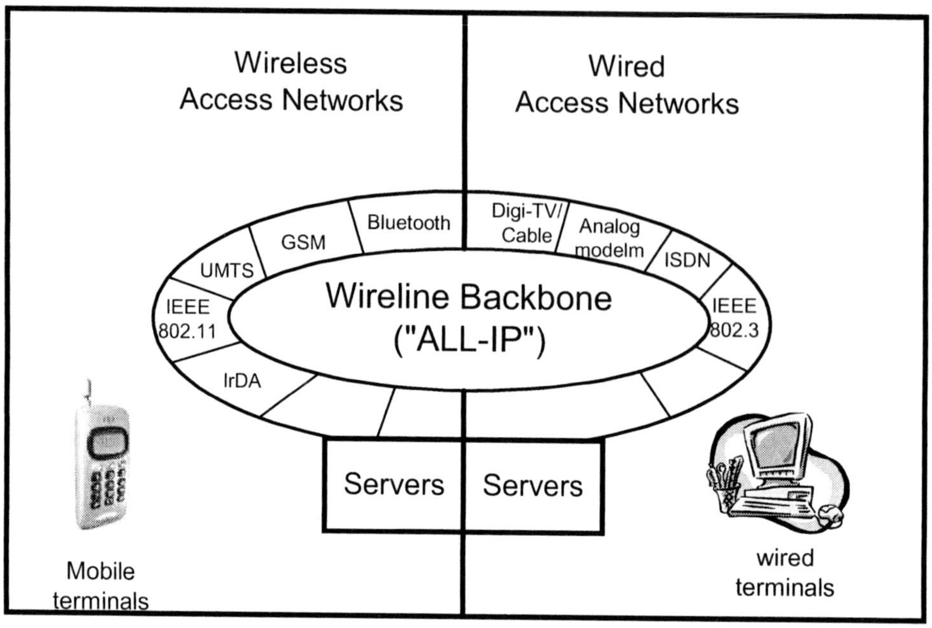

services. A typical example of a location-dependent service—and mobile commerce—is ordering a taxi in any city just by pressing a button on the PTD a traveler carries. The location-dependent services are those that make the mobile commerce different from other forms of e-commerce.

Personalization of the services is also possible, at least to the same extent as in the Internet, because PTDs are highly personal devices, and sophisticated user profiles can be maintained by the operators. In Devine and Holmqvist (2001), it was argued that the most promising M-services are those that belong to the categories: location-dependent, or personalized, or timely. The most sophisticated and promising services, however, belong to the intersection set, i.e., are location-dependent, personalized, and timely. These sophisticated services can be developed by collecting context information from the people, for instance, information on the profile and on the current physical location. But collecting private information compromises the privacy of the people. Thus, there is a trade-off between the quality of the services offered and the loss of privacy.

Services where the current location of the user device is crucial only make sense for a person on the move and are therefore a proper extension of Internet-based electronic commerce facilitated by terminals in fixed locations. Indeed, some reasonable but simple location-dependent services (e.g., simply displaying the current coordinates) can be provided by the wireless networks and/or by a GPS-enabled terminal, without the help of the Internet electronic commerce infrastructure. On the other hand, there are Internet electronic commerce services, which can be used from both wireline and wireless terminals. Typical examples are banking services in Scandinavia, as reported by Nordea Bank (2001). This shows that Internet electronic commerce infrastructure can be used to support mobile commerce. Technical, business, and legal issues become more complicated in m-commerce than in electronic commerce performed using stationary workstations and similar devices. See Veijalainen and Tsalgatidou (2000, 2001) for further discussion on this.

Given this technological background, the complexity and dynamic nature of technological as well as business advances in the electronic commerce area in general and in mobile commerce in particular overwhelms not only users but also business experts and information systems people. The general concept of m-commerce environments is rather imprecise, often even fuzzy, triggered by their highly dynamic nature: While current standards and technology are still not very well understood, new business models and applications are popping up almost on a weekly basis. In this situation, it is hard to identify the fundamental properties of m-commerce environments and to answer questions like, "What are the fundamental technological properties of mobile devices, independent of the current technology in place?", "What are the main players in m-commerce scenarios and how are they related by particular business models?" and "What are the interrelationships between properties of mobile devices on the one side and business models and the players involved on the other side?" "What is the impact of the user's willingness to use her time to interact with a wireless terminal?"

How to tackle the issues? First, we view mobile commerce to be a subset of electronic commerce. More precisely, as m-commerce we define any type of economic activity that is considered as electronic commerce by the legislation of some country or by the business community and that is performed using a mobile wireless terminal by at least one party. In most cases the mobile terminal is used by a customer (not, for example, by merchant or bank) and the wireless network used is a wireless telecommunications network, although any other wireless network, such as a wireless IP network or Bluetooth link, could be used, too.

The abovementioned definition of mobile commerce above is still vague and raises many questions. One reason for this is that the very concept of electronic commerce is currently not very precisely and uniquely laid down. Our view is thus that mobile commerce is a special case of electronic commerce, i.e., mobile commerce has all the opportunities and problems that Internet-based electronic commerce has, but it offers in addition some novel and very exciting possibilities—as well as new threats and challenges.

This chapter tries to remedy this fuzzy understanding of m-commerce environments by identifying the main fundamental and invariant structures and properties behind m-commerce application scenarios. This is aimed by organizing them using an object-oriented modeling approach. What we mean by an invariant here is a concept whose extension exists over a long period of time (years, tens of years in this case) before it vanishes entirely or changes so much that it becomes something else in quality. A typical example is a "terminal." We do believe that some kind of a portable device (or a set of portable devices) is needed at any point of time in order to perform m-commerce. This is because a human being is not able to directly exchange data and access services offered at a network but needs a technical device or devices in order to do it. This holds now and in the future. However, how these devices are constructed at a particular moment is another question; we have already seen during the last 5-10 years a tremendous development in the terminals; while the original size and weight was comparable to a voluminous book, they now are small like matchboxes—and have still more processing capability and memory than a PC 10 years ago. The newest telecom terminals already incorporate video cameras, and a person-to-person video service V-LIVE has been launched in May 2002 in Japan (NTT DoCoMo (2002)). The first really wearable terminals have also seen the light; they are integrated into the clothes, and the parts can use wireless Personal Area Network (PAN) or cables for communication (see, e.g., Kaario, 2000; MIT, 2002; Kahney and Leander, 2002). But they still have the same functionality as any terminal in our sense: they allow the people to access m-commerce services.

The other invariant concepts, like business model, are similar. Although at a certain moment a business model has one form and at another moment another form, the concept itself stays for long time. Examples of these kind of invariants in mobile commerce settings are the inherent mobility of users, limitations of user devices with respect to input and output capabilities and connectivity, where the latter is either due to the noncoverage of certain parts of the world by service providers, or because the user decides not to be reachable for some time. As shown by Veijalainen (1990) and Veijalainen, Eliassen, and Holtkamp (1992), one of the invariants at this level is that the terminal exhibits communication autonomy.

We are here not only interested in individual permanent concepts, but in a set of concepts that together are essential for the comprehension and development of m-commerce. We want to address the legal and organizational prerequisites of m-commerce, enabling technologies, and the actual global infrastructure in place. We thus identify four spheres of concern that refer to the above aspects. The most abstract but at the same time most pervasive sphere of concern addresses the patchwork of the *Regulatory Frameworks* emerging in different parts of the world. It influences the *Business Models*, *Enabling Technologies*, and the *Global Infrastructure*, the other spheres of concern. The essential criterion in establishing the spheres of concern is that they have a dynamics of their own that is relatively independent of the other spheres. Still, they are dependent on each other; for instance, even if the Regulatory Framework logically precedes m-commerce, deeming some forms of it "illegal" and others "legal," historically, m-commerce can be performed without a special legal framework, and in fact, the emergence of m-commerce is the reason for establishing a particular legislation.

This article is organized as follows: the next section, *Mobile Commerce Model*, introduces the object-oriented model for mobile commerce environments by presenting the individual spheres of concern, introduced above. This section concludes with an integration of the submodels to present an overall mobile commerce environment model. In the following section, *Sample Scenarios*, the applicability of the model is illustrated by sample mobile commerce scenarios. A section on related work and concluding remarks completes this chapter.

MOBILE COMMERCE MODEL

By analyzing the complex environment of m-commerce applications, it appears that a variety of diverse entities and organizations are playing important roles, ranging from standardization organizations and technical specifications of the network infrastructure to user terminals and particular business models and, finally, to revenues. The approach presented in this chapter takes into account a wide range of these static aspects. Rather than viewing the environment as an unstructured collection of diverse entities and concepts, we organize them in four *spheres of concern* and describe them using object-oriented modeling techniques. As noted above, the organization of the entities into spheres of concern is based on the relative independence of the dynamics within the spheres. The spheres classify the entities into four broad categories, like government rules, enabling technologies, network/terminal technology, and business aspects (they will be specified more precisely below). Since the spheres of concern represent parts of the overall model, they can also be regarded as submodels.

Object-oriented approaches have been proven well-suited to capture the static aspects of complex application scenarios. However, their use is not limited to specific low-level design models used for systems implementation. On the contrary, due to abstraction concepts provided by the object-orientation paradigm like classification, association, and inheritance, it is well suited for modeling complex application level scenarios. In this chapter, we use the de facto standard Unified Modeling Language (UML) introduced by Booch, Rumbaugh, and Jacobson (2001) to model the static aspects of mobile commerce environments.

Based on the object-oriented approach, we organize the entities involved in m-commerce environments in classes and represent their relationships by associations between the classes. While typically each class belongs to one sphere of concern, there may—and must—be overlapping classes, as well as associations between classes of different submodels that act as the glue between the respective spheres of concern. Hence, each sphere of concern is characterized by an object-oriented submodel, and these are interdependent through overlapping classes and associations. We do not model all possible known concepts and their relationships as explicit classes and associations in our model. In order to be explicitly present as a class in the model, the corresponding concept must be essential for the existence and evolution dynamics of the sphere of concern it occurs in. Organizing the static aspects of m-commerce environments in this way aims at clarifying the concepts, technologies, and players in this highly complex and dynamic area.

To organize the presentation in the remainder of this section, the submodels are addressed in turn; finally, these parts are integrated into an overall model for m-commerce environments and the relationships between the spheres of concern are discussed. The spheres of concern are listed as follows:

- *Regulatory Frameworks*: The organizational and technical aspects of laws, standards and recommendations, as well as the bodies involved in their definition or ratification.
- *Enabling Technologies*: This sphere includes technical aspects like user terminal and network technologies and cryptography, and organizations developing these technologies.
- *Business Models*: Business aspects, including business players, provided services, business protocols, revenue sharing, and code of conduct are important artifacts in this sphere.
- *Global Infrastructure*: The global infrastructure sphere deals with the global network and the concrete terminals that facilitate m-commerce.

The general approach for each sphere is as follows: We start by discussing in an informal way the concepts and artifacts specific to that particular sphere and those that should appear in the respective submodel. The artifacts are then classified using an object-oriented approach. Relationships between artifacts are represented by associations between the respective classes. As said above, we use structure diagrams provided by the Unified Modeling Language as notation.

It is important to stress that we are not focusing on specific technical or organizational properties of artifacts in mobile commerce as they have developed during their history, but—on the contrary—have general, long-lasting patterns and properties in mind. For instance, rather than modeling individual types of terminals, we introduce a class representing these terminals. This means that whenever a new terminal enters the market, conceptually, an object will be inserted in the terminal class. As a result, the model will not change in response to the introduction of a new terminal type. The same holds for a new standard, a new business model, a new network protocol, or a new modulation technology. Hence we regard the concepts of the model to be invariant against changes in mobile commerce environments in the foreseeable future. A schematic picture of the spheres of concern and their dependencies is presented in Figure 2.

Notice that the spheres are dependent on each other, as specified in Figure 2: the two up-most layers rely on Enabling Technologies; Business Models cannot be deployed without the Global Infrastructure that facilitates the concrete m-commerce transactions. The picture also alludes to the relative independence of the layers. One can change something in one layer without necessarily changing the other ones. For

Figure 2: The four spheres of concern

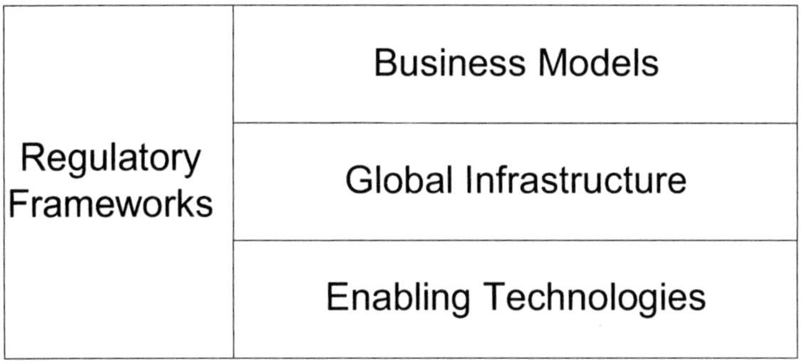

instance, offering GPRS data transfer service within the Global Infrastructure sphere does not necessitate changes in the existing concrete business models (albeit new business models might become feasible with the "always-on" functionality of the terminals). Business models can be developed and changed without the need to change the terminals or the network infrastructure in place. The limits for possible business models are still dictated by the infrastructure; think of selling video-on-demand services over 9.6 kbps wireless links to slow terminals. Putting it the other way round, certain business models set certain requirements for the infrastructure and new technologies. New enabling technologies can be developed based on the business requirements or for other reasons, but this does not have influence the Global Infrastructure as long as they are not deployed in the concrete networks; existing new enabling technologies are, however, a necessary condition for the changes in the global infrastructure. Therefore, Enabling Technologies is the lowest sphere of concern in the picture; the upper spheres stay on top it.

The difference between the Global Infrastructure and Enabling Technologies is exemplified by UMTS technology. The technology as such is ready and available since 2001, and the network components and terminals are in production. But only a tiny part of the wireless access networks are really currently (Fall, 2002) of this type, and there are much less than one million subscribers in the world.

Regulatory Frameworks put forward requirements for all other spheres of concern and restricts their form. It stands on its own but leans on the other spheres. This reflects the relative independence of the governments, international organizations, and companies and their leading role against the other spheres of concern. The Regulatory Frameworks sphere influences directly all other spheres, but there are also indirect influences. For instance, legislation can require certain privacy rules to be obeyed by the business actors. This can be reflected by the business model applied, and further in the Global Infrastructure sphere, where technical support for privacy maintenance must be in place. On the other hand, developments in the three other spheres of concern can require regulatory measures to be taken. Typically, a new body must be formed to take care of some aspect of global interoperability (see below), a new consumer protection law passed, or emerging technology otherwise regulated, etc.

Regulatory Frameworks

The Regulatory Frameworks sphere represents the global organizational level that is needed to facilitate the interoperability of the underlying network infrastructure, terminals, and the higher-level—m-commerce—services. This level is also needed to foster and guide the development of new technologies, and last but not least, to provide the legislative framework for the m-commerce. Interoperability is closely related with the autonomy of the organizations offering m-commerce services. As was shown in Veijalainen (1992b), in order to guarantee interoperability of systems, one needs to establish an organizational entity called *global designer*. It is responsible for specifying the rules for interoperability within the *technical domain* controlled by it. A global designer has to have an exclusive power to control the domain, otherwise there almost certainly will be mismatch and confusion about what is the right way of handling things (of course, sometimes this is also intended for reasons of competition). Should several global designers design parts of the same environment they must guarantee that the parts are interoperable. Thus, they must cooperate with each other—or establish a higher-level global designer, and so on, if they want to ascertain interoperability of their standards and recommendations.

We distinguish between three main groups of organizational entities, namely *legal bodies*, *standardization bodies*, and *interest groups* that have formed so far and that most probably will persist in the foreseeable future, because the reasons for their existence will persist in the future. They all play a role *within* global designers for various aspects of the global environment.

The legal bodies are national governments and other organizations with legislative powers, like the European Commission or the United Nations. These issue legislation that regulate business in general and mobile commerce in particular. In general, laws say which business models are legal, which are not, and what has to be taken into consideration when performing m-commerce transactions. Especially the privacy of customers is of concern at this level, as well as jurisdiction and applicable law addressing consumer protection, liability of merchants, and dispute resolution. The governments also determine which wireless technologies and access networks using them are allowed in their territory, thus creating the basis for mobile commerce. In the legal respect, Japan, EU and USA currently form three different areas of rules as concerns wireless networks and business models for mobile commerce. So the domains controlled by these three legal bodies are geo-spatially separated and nonoverlapping. The rest of the world is mostly a "gray zone" but will join the hopefully commonly established domain (or one of the three domains).

Legal bodies can enforce and ratify laws and other legally binding rules. In this context, the laws set forth constraints that mobile commerce systems have to meet. It is the responsibility of the business partners to guarantee that the legal constraints imposed by the legal bodies are met. Within the European Union, electronic commerce legislation in general and mobile commerce legislation in particular is to a large extent in the hands of the European legislative bodies. The European Commission and Parliament has issued Directives that are to be integrated into national legislations. The legislation has thus both a direct impact on technology, business models, and network infrastructure, and an indirect one through the adjacent spheres of concern. Interested readers are urged to visit the EU Information Society portal (2002) and EU e-Commerce legislation portal (2001) to learn more about the various activities performed and legislation issued by the EU.

Legislation can sometimes directly dictate a standard, but usually it just refers to an existing one that has been prepared by a standardization body or has become a *de-facto* standard. Agreements on common technologies and standards are the key to interoperability. This holds for a variety of domains, including wireless networks, software technologies, positioning methods, etc. Standards can be either international ones ratified by international standardization organizations, such as ISO (2002), national bodies, such as American National Standards Institute (ANSI, 2002), or international institutes like ETSI or IEEE (ETSI, 2002; IEEE, 2002). However, often industry best practices that act as *de facto* standards are eventually accepted as an official standard. This pattern has been prevalent in the past in various technology areas. It has also been observed in Internet network technology where TCP/IP became the standard transport and network layer protocol of the global network infrastructure, as well as the layers on top of it, and it does not comply with the OSI standard open protocols suggested by ISO. GSM technology as a wireless access technology has also conquered majority of the developed countries in the world, although Japan is an exception, as well as partly the USA. UMTS will most probably have a still wider applicability than GSM, because Japan and USA will be on board (UMTS Forum, 2002) and the rest of the world will follow, eventually.

While standardization bodies have—strictly speaking—no legal powers, they are an important instrument for the industry and governments to agree upon standards in a specific technical domain. Interest groups are a weak form of standardization body. In our sense they are an instance of the abstract global designer, that is, usually established for a limited period of time in order to solve a pressing technical issue. In mobile commerce, there are several very strong interest groups, each of which focuses on a specific technology, i.e., on a particular technical domain. The common behavioral pattern of interest groups is to get the key players of the field involved and define a *de facto* standard with the aim of pushing it into the market through their power. An ultimate case is NTT DoCoMo in Japan who led the specification of the i-Mode system, including the whole protocol stack, mark-up language, terminals and business model(s), and also deployed the core system.

In order to reach the true global interoperability in wireless communication systems at the application level, many companies are involved in various interest groups in diverse technical domains. These include manufacturing terminals, setting up wireless networks, and designing and implementing m-commerce services. The key factor for success in this context is indeed interoperability, meaning the ability of devices and network technology produced and provided by different manufacturers and network operators to communicate with ease and efficiency as desired by customers.

The division of the world into geographic zones or areas is an important issue for roaming customers who would like to use location-dependent services (like ordering taxi, see below). His or her PTD must be able to not only use the local wireless access network, but also the local services. 3G networks and compatible terminals might solve the first part of the problem, but the second problem area—the heterogeneity of the services—still remains. In 3G networks, ITU (2002) and ETSI (2002) played a central role as global designers, but as far as m-commerce services are concerned, there is no organization that has the power to dictate how they should look like. The m-commerce services are, first of all, regulated by the local legislation of the zones, and they are highly dependent on the local culture and rules, as represented, for instance, by specific business models issues and codes of conduct.

The companies involved in the wireless telecom business have realized that if the business models and m-commerce services become incompatible in different parts of the world, or indeed within a zone with the same legislation, roaming customers would suffer. This would have a negative effect on the whole industry. As a result, the world has now seen many global designers to emerge that try to tackle the problems. These include the WAP forum (WAP Forum, 2002), but especially the Mobile electronic Transaction Forum (MeT, 2002), Location Interoperability Forum (LIF, 2002), and last but not least, the Open Mobile Architecture Initiative (OMA, 2001, 2002). All these must closely follow the work of the W3C Consortium (W3C, 2002), as well as the work of many other similar bodies and standardization organizations. The situation raises many questions, like what guarantees that the diverse recommendations developed by all these bodies will be coherent and interoperable? Second, what chances does the work of these interest groups have to become global de facto standards or even a sort of guideline for concrete service design? The first problem is more or less satisfactorily addressed because the interest groups consist essentially of the same big players, and these coordinate the overall architecture development and the concrete work within the different forums. In general, however, there is no guarantee that many global designers specifying closely related recommendations would succeed in achieving interoperability.

The second question above is a tougher one. The OMA Initiative is the answer of the big players: "The objective of the Open Mobile Architecture (OMA) initiative is to create a non-fragmented, interoperable global market for the next generation of mobile services" (OMA, 2002). As the name suggests, the idea is to use only standard, openly available specifications and software. What is the controlled domain in this case? Formally software, but of course with software comes the functionality. To which extent does this consortium want to specify the functionality of the software? Theoretically, the interest group can indirectly specify even the "open" business protocols embodied in the "OMA" software. Whether it wants to go so far remains to be seen. Currently, the emphasis is rather on open industry standards, standard tools, and open platforms.

It is worth of noticing that a commonly agreed upon syntax, such as EDIFACT or a particular mark-up language in the XML family, such as XHTML, does not yet help to completely solve the interoperability problem; it remains to be decided by autonomous organizations, which concrete EDI or XHTML specifications to adopt and what the elements exchanged mean exactly. Furthermore, one must specify which protocol stack is used to exchange the content. The m-commerce protocols are application protocols that support the chosen business model, and these will most probably vary from zone to zone. The services should be paid in a uniform way so that the roaming customers are able to use their terminals smoothly. In any case, should this consortium be able to keep its promise, it will certainly have a strong impact on the way mobile commerce is conducted in the future.

Based on this general introduction to regulatory frameworks, their players and issues, we now take the step towards an object-oriented model of that sphere. This is shown in Figure 3. Starting with organizational aspects, regulatory bodies and companies are represented as specific classes of organizations, with inheritance relationships between them. Regulatory bodies can be either legal or standardization bodies, represented by subclasses of the regulatory body class. Companies join interest groups, which aim at defining *de facto* standards, represented by an aggregation association between these classes. If the interest group succeeds in defining a recommendation, the companies participating in the interest group have a considerable competitive advantage, since they already offer products and services complying with that new recommendation. The left side covers the legal bodies, like governments, the European Commission, and the United Nations. They regulate the technical or business domain in a specific geographic area, i.e., a country, or a larger economic region.

The model shown in Figure 3 provides a detailed, yet abstract view on the entities involved in the Regulatory Frameworks sphere. It is important to stress that it details the type level, not the instance level. In general, objects are the instances of classes. Thus, objects of the company class are Nokia, Ericsson and Siemens to name a few players in the mobile commerce arena. These companies participate in interest groups with the aim of setting up recommendations. For instance, the Mobile Electronic Transaction Forum is an interest group aiming at defining *de facto* standards for secure mobile transactions. Tens of big companies, including Nokia, HP, Sony Ericsson, Oracle, Sun, IBM, NTT DoCoMo, NEC, Siemens, Vodafone and AT&T Wireless have joined the OMA Initiative (see OMA, 2002 for further details).

Figure 3: Regulatory frameworks

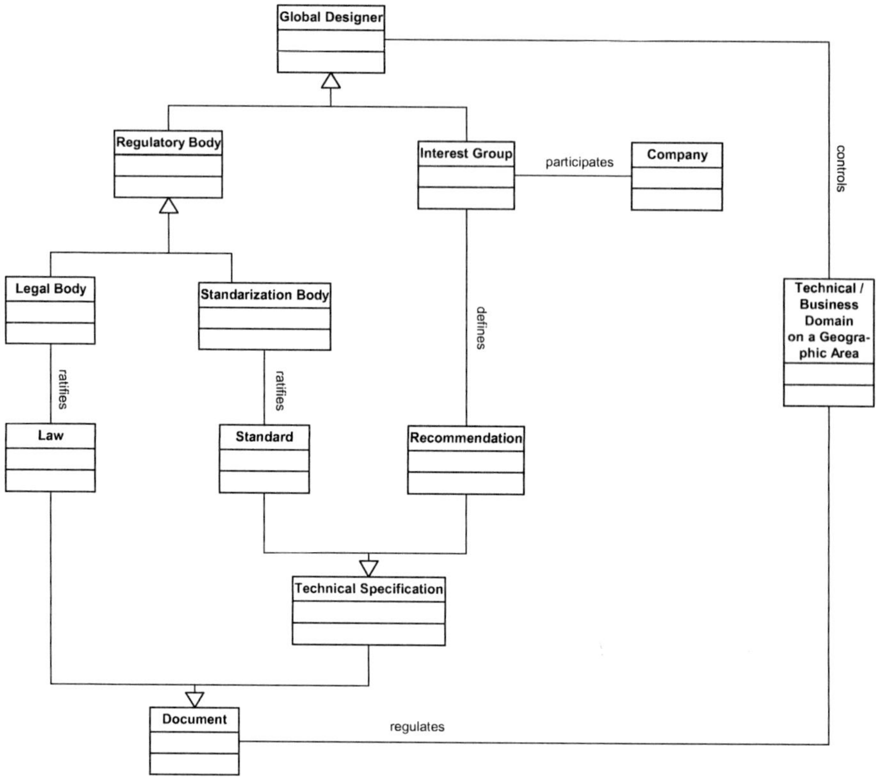

We believe that the class structure of the organizational entities as shown in Figure 3 will persist in the foreseeable future. We are still fully aware that a deeper analysis requires populating the instances and grasping the concrete dynamic relationships between interest groups, standardization bodies, and legal bodies, as well as the analysis of the individual organizations. For instance, the analysis of interest groups, the reasons for their emergence and dissolution, their impact on the technology and markets, their relationship to globalization, etc., could be studied in research projects. There is work done in this vein on the standardization of 1G, 2G, and 3G systems: see, e.g., Telecommunications Policy (2002).

Enabling Technologies

Advances in electronic commerce, in general, and in mobile electronic commerce, in particular, have, to a large extent been spawned by technological advances, both with respect to wireless network technology and user device or terminal technology. In this section we focus the second part of the overall mobile commerce model, i.e., Enabling Technologies.

It is not just one technology that makes mobile commerce possible. It must also be available for users through mobile terminals, access networks, and servers. Personal Trusted Devices (PTDs)—i.e., wireless terminals with personal flavor and security facilities as access devices in mobile commerce—are in a crucial position in this respect. This is due to the fact that the technologies incorporated in them (such as processors, protocol stacks, and local applications) largely determine which m-commerce services can be offered. Indeed, the current PTDs where a PDA and a GSM phone are integrated and run WWW and WAP browsers offer mobile access to the Internet and m-commerce services but use the basic GSM services offered since the beginning of 1990s. The GSM network technology was thus mature enough from the beginning for mobile commerce, but before the terminals had reached the current stage of development, mobile commerce could not be realized. For these reasons, PTD is in the center of the Enabling Technologies sphere in our model, as shown in Figure 4.

Figure 4: Enabling technologies

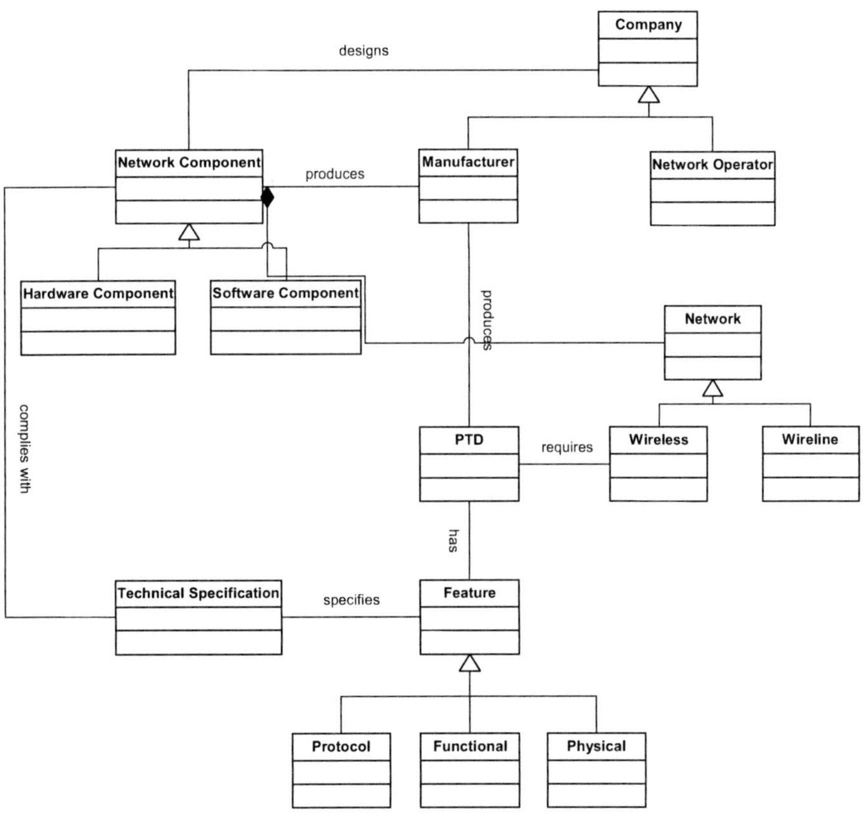

Each Personal Trusted Device is associated with a set of features, which can be classified as protocol features, functional features and physical features. Protocol features represent wireless network protocol stacks that the device supports (e.g., GSM and WAP stack). Functional features subsume a variety of different functional aspects, for example, positioning (like GPS receiver functionality), and security and privacy features. As indicated above, features of PTDs are defined in technical specifications that manufacturers have to adhere to. Notice that the technical specification class is also present in the Regulatory Frameworks part of the overall mobile commerce model. These overlapping classes will act as the bridge between the Enabling Technologies and Regulatory Frameworks spheres. That is, the impact of the Regulatory Frameworks on the Enabling Technologies comes, among other things, through the technical specifications.

Besides the manufacturer, the network operator is another subclass of company involved in Enabling Technologies. While manufacturers produce Personal Trusted Devices, network operators also take part in technology development either directly or through various interest groups. For instance, NTT DoCoMo specified the i-Mode technology and let the manufacturers produce the concrete network components and terminals. In the WAP Forum, network operators try to influence technology development through technical specifications.

Global Infrastructure

Enabling technologies are a necessary ingredient for the concrete deployment of wireless network technology. We keep the deployed networks separate from the base technologies for several reasons. First, there are many individual technologies (such as encryption and protocols) that must be joined coherently in order to form a functioning global infrastructure. Second, the individual technologies were developed rather independently from each other and are controlled by players other than telecom operators or network providers. Third, the deployment of the concrete networks is largely on the responsibility of the individual telecom and other operators, such as Internet Service Providers (ISPs). It follows that the global infrastructure is fragmented into hundreds of wireless access networks and a (few) backbone network(s). Each network is conformant with the legislation of the country or area where it is deployed. This causes many potential differences. Although the different network generations might be interoperable among each other, the concrete networks still might be, technically, at different stages (e.g., 2G, 2.5G (GPRS), 3G). In addition, a terminal can roam and can or cannot get access to the wireless networks of different operators based on the roaming contracts. But getting access to the wireless network at the levels 1-3 does not yet guarantee that the terminal is able to, in fact, access the local m-commerce services. This is because the business model or the technical level implementation might be different (e.g., markup languages in the home network and in the visited network may differ), or it simply cannot find the services in the local network, because it does not know the address of the service directory or cannot access it due to protocol differences. Finally, even if the terminal would be able to find the services and offer them to the user, it might not understand the local language used in the services in order to properly use them.

The abovementioned problems are largely independent from the Enabling Technologies and also, as it turns out, from the Business Models. They all manifest the same phenomenon at different system levels. The phenomenon is sometimes called *roaming heterogeneity*. It is a severe issue of the wireless global infrastructure that is currently not fully understood, albeit recognized or even solved. For instance, the OMA Initiative (2002) does not use the term, but it addresses exactly roaming heterogeneity at the service level, as was discussed in a previous section.

We envision that the PTDs and the wireless networks will persist far into the future and the networks are deployed and managed by (mobile) network operators. Irrespective of the technology generation, the structure of the networks is rather coarse, such that the wireless link is between the PTD and a base station or another entry point and the backbone network is wired. We also assume that there are hundreds or thousands of operators in the world that have their own customer base. Each wireless network occupies a restricted geographical area. Because competition is allowed and required in the telecom markets, there are several wireless networks operated in the same geographical area. From this basic organizational arrangement, it follows that customer roaming requires a patchwork of contracts between operators. It also requires technical support within the network that is at least as powerful as in GSM networks. In theory, a user can roam between different network operators even if she would not leave her home country, but because of competition this is not common. Roaming precludes, thus, that the customer moves to the operation area of a foreign operator in another country or region.

The submodel of the Global Infrastructure sphere is shown in Figure 5. As represented by the inheritance hierarchy shown, the networks managed by operators may be either wireless or wireline. By definition, Personal Trusted Devices require wireless networks in order to operate. As indicated above, these wireless networks can be used to access wireline networks and the services provided. For instance, second-generation wireless networks, such as GSM, can act as access networks to wireline networks, typically to the internet. WAP 2.0 specification assumes that the wireless terminal can indeed be an IP-enabled device, although it is compatible with the complete WAP stack solution introduced earlier, as specified by the WAP Forum (2002).

To sketch the main classes of the submodel shown below, we remark that manufacturers deliver network components that are being used to build and run networks. Networks should obey local legislation, i.e., legislation is pertinent to areas. PTDs may roam to areas and networks. In particular, area binds the validity of the business models and extension of the networks into a specific geographic area. The terminal must roam on this area whenever it wants services from a specific network. It is worth mentioning that there is also a direct relationship with Figure 1. The PTDs are a special case of the mobile terminals, and the wireless access networks are usually those deployed by network operators. To this end, Figure **1** can be seen as a more detailed representation of the different network types and their instances, as well as a rather abstract topology of the components of the overall global infrastructure.

Business Model

The abovementioned spheres of concern provide the environment for the business aspects of mobile commerce. After all, economic aspects are the driving force behind mobile commerce developments and applications, and are the source of revenues for the companies involved.

Timmers (2002) defines business models as follows: A business model consists of:

- an architecture for the product, service and information flows, including a description of the various business actors and their roles;
- a description of the potential benefits for the various business actors; and
- a description of the sources or revenues.

In brief, a business model thus describes the economic player categories, the products and services offered, how the players interact, the information and goods that flows, the sources of revenue, how the economic yield is shared among the players, and how they are related with the above flows. We add the code of conduct to the above list, because it has a considerable local impact on the form m-commerce can adopt.

Japan is probably the most advanced mobile commerce market today. Therefore, we look more closely at it. Devine and Holmqvist (2001) distinguish the

Figure 5: Global infrastructure

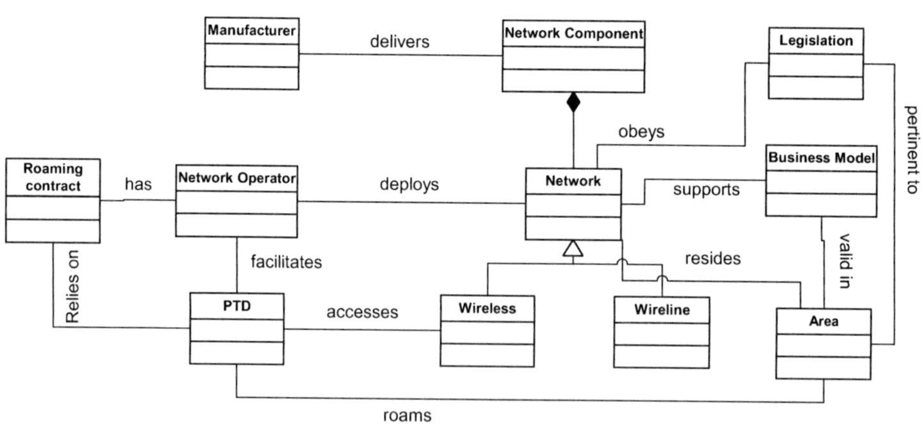

Figure 6: Business model and related concepts

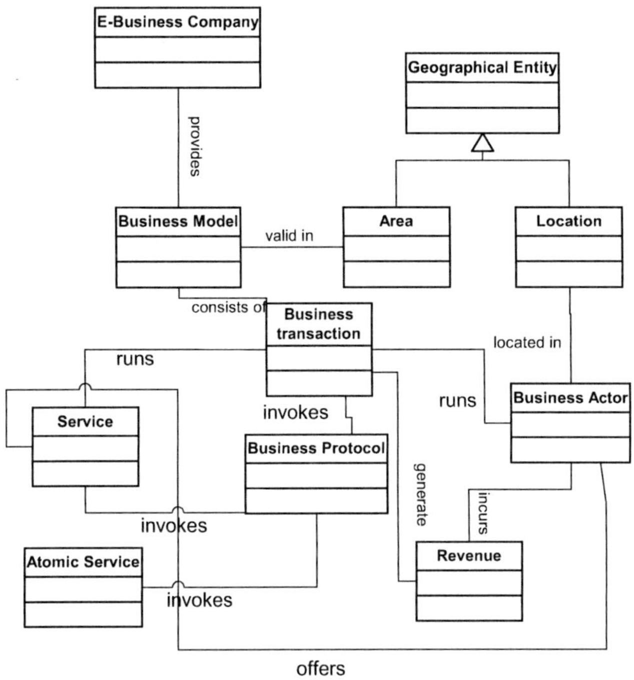

following players on that market: user, mobile network operator, telecom operator, application provider, facility supplier, information provider, contents holder, solution provider, financial institution, and terminal manufacturer. All these players get their revenues from the user, but only the mobile network operator, the information provider, the application provider, and the financial institution are directly involved in individual m-commerce business transactions. Thus only these directly participate in providing the mobile e-commerce services and in direct revenue sharing. The other players (like solution provider and contents holder) get their revenues indirectly. NTT DoCoMo's i-Mode service has currently over 30 million subscribers, and the average monthly revenue per subscriber is about 40-50 US $.

The Business Model sphere of the mobile commerce model, reflecting the above narrower view, is shown in Figure 6. The E-Business Company class contains the enterprises involved in electronic business. Business models are in the heart of the e-business companies. A business model consists of one or more business transaction specifications. A business transaction specification consists of a business protocol and service descriptions the protocol can access and handle. One can, for instance, imagine that the business protocol is an abstract protocol specification and a service is a WWW page that delivers the content needed or runs a program changing a database state at the requested organization, e.g., modifies a balance in a bank. The service can be atomic, in which case it does not invoke further business transactions. Notice that the business protocol can also, in this case, still invoke several request-response pairs towards the service. A service can further require other services to be invoked at different sites. Therefore, it runs a new business transaction. Typically, this happens, for instance, if somebody is ordering a book and paying with a credit card. The payment would be a new business transaction between Amazon and VISA. Upon running this business transaction, revenues are generated. We thus model revenue generation at this level of abstraction.

A simple request-reply PDU pair may access a service, but it may also be a complex of services accessed by many request-response pairs. This means the service consists of a set of other services, which are related by a Business Protocol. The concept of Business Protocol is treated in more depth in the next section when discussing an example. At the model level, Business Transaction has Business Protocol as a direct component, but this can invoke Service or Atomic Service.

A business protocol specifies the steps that occur when business partners cooperate, as well as the format of the data exchanged during the cooperation steps. Each step within such a business protocol is typically characterized by the received message and the response sent in another message. A simple one is where upon receiving an order, a mail order company responds by sending a message, acknowledging the order. We remark that there are a number of technologies for business-to-business data exchange, known as business protocols. There are also a number of standards for business protocols in place, e.g., Electronic Data Interchange (EDI). These technologies cover static as well as dynamic parts of e-commerce environments, because they provide data format standards (e.g., data structures of orders) and data exchange standards (messages sent during a successful cooperation). An Atomic Service is conceptually indivisible in that it does not require a further business transaction to be run. An example of a very simple atomic service instance is responding to a weather forecast request with the respective temperature forecasts. As these brief examples show, on the one hand business protocols are used to implement business models; on the other hand they make use of the global infrastructure to enact the services. They are thus the crucial link between these two worlds.

Service composition has recently emerged as an important aspect to develop Web applications based on so-called Web services, sponsored by the World Wide Web Consortium, W3C (2002). Web services are basically applications provided by a company that can be accessed via the HTTP network protocol. Using suitable wrappers, many services that are implemented by back-end systems, e.g., Enterprise Resource Planning systems, can be made available to a large market by Web services technology. While Web services are developed in the context of electronic commerce, they can also be used in mobile commerce environments to provide the base functionality required by a complex mobile commerce application. Consider, for example, a mobile commerce application, where customers can make reservations and book and pay for flights via WAP devices. While the interaction with the customer is provided by mobile commerce technology, the back-end functionality like flight reservation and payments has to be offered by a suitable back-end system.

Web services are a new technology based on accepted standards. While the key technologies are already in place, the composition of Web services is currently under research. The basic understanding of Web services is that the individual functionality accessed by individual Web services (e.g., making a flight reservation) can be combined with other services within a predefined order. It is in this way that application processes can be developed based on individual steps, covered by Web services. For example, a payment step can be followed by a successful flight reservation step. In the next level of abstraction, the new Web service processes, flight reservation and payment, can be made available. We model these aspects of service composition, in general, and Web service compositions, in particular, in the static structure model by a recursive federation between the Services class through Business Transaction and Business Protocol.

Business partners provide services to their customers, i.e., business actors. A business actor can either be an individual (e.g., a person ordering a weather forecast of southwestern France for next Friday), or it may be a corporate business actor. These can make use of services to offer value-added services to their respective customers. The goal of each of the business partners involved is to generate revenue from service's executions. Notice that the legislative bodies provide the legal environment in which the business models are developed.

The Business Models sphere also addresses the geographical validity region of a Business model, represented by the class Geographical Entity, associated with the Business Model class and with the Business Actor class. This is important, as the business actors and the models applied do have a location on earth where they are legal and accessible. It is interesting to notice that the association between business models and geographical location are quite stable: As defined by legislative rules (in the Regulatory Frameworks), a given business model is associated with a geographical location on earth. This may not only be a XYZ-coordinate, but it can also be an organizational entity associated with an area, such as a state or an association of states (for example, the European Union). Hence, the geographical entity class has subclasses geographical location and geographical area, representing the abovementioned concepts in the model.

Complete Model

The four spheres developed above are now integrated into the complete mobile commerce model. As discussed above, the glue between the submodels is provided by the associations between classes of the different spheres as well as by common classes. As an example, the Technical Specification class of the Regulatory

Framework submodel is associated with the Feature class of the Enabling Technology submodel, gluing the two submodels together. The same holds for the Business Model sphere and the Enabling Technology sphere, where the Service class is associated with the Feature class. Another form of glue between the spheres is characterized by inheritance: The E-Business Company class of the business submodel is a subclass of the Company class. The overall mobile commerce model is shown in Figure 7. For ease of convenience, each submodel is represented by a package in that figure. These relationships between the packages are represented by the dotted lines in Figure 7.

Additional associations between classes can be drawn. For instance, Regulatory frameworks have a validity region on earth; in particular, the three main regions (US, JP, EU) can be regarded as three geographical regulatory frameworks. Business models adhere to a certain legislation and thus to a certain region, and a terminal is in a certain location on earth that belongs to a certain validity region. The issues in this context are of two kinds: the terminal is able to interact with the rules of a certain region. These rules are characterized by the wireless access network in place, the business model that the services implement, and the protocol stack that is used to facilitate the m-commerce services. While roaming to another region, roaming heterogeneity can occur, i.e., the terminal (or user) is not anymore able to take advantage of the services. Second, and this is a general problem of electronic commerce, as well: the terminal can use services in another validity region than where it is located. If the rules applied to the services differ, whose rules should be

Figure 7: Complete m-commerce model

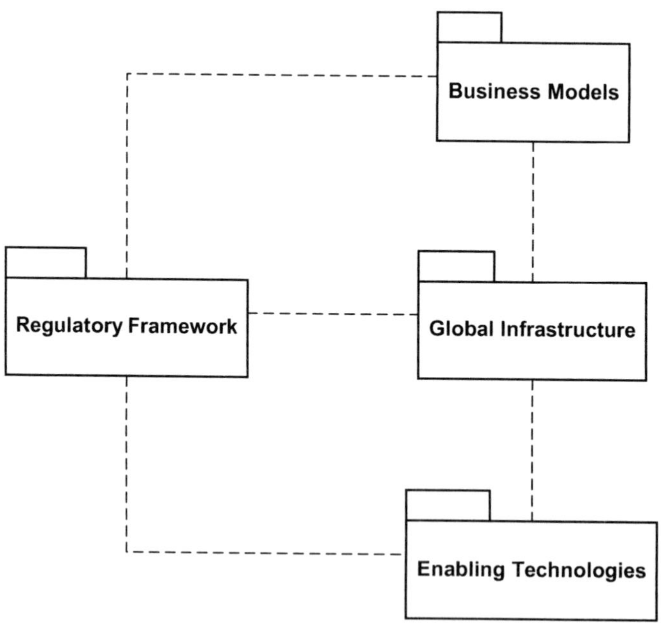

followed? Examples of these rules are legislation, governing taxation, duties, and the handling of disputes.

Some of the above classes have also a direct relationship with the entities presented in Figure 1. The PTD is a subclass of the wireless terminal and the wireless access networks are instances of wireless network in the Enabling Technologies sphere.

SAMPLE SCENARIOS

In this section we look closer at some mobile commerce scenarios in order to illustrate the feasibility of the model presented and in order to discuss location-dependent services —which are at the heart of mobile commerce—in more detail.

Ordering a Taxi

The first scenario is a typical situation in urban environments, where a traveler arrives in a city and wants to order a taxi for local transport. Different phone numbers of taxi agencies in different cities and—in case of foreign countries—language problems render traditional ordering by phone cumbersome, so that mobile commerce services will provide a more convenient solution for the customer.

The presence of location information on both the traveler and the taxi facilitates this new convenience. The basic idea of this service is that once the terminal (i.e., the traveler carrying the PTD) and a taxi can be positioned accurately enough, the taxi can be ordered based on that information without the need for the customer to know local phone numbers or local business practices.

At the business model level, it can be expected that a taxi that is closest to the terminal can offer the transportation service cheaper than a more distant taxi. The overall business protocol involved is represented by a simple sequence diagram shown in Figure 8. For each party involved in the scenario, there is a vertical line. The messages between different parties are represented by arrowed lines between the vertical lines associated with the parties. More elaborate techniques to specify business protocols are feasible, but for the purpose of this chapter, the notations based on sequence diagrams will suffice.

Briefly, the steps go as follows:

1. Positioning of the customer terminal (triggered by the terminal)
2. Sending the coordinates and other parameters to a global directory service server that determines an appropriate service provider based on the position and service type
3. Sending the coordinates and other parameters to a taxi server instance, selecting an appropriate taxi (terminal and server involved)
4. Selecting the taxi and guiding it to the customer (server, taxi involved)
5. Finding the customer and picking him/her up (taxi, terminal)

Figure 8: Business protocol, taxi scenario

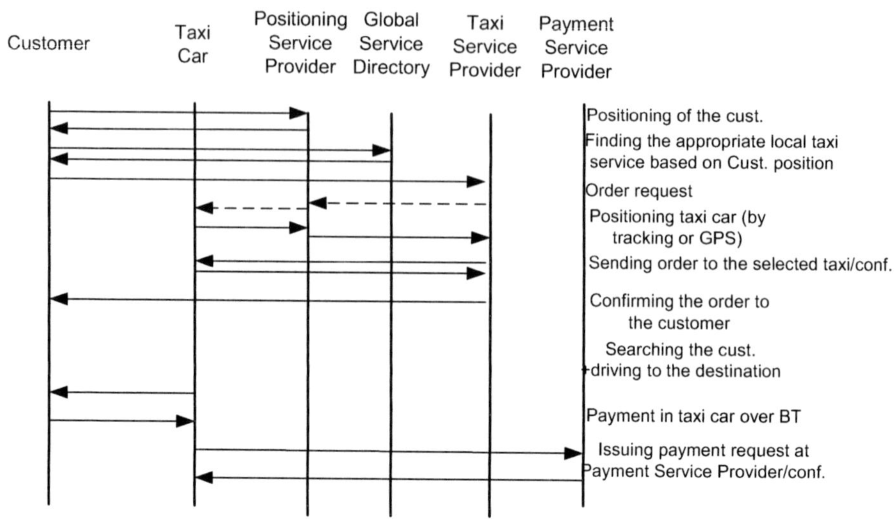

6. Transport as determined by the customer (taxi, physical step)
7. Taking care of payment (terminal, taxi)

In brief, the m-commerce infrastructure does the searching and possibly the negotiation for the customer. It is evident that in step 1 the terminal must be able to position itself and in step 3 subsequently send out the coordinates to a server instance that takes care of the taxi ordering. The first step can be performed using the GPS functionality of the terminal or asking the wireless telecom network to find out the coordinates (existing technology for this tasks includes E-OTD, a GSM-based location technology). In the latter case the terminal uses a special positioning service offered by the network.

In step 2, the main task is to find the address of the local taxi service instance to which the taxi order request should be send. This corresponds to the problem of finding the phone number of a local taxi service. In principle, the country, city, and suburb can be deduced from the coordinates of the terminal, but a rather heavy infrastructure service is needed in order to determine the appropriate taxi service center server in the network to which the request should be sent. Marketplace and auction mechanisms can be deployed to trade transportation resources.

Assuming that step 3 is successful, step 4 is taken. It is more complicated than the former ones. During this step, the specific taxi to serve the customer is selected. It is immediately clear that the location of the individual taxis must be tracked by the taxi service center in regular intervals so that the location is known with a reasonable accuracy. Which taxi to select? An obvious answer is "the closest one which is not occupied," but this is evidently not the only optimization criterion. The closest in Euclidean sense is not necessarily the most appropriate, because the street network might require the closest to drive a long way and maybe stop at many traffic lights before reaching the waiting traveler. Second, the customer can give as part of the order the destination and timing parameters. The latter might indicate that the taxi is needed in, e.g., 10 minutes at the latest. This gives possibilities to optimize the allocation. The destination is also important. Based on the parameters, the taxi service center can optimize the allocation of taxis ahead of time: It knows, at least partially, what will be the location of the fleet in the next 10, 20, or 30 minutes. Knowing the destination might also prune some allocations, because certain drivers would not want to go to the destination indicated by the customer for various reasons. Finally the number of people to be transported as well as specific constraints such as large luggage items are also important and could be provided by the customer. Based on this information the taxi service center can allocate an appropriate vehicle.

Step 5 is about finding the customer. This can be tricky, but ordering the taxi through a PTD helps in several ways. The customer could get as part of the response to his/her order the order number, the license number of the taxi (and even a color picture of it), and the driver's phone number. An even more sophisticated way is to let the taxi track the customer's position while approaching. This is would be handy, if the customer wants to move from the ordering location while waiting. With customer tracking, the need to order the taxi to a certain address and wait there is not necessary anymore. In step 6, the actual trip is performed. The customer can express the destination by giving the address or just confirming the destination given earlier. The point is that the driver and customer need not have a common spoken or written natural language in this phase. The trip is paid in step 7. This can be based on cash, physical credit card, or PTD with payment functionality. PTD can communicate with the payment infrastructure in the taxi using a Bluetooth link and the customer can confirm the payment on the PTD.

Of course, the scenario presented is simplified with respect to numerous aspects. There might be disputes about whether the taxi arrived on time or not, whether the shortest path was chosen, etc. Although important, location-related disputes are for further study (see Tang and Veijalainen (2000) for more on disputes in e-commerce). The commitment to an order and disputes are tricky aspects that are dependent on the business habits (code of conduct) in a country and contribute to the roaming heterogeneity, i.e., business model and technical heterogeneity between countries and mobile network operators. These become evident because people roam and need services from different local providers. Another not so trivial aspect is to map the (XYZ) coordinates to <address>. It is by no means easy in an arbitrary country that uses local coordinate systems. It requires, most probably, coordinate transformations and attachment of them to the local map. Using the XYZ coordinates of the customer directly (in WGS-84 format) is rather hopeless, because they do not tell the taxi driver how to reach the customer. Which player provides this mapping? It can be the taxi ordering company, each individual taxi owner (GPS car map), or an external service provider that upon getting a coordinates <XYZ> returns the <address>. These are actually issues that must be solved within the Global Infrastructure. These infrastructure issues are for further study.

From a business model point of view, it is interesting to investigate what types of players exist and how the revenues obtained from the taxi customer are divided among the parties in the above case. The customer pays both to a mobile network operator and to the taxi company after the trip has been done. If the automated service is not better than the current voice based, the customer is hardly interested in paying more than now. There are still many benefits both from the customer point of view, as well as from the taxi company point of view. Thus, it could be expected that these kinds of systems would become common in the future.

This simple specification of the taxi scenario can be extended in multiple ways. For instance, the ordering system could also support small-scale auctioning; who takes this customer to the destination, how fast and for what price? The result of the auction could then be returned to the customer as a response. These are business model ideas made possible by technology, but the local taxi companies and legislators should endorse them before they become reality.

The above service requires privacy protection measures and customer trust. For instance, the tracking of the phone number should only be enabled for this particular purpose, this taxi-order here and now. In general, m-commerce transaction security can probably be improved through location-based authentication, as it is shown by Denning and MacDoran (1996). Location, as calculated from a location signature, adds a new dimension to user authentication and access control. It can be used to determine whether a person is attempting to log in from an approved location and using approved services. It is for further study, how this kind of authentication can be combined with the m-commerce infrastructure services.

From the discussion of this sample scenario the mapping of the artifacts involved to the mobile commerce model, as presented above, is quite clear. Rather than discussing this mapping in detail, we state the main entities for each submodel:

- *Regulatory Framework:* The Regulatory Framework submodel determines the form of the overall infrastructure and facilitates the terminal and the servers to communicate. While the sample scenario does not rely on any particular wireless network (e.g., no video streams have to be passed), the terminal and the servers have to be connected in a way that facilitates their communication. But technical specifications are not only necessary for wireless networks, but also for the terminal itself and the positioning functionality provided. In some cases the network allows a positioning provided by cell information, unless the terminal can provide the coordinates itself by using a satellite positioning method (GPS, or GALILEO). To provide a more detailed mapping from the entities of the example to the classes of the model, we mention that the device that the traveler carries must conform to some standards (class Standard) set by an organization (Standardization Body). Standards define technical specifications (Technical Specification) consisting of a set of features (Feature) that the personal trusted device (PTD) of the customer offers.

- *Enabling Technology:* The Enabling Technology submodel concentrates on the terminal and its features. In particular, the personal trusted device was built by a manufacturer that is a company, represented by a super-class relationship between the Manufacturer and the Company class in the model. In order to operate in a particular network environment, the PTD requires a wireless network (Wireless), which is a sub-class of the Network class. Networks are provided by network manufacturers, as specified by the association in the class diagram. In this sample scenario, the PTD has to offer specific features, mainly the ability for positioning to make use of the taxi-ordering service. In the class diagram, this feature is represented by the named association called *Requires* between the Feature and the Service class.
- *Global Infrastructure:* The Global Infrastructure models entities that are relevant for facilitating mobile commerce applications from an infrastructure point of view. In the example at hand, this submodel is required for defining the network communication standards as well as specific PTD properties that, e.g., enables the traveler's PTD to connect to the local mobile network. In addition, roaming contracts between the network operators involved have to be in place to allow the traveler to use services in his or her destination city. Appropriate business models, as well as the legal context, have to be present at that location. As this brief discussion shows, the main classes of the global infrastructure diagram shown in Figure 5 are required in this particular example.
- *Business Model:* The business model is implemented by a set of services, which are executed according to some business protocol. While an exact specification of the protocol is outside the scope of this chapter, we mention that the customer requests the service using his or her terminal. The service is transferred to a service provider, which now performs a series of activities in order to serve the customer well. In particular, location information has to be tracked and an appropriate taxi has to be triggered to serve the customer. The PTD must thus be able to run the business protocol in Figure 8. This is a tricky requirement, unless there is a global standard that the terminal can rely on. The global directory service is an easy issue, because one can assume a globally specified protocol. Particular questions arising from step 3 are: in which format are the coordinates presented to the local taxi service? How are the parameters encoded? What parameters are allowed? How is the result to be interpreted? The taxi service is also a service in our sense, but the special feature is that this service uses a physically moving object (a car) to realize the service. In the business protocol sense a taxi is a service provider and is accessible through the wireless network.

As indicated above, complex positioning issues may arise, taking into account physical aspects such as streets, and maybe even traffic. However, the geographical entity class represents the location-dependent properties of the example quite well, including the location of the taxi and the customer (by the Location sub-class of the Geographical Entity class) and the area specification where the taxi service is provided (by Area class).

Wireless Payment

Just like fixed payment terminals, PTDs can be used to pay for goods or services, but there are new possibilities, too. A PTD can contain the private key and credit card information of the customer. This information can be used to build services that are based on electronic commerce infrastructures, on the one hand, and on a wireless short-range link on the other. Concretely, the PTD can communicate with the cash register over Bluetooth in order to pay for the (physical) goods in the shopping cart.

In a standardization effort, the Mobile Electronic Transactions Forum (MeT, 2002) has developed a protocol to facilitate wireless payment. As a result, MeT set up recommendations that service providers have to adhere to in order to provide interoperable and efficient m-commerce solutions, e.g., in wireless payment. But wireless payment also requires Enabling Technology and Business Model aspects. For instance, the Enabling Technology includes a public key infrastructure, which can be modeled as a feature of PTDs. Just like previously, these features are defined by Technical Specifications that in the context of payments may also be specified by law (cf., privacy issues related to payment). In addition, wireless payment is organized as a business model of financial institutions. These payment services are provided to the customers, and they consist of a number of services, which are executed, in some order, according to the business protocol in place. Business actors (for instance, the traveler using a taxi) use that service for convenient payment. The financial institution maintains an infrastructure and business relationships that enables it to organize the payment between the parties involved.

Ticketing

The idea of ticketing applications is structurally similar to the location-based services where the customer first determines the location and then uses it to access some service. In the case of ticketing, however, the customer gets the ticket from a merchant, similarly to the goods in the wireless internet e-commerce. The payment scheme can be a credit card, online banking, and also billing. Goods are, of course, in this case intangible, i.e., bit strings that are stored into the memory of the PTD, encoding the ticket information. Days or even months later after loading the digital ticket into the memory of the PTD, the ticket is used when entering the event, for instance a concert hall or an aircraft in case of a travel ticket. The reasonable usage of this form of electronic tickets requires that the PTDs are capable of communicating over short distances with the cash registers or ticketing devices in the busses and concert houses, etc. This can currently best happen over Bluetooth, although Infrared connections might also be feasible in some cases. The general scheme looks as follows:

Tickets can often be cancelled or modified before they are used. If cancelled, the money charged is typically paid back. The cancellation transaction assumes that paying back is possible by knowing, for example, a terminal owner's identity. In practice, money return is possible by using credit card or bank account number stored at merchant's database for customer with the recorded identity. Notice that the ticket could also be modified (typically, an airline ticket is often changed before it is used) and even transferred to other person's PTD, using a Bluetooth or infrared link. We stress that MeT develops schemes for standard ticket formats. In this scenario we refrain from discussing the various classes of the mobile commerce model but leave these considerations to the interested reader.

RELATED WORK

As concerns related work, we are not aware of as extensive approaches covering wide aspects of mobile commerce environments as ours. A work somewhat relevant to the one presented in this paper is Siau and Lim (2001), where the authors provide an overview of mobile commerce and a research agenda. A clear separation between Enabling Technologies and Business Models is drawn, but no particular modeling incentive is applied. That paper concentrates on Enabling Technologies and Business Models spheres of concern in our sense. The Regulatory Frameworks sphere is hardly touched, and it is not treated as a special theme requiring attention.

A somewhat similar framework for m-commerce applications, as ours is presented by Varshney and Vetter (2001). The framework also uses a layered approach, but it focuses strictly on technological layers. The main purpose of that work is to assist in the development of m-commerce applications. It analyzes m-commerce application types, such as mobile financial applications. The paper shows, in a pragmatic way, that the properties of wireless network technology can be reflected to the application under development. Thus, the paper is well equipped to improve the development of m-commerce applications. The focus of our chapter is different: it is broader and more formal. It is broader since not only technical but also organizational and legal aspects are covered in the four spheres of concern. It is more formal, since for each layer the main artifacts are modeled using an object-oriented approach. Finally, we seek to find invariant properties of mobile commerce environments, which are not covered in Varshney and Vetter (2001).

CONCLUSIONS

This chapter discusses mobile commerce environments from a bird's-eye perspective. The focus is on organizing the main concepts and players in mobile commerce into four spheres of concern with the aim of a better understanding of this complex and dynamic area. The fundamental basis is the Global Infrastructure. It consists of wireless and wireline access networks and a global backbone, able to transmit huge amounts of data between terminals and servers. The Global Infrastructure covers a huge number of wireless terminals, soon more than 1 billion. This basic structure facilitates mobile commerce and will persist a long time into the future, although the concrete wireless and wireline access technologies and those of the backbone itself will change over time. We expect the digital convergence to homogenize the backbone technology, and it is also rather probable that all wireless and wireline terminals will have a unique global network address in the future. The access technologies will show more variety and at least the big division into wireless and wireline technologies will persist.

From an m-commerce point of view, transmission speeds have an important role, because faster—and cheaper—wireless data transfer enhances the set of m-commerce services that can be used by the customers. Transmission speeds on individual channels will increase in both categories over time. A curious exception seems to be the telecom voice traffic, where the 4kHz bandwidth and 64 kbps uncompressed capacity requirement are expected to persist years. Wireless channel data transfer rates will have to grow but will in general be much smaller than wireline ones. Still, already GPRS offers transmission speeds that are essentially the same as typical wireline modems, and 3G should mean a considerable improvement in this respect. In overall traffic volume, wireless traffic will increase relatively and absolutely. Within the wireless access networks, voice traffic will form a smaller portion than currently and data traffic a greater portion. Still, the wireless voice traffic will grow many years to come, because the rapid proliferation of the wireless phone customer base increases the traffic volume.

Global Infrastructure will consist of hundreds or thousands of different networks, run by different operators. The terminal owner has a home network, and if that cannot service her while roaming, foreign networks have to offer wireless access and network services. The nasty problem is roaming heterogeneity that is encountered at each system level. The Open Mobile Architecture Initiative tries to tackle this problem at the m-commerce service level.

It remains to be seen how this will succeed, because it is not only a technical and organizational issue, but also a legal and business issue.

Another fundamental sphere is Enabling Technologies. Their development was a necessary condition for the mobile commerce to emerge. The Personal Trusted Devices are in a crucial position in this respect, because the technologies incorporated in them largely determine which m-commerce services can be offered. The advances in enabling technologies, such as integrated circuits, processor technology, wireless communication technologies, software technologies, cryptography, battery technologies, positioning technologies, etc., and their implementation in PTDs are a crucial condition for mobile commerce to advance qualitatively. Through increasing m-commerce functionality more sophisticated services are possible and, consequently, higher revenues become possible.

A further relatively independent sphere of concern is Business Models. For the concrete business models the Enabling Technologies and the Global Infrastructure are a necessary basis. Our modeling effort concentrates on the basic business actors and their interactions over the network that form the concrete instantiation of the mobile commerce. We exclude the infrastructure providers that do not explicitly take part in performing m-commerce transactions. The main source of revenues is the customer accessing the m-commerce services over the network infrastructure, but also advertisement or other similar sources are possible. The Business Models sphere also addresses the geographical validity region of a business model. This is important, as the business actors and the models applied always have a place or area on the earth where they are legal and/or accessible. How high can the revenues of individual actors be? What kind of business actors and models can exist? These kinds of questions are not answered at the level of our model, but they are interesting and must be tackled in the future. Our modeling incentive helps in focusing on the right issues.

Finally, we discuss Regulatory Frameworks. These include the legal bodies that issue legislation, as well as standardization bodies and interest groups. They all can be seen as global designers specifying a common interoperability area. Some standardization bodies, like ISO or ETSI have basically global coverage. They do not, however, have any legal means to enforce the standards into usage. A typical problem with the current truly global standardization bodies is that they are rather slow in developing the standards, in comparison to the emerging needs. This has led to the establishment of industry-led unofficial bodies that we call interest groups. For mobile commerce area the most important bodies are the WAP Forum, the Mobile Electronic Transactions Forum, and the Location Interoperability Forum. A still less cohesive conglomerate is the Open Mobile Architecture Initiative (OMA), but it has the most ambitious goal: "to create a nonfragmented, interoperable global market for the next generation of mobile services." Should this initiative be successful, it will have crucial impact on the future of m-commerce.

Common to mobile commerce and electronic commerce, as they exist currently, are technologies like HyperText Transfer Protocol (HTTP) for communication, HyperText Markup Language (HTML) or a member of the eXtensible Markup Language (XML) family for content, and the Java programming language for functionality. Regulatory bodies and especially interest groups also specify these. Furthermore, they play an important role also in business aspects. For instance, there are a couple of interest groups involved in setting up XML-based recommendations for business to business integration; the ebXML approach is put forward by OASIS and UN/CEFACT, a United Nations body. Naturally, there are a couple of differences in the Enabling Technology sphere between electronic commerce and mobile commerce. There is little need for defining specific devices and their properties, since electronic commerce applications are typically run by a computer connected to the Internet. Business aspects, however, are quite similar. There are electronic business companies, which offer services and business protocols to implement their business models. The Mobile Electronic Transaction Forum, for instance, is actually defining essential parts of the business model (services and protocols), but also lower-level technical specifications. Whether this is feasible, globally, remains to be seen.

In this chapter we propose a conceptual, object-oriented model to describe different spheres of concerns in mobile commerce environments. The framework seems promising. We are confident that the spheres of concern introduced here are fruitful in the forthcoming research. At the same time we are aware of the limitations of the work as concerns the analysis of the dynamics of the m-commerce field. This is for further study.

ACKNOWLEDGEMENTS

The work of the first author was performed to a considerable extent in summer 2001 while he was visiting FhG-FIT in Sankt Augustin, Germany, during his leave of absence. The support of FhG-FIT, as well as the support of the National Technology Agency of Finland (TEKES), Nokia, HP Finland, and Yomi Solutions under contract 40599/99 (MultiMeetMobile) are highly appreciated. The valuable comments of Jukka Heikkilä, Antti Aarnio, and Aki Enkenberg are also highly appreciated.

REFERENCES

ANSI (2002). *American National Standards Institute* [online]. Accessible at http://www.ansi.org.

Booch, G., & Rumbaugh, J., & Jacobson, I. (2001). *The Unified Modeling Language User Guide.* Boston: Addison-Wesley.

Bluetooth Consortium [online]. Accessible at http://www.bluetooth.com.

Denning, D. E., & MacDoran, P. F. (1996). *Location-Based Authentication: Grounding Cyberspace for Better Security, Computer Fraud & Security.* Elsevier Science Ltd. Accessible at http://www.cosc.georgetown.edu/ denning/infosec/Grounding.txt.

Devine, A., & Holmqvist, S. (2001). *Mobile Internet Content Providers and Their Business Models.* Master's thesis, Stockholm Kungl Tekniska Högskolan, January 2001. Accessible at http://www.japaninc.net/online/sc/master_thesis_as1.pdf.

Durchlacher Research, Ltd. (2000). *Mobile Commerce Report* [online]. Accessible at http://www.durchlacher.com.

EU Information Society portal (2002). Accessible at http://europa.eu.int/information_society/services/sitemap/index_en.htm.

EU E-Commerce legislation portal (2001). Accessible (archived) at http://europa.eu.int/ISPO/ecommerce/legal/legal.html.

ETSI (2002), European Telecommunications Standards Institute (ETSI) [online]. Accessible at http://www.etsi.org.

Oasis, UN/CEFACT: *Electronic Business Exchange Jointly Sponsored by OASIS and United Nations Centre for Trade Facilitation and Electronic Business* [online]. Technical Documentation accessible at http://www.ebxml.org.

Helal, A., Haskell, B., Carter, J. L., Brice, R., Woelk, D., & Rusinkiewicz, M. (1999). *Any Time, Anywhere Computing; Mobile Computing Concepts and Technology.* Kluwer Academic Publishers.

IEEE (2002). The Institute of Electrical and Electronics Engineers, Inc. [online]. Accessible at http://www.ieee.org.

IEEE LAN/MAN Standards (2002). LAN/MAN Standards Committee 802 [online]. Standards available at http://grouper.ieee.org/groups/802/.

ISO (2002). *International Organization for Standardization (ISO)* [online]. Accessible at http://www.iso.org.

ITU (2002). International Telecommunication Union [online]. Accessible at http://www.itu.int/home/index.html.

Kaario (2000). A wearable project presentation [online]. Accessible at www.uta.fi/hyper/projektit/iti/esitykset/kaario_clothes.pdf.

Kahney & Leander (2002). *Video Clothes: "Brand" New Idea.* [online]. Accessible at http://www.wired.com/news/technology/0,1282,36698,00.html.

Location Interoperability Forum (LIF) (2002). Accessible at http://www.locationforum.org.

MeT (2001). *MeT overview White Paper: The MeT Initiative—Enabling Mobile E-Commerce* version 2.0, January 2001 [online]. Accessible at http://www.mobiletransaction.org.

MeT (2002). Mobile electronic transactions forum [online]. Accessible at http://www.mobiletransaction.org.

MIT (2002). The MIT wearable computing Web page [online]. Accessible at http://wearcam.org/computing/.

Nordea Bank (2002). *See Especially Solo Services* [online]. Accessible at http://www.nordea.com.

NTT DoCoMo (2002). FOMA Services/V-LIVE [online]. Accessible at http://foma.nttdocomo.co.jp/english/.

OMA (2001). *Open Mobile Architecture Initiative* [online]. Press releases available at http://press.nokia.com/PR/200111/840158_5.html, http://www.hp.com/communications/nokia_mobile.html.

OMA (2002). *Open Mobile Architecture Initiative* [online]. Accessible at http://www.nokia.com/oma/.

Siau, K., & Lim, E. (2001). Mobile commerce: Promises, challenges, and research agenda. *Journal of Data Management, 12* (3).

Tanenbaum, A. S. (1996). *Computer Networks,* 3rd ed. Englewood Cliffs: Prentice Hall.

Tang, J., & Veijalainen, J. (2000). On e-commerce transaction protocols that support atomicity based dispute handling with untrustworthy players. *Proceedings of the 3rd International Conference on Telecommunications and Electronic Commerce* (ICTEC 2000), Dallas, TX, USA, Nov. 16-19, 2000, pp. 299-314.

Tang, J., Terziyan, V., and Veijalainen, J. Distributed PIN verification scheme for improvement security of mobile devices. *ACM MONET*, Special issue on security in mobile computing environments (forthcoming).

Telecommunications Policy (2002). 26 (3-4). List of contents accessible at http://www.elsevier.com/cdweb/journals/03085961/viewer.htt?vol=26&viewtype=issue&iss=3-4.

Timmers, P. (2002). Business models for electronic markets. Accessible at http://www.electronicmarkets.org.

UMTS forum (2002). Accessible at http://www.umtsforum.org.

Varshney, U., & Vetter, R. (2001). A framework for emerging mobile commerce applications. In R. Sprague (ed.): *Proceedings of the 34th Hawaii International Conference on System Sciences.*

Veijalainen, J. (1990). *Transaction concepts in autonomous database environments.* Ph.D. thesis. Berichte der GMD Nr. 183, Oldenbourg Verlag, 1990.

Veijalainen, J., Eliassen, F., & Holtkamp, B. (1992). The S-transaction model. In A. Elmagarmid (ed.), *Database transaction models for advanced database applications*. San Mateo: Morgan Kaufmann.

Veijalainen, J. (1992). Issues in Open EDI. In P. Ng, C. Ramamoorthy, L. Seifert, & R. Yeh (eds.), *Proceedings of the Second International Conference on Systems Integration* (ICSI'92), Morristown, NJ, June 15-18, 1992, pp. 401-412.

Veijalainen, J., & Tsalgatidou, A. (2000). Electronic commerce transactions in mobile computing environment. In Q. Jin, J. Li, N. Zhang, J. Cheng, C. Yu, S. Nogushi (eds.), *Proceedings of IS2000 Conference*, Fukushima, Japan, November 5-8, 2000, (pp. 37-45). (Best Paper Award). Accessible at http://www.cs.jyu.fi/~mmm.

Veijalainen, J., & Tsalgatidou, A. (2001). Electronic commerce transactions in a mobile computing environment. In Q. Jin, J. Li, N. Zhang, J. Cheng, C. Yu, & S. Noguchi (eds.), *Enabling Society with Information Technology* (pp. 131-140). Springer Verlag.

WAP forum (2002). Wireless Application Protocol Forum [online]. Accessible at http://www.wapforum.org.

W3C (2002). The World Wide Web Consortium [online]. Accessible at http://w3c.org.

Chapter VIII

Known By the Network: The Emergence of Location-Based Mobile Commerce

Stuart J. Barnes
Victoria University of Wellington, New Zealand

ABSTRACT

The use of mobile telecommunications devices for commercial transactions, coined mobile (m-) commerce, has been an emerging trend since the late 1990s. As the phenomenal growth of the Internet and mobile devices has continued unabated, the inevitable convergence of these two streams of technologies has occurred, promising a plethora of mobile data services to the handset user. Although these services have been considerably hyped in the media, and adoption has been somewhat patchy and limited, it does signal the emergence of a range of innovative value added services. With further developments in technology and markets, further services will appear, bringing new revenue streams. One potential area of m-commerce development is in location-based services (LBS). LBS are heralded as the next major class of value added services that mobile network operators can offer their customers. Using a range of network- and handset-based positioning techniques, operators will be able to offer entirely new services and improvements on current ones. Popular examples cited include emergency caller location, people or asset tracking, navigation, location-based information, or geographically sensitive billing. The purpose of this chapter is to examine the technologies, applications and strategic issues associated with the commercialisation of LBS. The chapter concludes with some predictions on the role of LBS in m-commerce.

INTRODUCTION

Since the technological convergence of the Internet and mobile telecommunications networks in the 1990s, the mobile Internet has brought the promise of significant changes in data communications. Separately, the Internet and mobile phone have witnessed extraordinary market penetration, and each is predicted to grow to 1 billion users by 2003-4 (IDC Research, 2001). Together, these technologies have created the platform for a raft of mobile data services, including business-to-consumer (B2C) applications for financial services, gaming, email and news, and business-to-business (B2B) applications for teleworking, logistics, field sales and after-sales servicing. Worldwide, revenues from mobile (m-) commerce—i.e., transactions over wireless telecommunications networks—are expected to exceed $200 billion by 2004 (Strategy Analytics, 2000).

In the emerging m-commerce economy, the knowledge of the position of a given service subscriber making a call is gaining particular interest among mobile operators who can, in turn, provide innovative location-based services (LBS), typically with the assistance of third parties such as service or content providers (see Barnes, 2002). Such ideas are not new. Location (l-) commerce has existed in a limited form for more than twenty years. The pioneers of location-based services were basic tracking services and automated vehicle location (AVL). In 2000, more than 100 companies were providing AVL products and services in the US (Airbiquity, 2000a). However, until recently, the specialised location-based industry survived as a niche market to both high-end businesses (such as trucking and freight) and well-to-do customers (via automobiles such as Lexus and BMW). Typically, high-priced devices required subscriptions to special location services, suppressing demand.

Large-scale commercialisation of location-aware services has only been recognized in the early 21st century, as a series of events and trends have begun to provide an environment that is conducive. Underlying the growth in commercial LBS markets are recent technological advancements (in handsets, networks and positioning technologies), regulatory change (including the removal of restrictions of satellite positioning technologies and mandates for emergency services), industry trends (particularly the need for new value-added services, mergers/acquisitions, and call-centre development) and emerging business opportunities (as a result of converging market conditions, e.g., the growth of LBS in Japan driven by the popular i-mode service). As a result, the door has been opened to a vast array of commercial applications, including those for emergency services, asset tracking, navigation, location-sensitive billing, and location-based information services. Indeed, the Strategis Group (2000) estimates that LBS could be worth $3.9 billion by 2004.

The purpose of this chapter is to examine the emerging l-commerce phenomenon. To this end, it analyses the technologies and applications involved with introducing the new wave of LBS (in sections 2 and 3, respectively). The chapter continues by exploring a value proposition model for services (section 4) and some of the core inhibitors (section 5). It also briefly explores some of the strategic business implications of these services (in section 6). Finally, the chapter rounds off with some conclusions and predictions regarding the future of l-commerce.

Location Technologies for Mobile Commerce

One or more location methods can be used to determine the position of user equipment for LBS. It is also possible to distinguish between methods that are most useful inside and outside buildings. Leading candidates for indoor location identification include short-range radio, such as Bluetooth, and infrared (IR) sensors (Barnes, 2002). For example, developers could use Bluetooth or IR to build an automatic tour-guide system, such as for an art gallery; as the tourist with a suitably enabled PDA device moves into range of a piece of artwork, it could send out a signal that automatically displays information related to the artwork on the screen (Tseng, Wu, Liao, and Chao, 2001). However, interesting though this is, the focus of this chapter is on roaming, location-aware technology used largely outside buildings. For a detailed examination of the benefits and applications of short-range wireless technologies, see Barnes (2002).

Location techniques operate in two steps—signal measurements and location estimate computation based on the measurements—which may be carried out by the user equipment or the telecommunications network (Lavroff, 2000). Subsequently, positioning techniques can be categorized into several varieties, each with its advantages and disadvantages. The main types are cell-location, advanced network-based, and satellite-based positioning. Three of the main categories of positioning methods are shown in Table 1, in order of increasing accuracy.

Table 1: Three methods of location positioning

Location Service Category	Explanation	Typical Methods of Positioning	Accuracy	Response Time	Key Limitations	Market Requirements
Category LS1: (Basic service level)	Location of all handsets with at least cell accuracy	Cell of Origin (COO) or Cell-ID, including Service Area Identity (SAI), LocWAP and enhanced Cell-ID. May also include enhancements with propagation time measurements	*Low.* Depends on cell size and enhancements; typically 150m to 10,000m	*Very Fast.* Typically around 3 seconds	Very limited accuracy in areas with low cell radius	No modifications needed to networks or handsets
Category LS2: (Enhanced service level)	Location of all new handsets with reasonable cost and improved accuracy	Estimated Time of Arrival (EOTD) for GSM, and its variations such as Advanced Forward Link Triangulation (AF-LT) and Idle Period Downlink (IP-DL) for CDMA and WCDMA respectively	*Medium.* Typically around 50m to 125m	*Fast.* EOTD takes around 5 seconds	Dependent on visibility of base stations for signal measurement and number of location measuring units (LMUs)	Software modified handsets needed for positioning
Category LS3: (Extended service level)	Location of new handsets with high accuracy and higher costs than LS2	Global Positioning System (GPS) and Assisted Global Positioning System (GPS)	*High.* Outside buildings, approx. 10-20m; inside buildings, approx. 50m	*Variable.* GPS takes around 10-60 seconds, but AGPS around 5 seconds	Signal degradation and reduced accuracy in certain environments, e.g., inside buildings or "urban canyons"	New handsets needed for positioning

Cell-location Positioning Techniques

This technique works by identification of the cell of the network in which the handset is operating (the "cell of origin"). Cell of origin (COO), sometimes called Cell-ID, is the main technology that is widely deployed in wireless networks today. It requires no modification to handsets or networks since it uses the mobile network base station as the location of the caller. However, although locating the caller is fast—typically around three seconds—accuracy is limited. Positioning accuracy depends on the size of the cell and techniques used for enhancing location calculation, such as user self-locating (whereby end-users use landmarks and addresses to improve their positioning precision) and propagation time measurements. Position accuracy down to 150 metres in urban areas is not uncommon, growing very significantly outside major areas of population.

Advanced Network-based Positioning Techniques

Advanced network-based techniques rely on the measurement of signals from nearby base stations via the user's equipment. The position of the user is derived by triangulation, using techniques such as Enhanced Observed Time Difference (E-OTD) and Observed Time Difference of Arrival (OTDOA). The E-OTD method works on the GSM network. Variations of E-OTD such as Advanced Forward Link Triangulation (AF-LT) and Idle Period Downlink (IP-DL) have been developed for CDMA and WCDMA networks. The positional information is based on relative times of arrival of signals at the handset and fixed receivers as sent by base stations. Location receivers or reference beacons (referred to as Location Measuring Units or LMUs) are overlaid on the cellular network at a number of geographically dispersed sites. Location is then calculated using the time differences of arrival of the signal from each base station at the specially enabled handset and LMU (via time stamps and intersecting hyperbolic lines). E-OTD is typically accurate to approximately 50 to 125 metres, with a response time of around 5 seconds. In a manner similar to E-OTD, OTDOA location works by calculating the time difference of the arrival of a signal from a mobile device and three network base stations. The large cost of network synchronisation affords only small improvements over COO in urban areas, and the response time is much higher at around 10 seconds.

Satellite-based Positioning Techniques

In some cases, a global navigation satellite system such as the Global Positioning System (GPS) can be used to enhance the accuracy of radio positioning. GPS has been available for general use since the early 1990s. Operating in the L-band frequencies GPS can be used anywhere in the world. The system's satellites transmit navigation messages that contain their orbital elements, clocks, and statuses, which a GPS receiver uses to determine its position and thus its roaming velocity (Tseng et al., 2001). Determining the receiver's longitude and latitude requires three satellites, and adding a fourth can determine the user's altitude. However, only recently (in May 2000) has the US Army removed restrictions upon outdoor positioning to a sufficiently high resolution for advanced use—currently 10-20 metres.

Stand-alone GPS have the key problems of no indoor coverage and a relatively long time to first fix, usually 10-60 seconds. It also fails in radio shadows and requires considerable cost, complexity and battery consumption in handsets (Djuknic and Richton, 2001).

The same issues are also involved in the use of GPS for mobile ad hoc networks (MANETs). A MANET consists of a set of mobile hosts that roam at will and communicate with one another. Typical examples of MANET applications include battlefields, festival grounds, assemblies, outdoor activities, rescue actions and major disaster areas, where communications are needed immediately without core network infrastructure (Tseng et al., 2001). However, the flexibility of these systems could lead towards more developed forms being further used commercially. Communication in MANETs takes place through wireless links among mobile hosts, using their antennae, but no base stations are involved. Transmission limitation means that several hosts may be needed to relay a packet between sender and receiver (Tseng et al., 2001). In this environment, location-positioning technologies are needed that do not require traditional network infrastructure. GPS is the prime enabler for this type of outdoor positioning.

Using GPS in addition to a wireless network—often referred to as assisted-GPS (AGPS)—can provide significant extra benefits. Embedding a GPS receiver into the user's handset can directly provide positioning fixes in less than 5 seconds; the network may assist the user equipment by reducing the power consumed of the handset, by optimising the start-up and acquisition time and by increasing the sensitivity of the GPS device (Lavroff, 2000). AGPS can also be used indoors, where it is accurate to within 50 metres. In the future, technologies such as Bluetooth and IEEE 802.11 may enable assisted location positioning within building to even higher resolutions, suggested at around 10 metres (Nokia, 2001).

Overall, the various positioning technologies are complementary—there is no single universal solution. Where both accuracy and coverage are important, hybrid technologies may provide an optimum solution. Cellular and advanced network-based technologies can be used to fill in the gaps in coverage from satellite-based systems, like GPS. The basic positioning accuracy category is focused on market penetration and should be available for all phones, enabling a fast time-to-market. The intermediate category will have a software impact on handsets, whilst the high accuracy category will have a hardware impact on handsets. All three levels of accuracy will exist in parallel in the future (Nokia, 2001).

APPLICATIONS OF L-COMMERCE

The kinds of location-based technologies described above enable many advanced forms of data services based on the position of the user, in both personal and business markets. Typically, services can be categorized into four main areas, as demonstrated by Figure 1. Let us examine each of these areas in turn.

Safety

The prime driver for the implementation of l-commerce infrastructure in the US is safety. Emergency and rescue services have a vital need to know the current location of any host that sends an emergency message. The US government has mandated that providers of personal communication systems must, in the near future, add location-identification capability to their emergency 911 services. Specifically, handset-based solutions must, by 1 October 2001, locate an emergency caller to within 50 metres for 67% of calls and within 150 metres for 95% of calls. Alternatively, the carriers relying upon network-based technology must achieve location accuracy of 100 metres for 67% of calls and 300 metres for 95% of calls. Location platforms such as Xypoint, the largest provider of such services to operators, are pioneers in this area (Xypoint, 2001a; 2001b). Carriers must also undertake reasonable efforts to achieve 100% penetration of handsets allowing location services by 31 December 2004 (Lavroff, 2000). Europe has no such mandate, although during 2001 the European Commission is rewriting the telecommunication regulatory framework in an attempt to make appropriate location

Figure 1: Key areas of application of location-based services

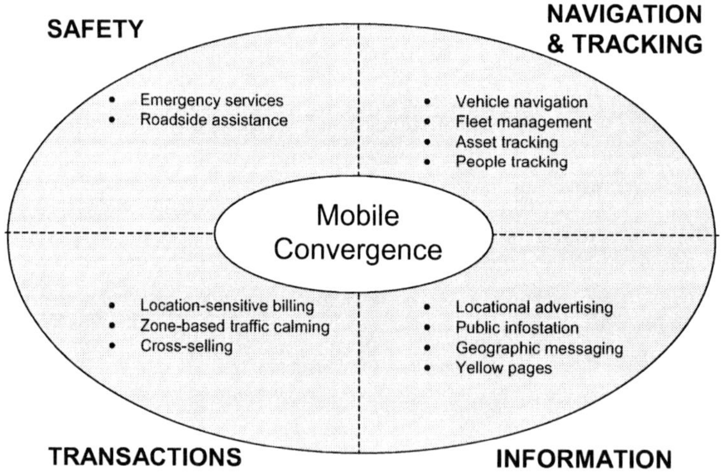

information available to emergency authorities by 1 January 2003 (Wieland, 2000). The same technology used for emergency "911" services also has value for other, related aspects of personal safety, particularly roadside assistance. In the event of an emergency breakdown or accident, the consumer's mobile device could be used to assist in getting roadside assistance to the right location. Accuracy requirements are likely to be around 125 metres or less for mass acceptance, although 500 metres is widely regarded as the entry level requirement for such services (Buckingham, 1999).

Navigation and Tracking

Driving directions and the tracking of fleet, packages and people are a core segment of the emerging LBS market. In the US, the average person spends 500 hours/year in an automobile. Interestingly, though, only 100,000 of the 146 million registered cars in the US and 20 percent of fleet trucks are equipped for telematics (i.e., wireless telecommunications for automobiles). Key players include ATX Technologies/Protection One, AAA/Response, Signature, Cross Country, and OnStar commercial call centres (Airbiquity, 2000a). With the widespread deployment of services, the telematics market could expand considerably in the next few years.

Location technologies can play an important part in logistics. Intelligent transportation systems are being introduced around the world, and location technology plays a key part in almost every solution. Taxis are being equipped with automatic vehicle location devices, allowing the fleet dispatch system to automatically select the taxi closest to the pickup location (Research in Motion, 2000). Similarly, fleet management systems are helping freight companies to monitor the status of deliveries and other logistics activities (Little, 2000). Wireless transceivers let portable terminals—such as PDAs—communicate with a central database. Terminals can log in shipments of materials from vendors and track those materials in inventory, as they are needed (Zeus Wireless, 2000). In this situation it is even conceivable to know all inventory in transit—or "rolling" inventory—allowing an efficient method of selecting a source of components based on their known location. By knowing the location of "rolling" inventory, times between transaction, manufacture and delivery can be further reduced (Varshney, 2000).

In Israel, mobile operator Orange offers a wireless workforce application, using CT Motion's Cellebrity platform, which includes location technology based on EOTD. Orange can offer companies the ability to monitor the movements of their workforce throughout the country to an accuracy of up to 100 metres. The central coordinator, or dispatcher, can see where each of the workers is on a map of the country and so can allocate tasks more efficiently (Wieland, 2000).

Tourists are a key customer segment requiring location-based information, since they are most often found in unfamiliar geographic areas. Some services, such as Bluesigns, are aimed at these consumers (Russell, 2001). Bluesigns works by the tourist phoning the Bluesigns tourist information centre via a telephone access number. During the tourist's call, location is determined, and location-sensitive information is generated from a database. Location can be determined either by GPS or verbal communication with an operator at the tourist information centre. Based on the customer's location, the tourist can be guided along highways to a particular destination, such as a petrol station, restaurant, hotel or tourist attraction.

Similarly, Webraska—a mapping and navigation company—offers a number of services through mobile operators such as KPN (Holland), AirTel (Spain), SFR (France) and Proximus (Belgium). Based on COO technology, the service requires users to type in abbreviations of their location (e.g., L-O-N for London, then B-A-K for Baker Street) before the network can locate roughly where they are and provide location-specific information, such as directions (Wieland, 2000).

Transactions

Location-based transactions are perhaps the most complex set of services. The main thrust of the business model is billing based on the customer's location. For example, a number of countries, such as Singapore, use road pricing as part of their traffic calming and environmental policy. Payments are made electronically through a special, in-car device on entry into a particular geographic area requiring payment. Given the availability of secure payment mechanisms through the mobile phone, such as electronic cash, this could present a convenient replacement.

Location-sensitive transactions open the way to new forms of price differentiation based on the location of the user. On one level, BellSouth Wireless Data is offering "distressed items" such as discounted last-minute tickets to Broadway shows for people that are near enough to pick them up before curtain (Bourrie, 2000) —a form of price discrimination. On another level, individuals could be charged and taxed according to geographic region, such as a US state or country, or proximity to outlets selling goods that the consumer wishes to purchase.

Location-based cross-selling is another possible stream of transaction revenue. For example, the mobile user who has just seen a film at the cinema could immediately be offered a CD or DVD of the soundtrack or film. Similarly, in addition to charging for information requests, such as a query for a restaurant address, service providers could earn additional revenue by asking subscribers whether, for another 10 cents, they would like directions to the restaurant (Bourrie, 2000). Ultimately, retailers, such as restaurants, could share in service costs to encourage customer interest.

In terms of payment systems, the mobile Internet has some way to go towards maturity. However, a number of solutions are underway. Mobile electronic cash refers to cash stored—via subscriber identity module (SIM) or credit-sized card—and transferred via the wireless network. In the UK, Visa piloted a debit smartcard system called Visa Cash in 1999, while France Telecom launched a similar service called Iti Achat. Such a system relies on a dual-slot phone such as Motorola's StarTac D model that can accept a credit-sized card. However, the extra size and weight of devices favour a more SIM-based approach. For example, in Finland, Sonera's Pay-by-GSM enables the user to dial a number to receive a charge to a prepaid phone or for a deduction from a mobile account. Similarly, KLELine (part of Paribas) allows a virtual wallet application that can be loaded from major credit cards. Visa, Nokia, and Merita-Nordbanken are piloting the dual SIM concept for the Nokia 7110 phone, where a second SIM is a Visa credit, debit and bankcard.

Information

The roaming user can be provided with information, alerts or even advertisements based on their locale. Typically, advertisements depend on location. For example, a particular sale may interest only people within a certain distance of a merchant's store. Thus, the sender will only need to transmit the advertisement—which can be regarded as a broadcast message—to users within a set distance (Tseng et al., 2001). For example, walking down the street in an urban area could set off a plethora of messages from retailers eager to tempt clientele inside. Bell Mobility is currently piloting such ideas via digital couponing, offering discounted products and services to subscribers within a certain radius of participating merchants (Bourrie, 2000). Similarly, GeePS is beta-testing its location-based wireless online shopping portal in New York and San Francisco, using couponing and other strategies.

Similar to advertising, geographic messaging is another useful application of location technologies. For example, an alert could inform the user of a security threat in a certain part of the city, such as a train station, stadium or shopping mall. Other types of public localised information can also be broadcast in a particular area—a public infostation; for example, the opening times of a public library, movie theatre listings, city phone directories, the schedule of bus services, or the availability of parking spaces could all be public broadcast information.

One of the most basic LBS offered by mobile operators is the mobile Yellow Pages. Indeed, many European operators are reluctant or unable to go beyond this sort of service (Hamilton, 2000), offered, for example, by Sonera, diAx and Telia. In this type of service, the roaming user asks the question: "what's near me?" For example, items such as locations of restaurants, shops, public transport or nearby ATMs may be useful to users as they move through an unfamiliar city. Weather or traffic information can also prove useful; Bell Mobility's Book4golf service allows the user to locate a North American golf course, book a tee time, and get a location-specific weather forecast.

An extension of the "what's around" type of service is the "who's around" service. Such an application determines who currently occupies a specific geographic area. These services are useful for planned or unplanned rendezvous between individuals, such as business colleagues or social friends. Meeting (or possibly avoiding) people becomes much simpler if individuals are enabled for such LBS.

CREATING RELEVANT USER SERVICES— A VALUE PROPOSITION MODEL

While geo-location technologies open the door to a variety of services in consumer and business markets, location in isolation provides a bounded set of opportunities, and the potential for developing relevant services for the user goes much farther in its scope. Indeed, some of the LBS discussed in the section above have hinted at some other important value propositions for wireless services. Overall, wireless devices have a number of features that together form a fertile bed for advanced value-added services (Kannan, Chang, and Whinston, 2001). Typically, devices are very personal to the user and carried on the person; aspects of the context of the user, such as time and place, can be measured and interpreted; services can be provided at the point of need; and, applications can be highly interactive, portable and engaging. For the consumer, this means that wireless services can be potentially very personal, timely and relevant, or even integrated with other services in a near-seamless way (Katz-Stone, 2001).

In terms of the value proposition, we can untangle three important aspects that influence the nature of service relevance: time, location and personal characteristics of the user. An individual's behavior is likely to be influenced by his or her location, time of day, day of week, week of year, and so on. Individuals may have a routine that takes them to certain places at certain times, which may be pertinent for mobile services. If so, marketers can pinpoint location and attempt to provide content at the right time and point of need, which may, for example, influence impulse purchases (Kannan et al., 2001). Feedback at the point of usage or purchase is also likely to be valuable in building a picture of time-space consumer behavior. Further, the nature of the user, in terms of a plethora of personal characteristics such as age, education, socioeconomic group, cultural background, residence, memberships, and so on is likely to be an important influence on how information is processed. Some of these aspects have already proven to be important influences on Internet use (OECD, 2001), and, as indicative evidence has shown, elements such as user age are proving an important influence on data communications via mobile devices (Funk, 2000; Puca, 2001). The wireless medium has a number of useful means for building customer relationships. Ubiquitous interactivity can give the customer ever more control over what they see, read and hear. Personalization of content is possible by tracking personal identity and capturing customer data; the ultimate goal is for the user to feel understood and to simulate a one-to-one personal relationship.

Let us examine a simple example, where the user wants to catch the next train from work to home. Figure 2 presents the value proposition model and analyses the value (V) attributed by the user to a number of service options. The lowest value proposition involves the traditional provision of train schedules to the user, devoid of context (V = a). This is the situation at the origin of Figure 2. By considering further value functions (f), we may be able to create additional value associated with the service options. Adding personal characteristics (P) to the situation would enable a

Figure 2: The value proposition model—value-added services for the train commuter

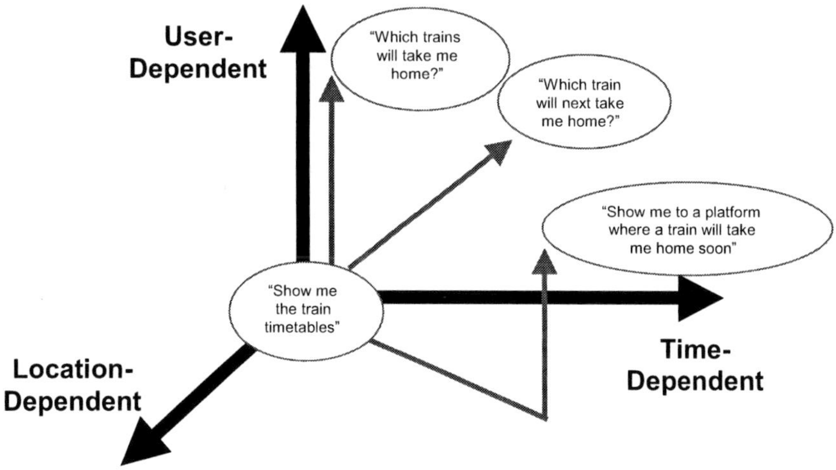

higher value proposition, allowing the user to ask for trains that pass the individual's home station (V = a + f(P)). Adding time criticality (T) to the user's options can generate an even higher value proposition (V = a + f(P, T)). Here, the user could, for example, ask for trains traveling home soon. Finally, adding location-dependence (L) to the user's request can create the highest value proposition (V = a + f(P, T, L)). Thus, the individual could ask for directions to a platform where they could catch the next train home.

Figure 3: Catching the next train home using the Kizoom mobile site

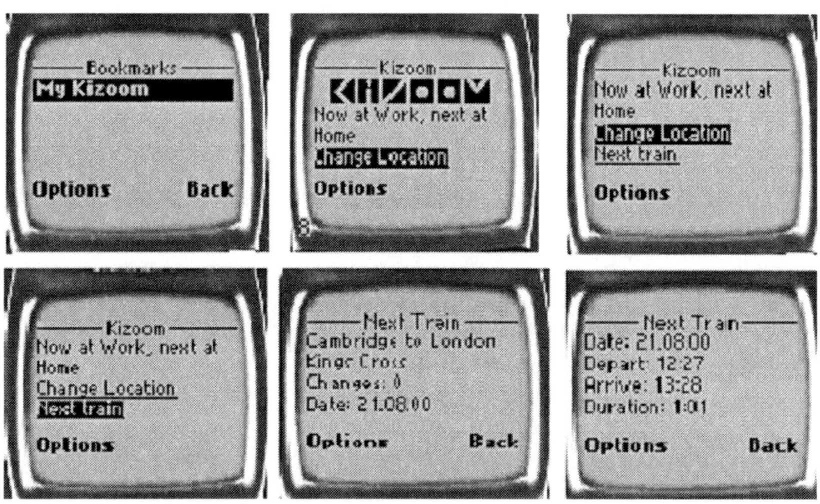

Various similar value propositions are available in services on the mobile Internet. Kizoom, the online travel service, provides a good example. The WAP service allows customised, time- and location-sensitive planning of travel. Figure 3 gives an example of how a user of the site might find details of the next train from his or her workplace to home, based on a known personal profile. The user may also go on to buy a train ticket. Essentially, the site relies on content provided from national timetables, journey planners, location services, spatial database maps, real-time travel information feeders, personal alert services, advertisers, transport companies (involved in m-commerce ticket sales), operators and m-commerce portals (Kizoom, 2000).

Other mobile services can provide similar value propositions, varying in the value-added functions of time, space and personal characteristics. Figure 4 provides a simple categorization of services using the value proposition model. Focusing on LBS, there are a variety of services that are largely based around the value proposition of location dependence. Such services include mobile yellow pages, for example, restaurant guides or nearest ATM, and location-based messaging, for example, security alerts or broadcast advertising. Adding time dependence into the value function broadens the scope to include more sophisticated services such as emergency E911, real-time traffic information, roadside assistance and parking information. Finally, personalization opens the door to further, bespoke services. The opportunities here are very large and include targeted advertisement, personal navigation aids, personal scheduling, and many other personalized services and

Figure 4: Categorization of mobile services using the value proposition model

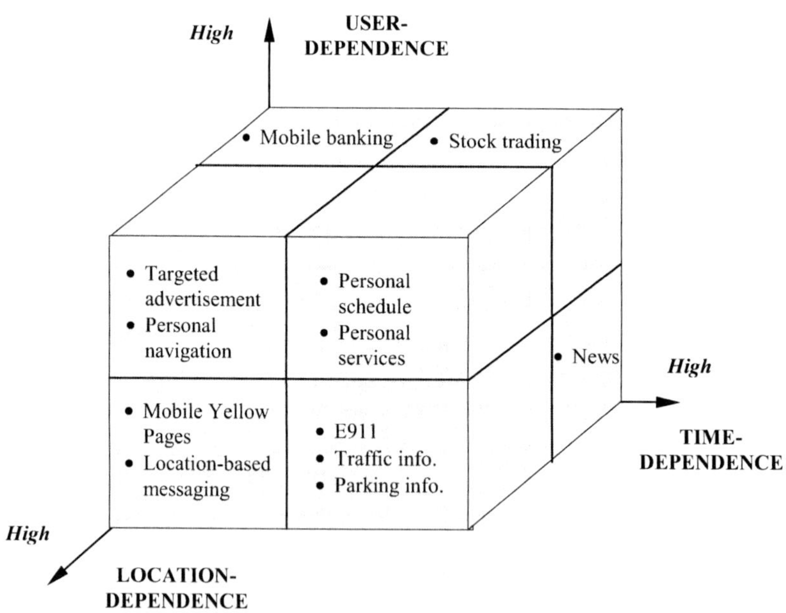

applications. It is important to bear in mind that other services exist outside of LBS that may capitalize on other aspects of the value proposition model. For example, news is largely time-dependent, mobile banking is dependent on the user and account details, and trading on a stock portfolio is dependent on both of these aspects.

ISSUES IN THE COMMERCIALISATION OF LBS—PRIVACY AND STANDARDS

Although considerable progress is being made in the commercialisation of LBS, they are still very much at an embryonic stage of development. Further advancement of l-commerce requires overcoming a significant number of obstacles in technology, markets and policy. Even before companies begin to examine whether customers are willing to pay for these new services, they need to establish a technological and legal platform for service provision. Key areas of discussion among industry players are privacy (Airbiquity, 2000b) and technology standards (Buckingham, 2000a).

The location industry is currently in the ridiculous and destructive situation that every location finding system and positioning vendor has a different proprietary location finding technology (Buckingham, 2000a). Recently, a number of companies have come together to form industry associations aimed at establishing standards and discussing other important industry-wide issues. In September 2000, Nokia, Ericsson and Motorola announced the formation of the Location Interoperability Forum (LIF), a forum to establish global interoperability standards for mobile positioning systems and solutions (Buckingham, 2000b). LIF members represent a mix of network operators, equipment manufacturers and service providers responsible for deploying equipment. Prominent members include Cambridge Positioning Systems, CellPoint and Airflash (LIF, 2001).

In December 2000, eight leading companies involved in the wireless location industry in the US, Canada and Europe—Cell-Loc, SignalSoft, GoAmerica Communications, Cambridge Positioning Systems, Zero Knowledge Systems, Indexonly Technologies, iProx and ViaVis Mobile Solutions—established another industry group, the Wireless Location Industry Association (WLIA). WLIA will interface with government, administrative and regulatory bodies on behalf of the industry and create a forum to develop self-regulating policies, network and share information among members in the industry (Buckingham, 2000c). It will also provide references and information about the industry to the public and policymakers, both in the US and elsewhere.

High on the agenda for the WLIA is the issue of privacy, which promises to be a considerable challenge on wireless devices. In the US, the Fair Location Information Practices (FLIP) dictate that companies must: (1) inform customers about collection practices; (2) give the customer choice regarding any uses of the information; (3) allow for access to the data so that customers can ensure that it is correct; (4) maintain the data securely; and (5) comply with enforcement and auditing of the FLIP policies (Airbiquity, 2000b). Recent attempts to introduce further wireless privacy policies, such as the Wireless Telephone Spam Protection Act (aimed particularly at unsolicited wireless advertising), could further limit the use of location-based technology to market mobile users (Bassuener, 2001). Overcoming such significant hurdles in the US in order to reach the commercial market will prove difficult. In Europe, no comparable policies exist yet, but they are currently under discussion at the European Union in Brussels.

STRATEGIC IMPLICATIONS OF LBS

Ultimately, the implementation of LBS technologies and applications has significant implications at the strategic business level. One strategic issue is the selection of an appropriate technology for a particular service, both in terms of location positioning infrastructure and speed of network. Most LBS, except those for safety and navigation, can begin with cell-level accuracy and the current second generation of networks. However, for mass acceptance, the technological platform for specific LBS must go beyond this. To this end, Table 2 shows a variety of LBS applications and indicates some of the appropriate platforms. Typically, packet-

Table 2: Typical technology requirements for location services

Application	Typical accuracy requirement (typical technology)	Typical network type (typical technology)
Emergency services	High (AGPS)	Any
Roadside assistance	Medium (EOTD)	Any
Vehicle navigation	High (AGPS)	Packet (GPRS+)
Fleet management	High/Medium (AGPS/EOTD)	Packet (GPRS+)
Asset tracking, e.g., packages	Low (COO)	Packet (GPRS+)
People tracking, e.g., workers	Medium/Low (EOTD/COO)	Packet (GPRS+)
Location-based advertising	Medium/Low (EOTD/COO)	Packet (GPRS+)
Public infostation	Medium/Low (EOTD/COO)	Packet (GPRS+)
Geographic messaging	Medium/Low (EOTD/COO)	Packet (GPRS+)
Yellow pages	Medium/Low (EOTD/COO)	Any
Location-sensitive billing	Medium/Low (EOTD/COO)	Any
Road pricing	Medium (EOTD)	Any
Cross-selling	High/Medium (AGPS/EOTD)	Any

switched networks—such as the Global Packet Radio Service (GPRS) —are more suitable for applications where short, intermittent bursts of data are required, such as navigation, tracking or intermittent messaging. Other applications can use any network, although the requirements for network speed in areas such yellow pages and cross-selling will rise with consumer demand for multimedia. Accuracy depends on the criticality of location in an application. This is highest where the exact location of an individual's handset needs to be known, e.g., emergency services or navigation. More general requirements for zone targeting reduce the required accuracy.

Once a technological platform is in place that supports the strategic objectives of the firm, the benefits that accrue in developing and using LBS can occur in many parts of the organization. For example, existing services can become more efficient (e.g., emergency calls), processes can be transformed (e.g., logistics), and new services can be developed (e.g., location-based products). Figure 5 analyses the strategic benefits from LBS using the Index Matrix, which categorizes the benefits in terms of efficiency, effectiveness and transformation, in the areas of the individual, function and organization (Farbey, Land, and Targett, 1992). Note that, where an LBS application could be placed into several cells, it has been placed into the cell of best fit for illustration. As the matrix demonstrates, the strategic impact of LBS is very deep, and applications can provide significant strategic benefits ranging from basic efficiency improvement to complex redefinition and redesign of organizational aspects.

Large transformation benefits are proving difficult to achieve (Hamilton, 2000; Wieland, 2000), requiring a significant amount of risk, investment and thinking "outside of the box." An incentive to operators is that the more business transformation that occurs the more benefits can be achieved (Venkatraman, 1994), thereby generating lucrative new revenue streams. As such, companies that are willing to take risks in adopting wireless location technology and using it creatively have

Figure 5: Benefits of LBS applications – Index Matrix

Benefits \ Area of impact	INDIVIDUAL/GROUP	FUNCTION	ORGANIZATION
EFFICIENCY	*Task mechanization* e.g., emergency call	*Process automation* e.g., package tracking	*Boundary extension* e.g., fleet management
EFFECTIVENESS	*Work improvement* e.g., navigation	*Functional enhancement* e.g., rolling inventory	*Service enhancement* e.g., infostation, messaging
TRANSFORMATION	*Role expansion* e.g., workforce coordination	*Functional redefinition* e.g., supply-chain redesign	*Product innovation* e.g., location-based products

tremendous possibility for achieving strategic advantage in the marketplace. However, the novelty, risk, complexity and cost of the largest transformations and innovations will certainly prove elusive to many firms.

SUMMARY AND CONCLUSIONS

Consumers and mobile professionals are demanding access to location-specific information on a whereever, whenever basis (Kivera, 2001). This trend, fuelled by the increasing ubiquity of low-cost wireless services, growing use of GPS and other location technologies, and acceptance of the Web as a primary source of information, is compelling operators and other companies to deliver location-based services to their customers. For operators, this offers a new set of revenue enhancing and differentiating value-added services.

This chapter has examined the technologies, applications and strategic issues surrounding the development of commercial LBS. As we have seen, a variety of technologies based on the handset, mobile network and satellite positioning systems are available, such as COO, EOTD and AGPS. These have created the platform for a plethora of services in areas such as safety, navigation and tracking, information, and location-based transactions. Of these, safety is the key market driver, where US policy has mandated emergency service caller location. Advertising, roadside assistance, fleet management, people tracking, road pricing, and location-based products are some of the other possible LBS under development. By understanding the needs of the user, a suitable value proposition can be created that combines appropriate aspects in time and space with personal characteristics.

Notwithstanding, mobile operators face considerable obstacles in large-scale commercialisation of LBS. Key problems include standards and privacy. Before a realistic technological platform can be created, fragmented location solutions require an integrative framework. Further, privacy policies pose a significant challenge to many types of LBS. However, if such obstacles can be overcome, the strategic benefits of LBS are potentially enormous, not just in improving efficiency and effectiveness of current services, but in developing new services and transforming core aspects of business.

In terms of research, considerable future work is needed to better understand how to leverage the value of location-based services in both personal and business markets. This chapter has merely scratched the surface of an emerging and growing phenomenon. Questions for research include:

- How can user value (or utility) best be modelled and created in new services?
- What business models are supported by new LBS? What are the strategic benefits of LBS for firms?
- How can the fragmented standards for location positioning be reconciled to provide the best service for mobile users?

- Given the current relative anonymity of mobile devices, do consumers want to be identified? How can privacy be protected?

Clearly, commercial LBS for mobile consumers are still in the embryonic stages of development and use. The next few years will be fundamental in the advancement and adoption of services. Although basic services are beginning to emerge, telecommunications network advancement will enable more sophisticated location positioning services. Network standards such as GPRS, currently being rolled-out in many European markets, are eminently suitable for LBS. Beyond GPRS, third-generation (3G) network standards, such as the Universal Mobile Telephone System (UMTS) in Europe, offer greater flexibility; with speeds of up to 2 Megabits per second, UMTS offers the ability to have simultaneous voice and data calls. With such infrastructure in place, the possibilities for new and improved services become ever larger.

REFERENCES

Airbiquity (2000a). The emergence of the location-commerce market. Paper presented at L-Commerce 2000—The Location Services and GPS Technology Summit, Washington DC, May.

Airbiquity (2000b). No l-commerce without l-privacy. Paper presented at L-Commerce 2000 —The Location Services and GPS Technology Summit, Washington DC, May.

Arthur D. L. (2000). Serving the mobile customer [online]. Available: http://www.arthurdlittle.com/ebusiness/ebusiness.html.

Barnes, S. J. (2002). Under the skin: Short-range embedded wireless technologies. *International Journal of Information Management, 22* (3), pp. 165-179.

Bassuener, K. (2001). Wireless privacy protection act targets location-based spam. Wireless Week [online]. Available: http://www.wirelessweek.com/index.asp?layour=print_page&doc_id=14256.

Bergeron, B. (2001). *The Wireless Web: How to Develop and Execute a Winning Wireless Strategy.* New York: McGraw-Hill.

Bourrie, S. R. (2000). A sense of place: Getting there from here [online]. Available: http://www.bluesigns/Press/bluesigns_in_news/wirelessweek_05152000/.

Buckingham, S. (1999). An introduction to mobile positioning. Newbury: Mobile Lifestreams.

Buckingham, S. (2000a). Nokia, Ericsson and Motorola announce the formation of the Location Interoperability Forum [online]. Available: http://www.mobilepositioning.com/.

Buckingham, S. (2000b). LIF shakes up mobile location standards [online]. Available: http://www.mobilepositioning.com/.

Buckingham, S. (2000c). Wireless Location Industry Association founded [online].

Available: http://www.mobilepositioning.com/.

Djuknic, G. M., and Richton, R. E. (2001). Geolocation and assisted GPS. *IEEE Computer, 34* (2), 123-125.

Farbey, B., Targett, D., and Land, F. (1992). Evaluating investments in IT. *Journal of Information Technology, 7,* 109-122.

Funk, J. (2000). The Internet market: Lessons from Japan's I-Mode system. Unpublished white paper, Kobe University, Japan.

Hamilton, T. (2000). The mobile concierge [online]. Available: http://www.bluesigns.com/Press/Industry_Insights/The_mobile_concierge.htm.

IDC Research (2001). A billion users will drive e-commerce [online]. Available: http://www.nua.ie/surveys/index.cgi?f=VS&art_id=905356808&rel=true.

Kannan, P., Chang, A., and Whinston, A. (2001). Wireless commerce: Marketing issues and possibilities. Paper presented at the *34th Hawaii International Conference on System Sciences,* Maui, Hawaii.

Katz-Stone, A. (2001). Wireless revenue: Ads can work [online]. Available: http://www.wirelessauthority.com.au/r/article/jsp/sid/445080.

Kivera (2001). Kivera Spatial Suite—data sheet [online]. Available: http://www.kivera.com/pdf/kls_data_sheet/pdf.

Kizoom. (2000). Building a WAP application: Software engineering for the mobile Internet. Paper presented at WAP Wednesday, London, April.

Lavroff, J. L. (2000). Location services: How to enhance personal safety and to stimulate lucrative business opportunities. Brussels: European Commission.

LIF (Location Interoperability Forum) (2001). LIF statement, version 5 [online]. Available: http://www.locationforum.org/.

Nokia (2001). Mobile location services. Helsinki: Nokia Corporation.

OECD (2001). Understanding the digital divide. Paris: OECD Publications.

Puca (2001). Booty call: How marketers can cross into wireless space [online]. Available: http://www.puca.ie/puc_0305.html.

Research in Motion. (2000). The wireless workforce [online]. Available: http://www.rim.net/.

Russell, B. (2001). *Making Mobile Commerce Reality Through Concept, Convergence and Cost-Effectiveness—Bluesigns.* Anniston: Bluesigns, LLC.

Strategy Analytics (2000). Strategy analytics forecasts $200 billion mobile commerce market by 2004 [online]. Available: http://www.wow-com.com/newsline/press_release.cfm?press_id=862.

Strategic Group (2000). European wireless location services: Strategies and outlook [online]. Available: http://www.strategisgroup.com/press/pub/wlocate.htm.

Tseng, Y. C., Wu, S. L., Liao, W .H., and Chao, C. M. (2001). Location awareness in ad hoc wireless mobile networks. *IEEE Computer, 34* (6), 46-52.

Varshney, U. (2000). Recent advances in wireless networking. *IEEE Computer, 33* (6), 100-103.

Venkatraman, N. (1994). IT-enabled business transformation: From automation to business scope redefinition. *Sloan Management Review, 35,* 73-87.

Wieland, K. (2000). Where are the location-based services? [online]. Available: http://208.220.133.42/issues/200009/where_are_the.html.

WireFree-Solutions (2000). *Wireless Applications: Where is Your Opportunity?* Madrid: WireFree Solutions.

Xypoint (2001a). Platform white paper. [online]. Available: http://www.xypoint.com/platform/tech/index.html.

Xypoint (2001b). Location management [online]. Available: http://www.xypoint.com/platform/location.html.

Zeus Wireless (2000). *Wireless Data Telemetry.* Maryland: Zeus Wireless, Inc.

Chapter IX

Usable M-Commerce Systems: The Need for Model-Based Approaches

John Krogstie
Norwegian University of Science and Technology, Norway

Petter Bae Brnadtzæg
SINTEF Telecom and Informatics, Norway

Jan Heim
SINTEF Telecom and Informatics, Norway

Andreas L. Opdahl
University of Bergen, Norway

ABSTRACT

Mobile solutions are becoming more and more common in e-commerce and are giving rise to a new breed of m-commerce systems, *which are characterized, among other things, by strong demands on* usability. *This chapter discusses new challenges and possible solutions for developing and evolving* usable *m-commerce systems, with focus on* model-based approaches. *We have experienced these new challenges through several research and industrial projects on mobile solutions, usability and model-based approaches over the last years. In this chapter, we apply our experience to the emerging m-commerce field. We summarize the main challenges on how model-based approaches can support the development of usable m-commerce systems and indicate upcoming research issues in this very dynamic area. We argue that this research area is also timely, because the underlying technological infrastructure are just becoming sufficiently mature to make feasible research on personal, group and organizational issues, and not only on technical issues.*

INTRODUCTION

Today, the PC is only one of many ways to access information resources. On one hand, traditional computing technology is becoming more mobile and ubiquitous and, on the other hand, traditional mass media are becoming richer as in interactive TV. Whereas information services related to interactive TV (iTV) and ubiquitous computing are projected to become prominent in a few years, mobile computing is the most important *current* market and technological trend within information and communication technology (ICT). For instance, it is projected by IDC that in 2004 there will be more mobile devices than PCs connected to the Internet. With the advent of new mobile infrastructures that provide higher bandwidth and constant connection to the network from virtually everywhere, it is predicted that the way people use information resources will be radically transformed.

According to Siau (2001), the essence of m-commerce is *to reach customers, suppliers and employees regardless of where they are located and to deliver the right information to the right person(s) at the right time.* To achieve this, a new breed of mobile information systems (Krogstie, 2001) must be developed. A lot of work has been reported on the technical aspects of mobile computing and thus on the technical aspects of m-commerce. Work has also been reported on the societal aspects of e-commerce and some of it directly on m-commerce. However, relatively little attention has so far been paid to the *user side* of m-commerce. The ability to develop and evolve *usable* m-commerce systems may become an even more critical success factor for enterprises in the next few years than is their ability to develop and evolve usable e-commerce systems today.

Research on business (information) systems has so far primarily dealt with the development and evolution of systems accessed through PCs and workstations. As mobile solutions are applied in more and more situations, new challenges will face those who develop and evolve business applications. This chapter summarizes some of the corresponding research challenges and points to how they can be addressed by model-based approaches.

The structure of this chapter is as follows: In the next section, we describe some of the specific characteristics of mobility. In section 3, we highlight some of the main differences between m-commerce systems and traditional information systems on the user, group, organizational, and inter-organizational level. In section 4 we summarize the resulting issues, research opportunities and future trends in how model-based approaches can be used to develop and evolve usable m-commerce systems. The discussion is structured according to the same levels used in section 3. Section 5 then concludes the chapter.

NOVEL ASPECTS OF M-COMMERCE

M-commerce systems run on mobile devices that can be moved around easily in space and that operate independently of particular locations.

People in general are getting increasingly mobile in relation to both their professional and private tasks. The user of m-commerce systems is characterized by frequent changes in context:

- The *spatio-temporal context* describes aspects related to the time and space. It contains attributes like time, location, direction, speed, track, and place. This context is particularly important and will be discussed in more detail below.
- The *environment context* captures the entities that surround the user, e.g., physical objects, services, temperature, brightness, humidity, and noise.
- The *personal context* describes the user state. It consists of the physiological context and the mental contexts. The physiological context may contain information like pulse, blood pressure, and weight. The mental context may describe things like mood, expertise, anger, and stress.
- The *task context* describes what the user is meant to be doing. The task context may be described with explicit goals or the tasks themselves.
- The *social context* describes the social aspects of the user context. It may, for instance, contain information about friends, neighbors, coworkers and relatives. The role that the user plays is an important aspect of the social context. A role may describe the user's status in this role and the tasks that the user may perform in this role. The term *social mobility* refers to the ways in which individuals can move across different social contexts and social roles and still be supported by technology and services.
- The *information context* describes that part of the global and personal information space that is available at the time.

According to Kakihara and Sørensen (2001), *spatial mobility* is primarily about *people* moving about in space because they have wireless access to information and services. However, spatial mobility is also about all sorts of ICT-enhanced things (parcels, cars, etc.) that can move about in the environment and that can interact with other devices and in this way influence people. In *ubiquitous computing*, the environment itself has become ICT-enhanced by a large number of interacting, stationary or mobile computing devices that are effectively invisible to the user (Weiser, 1993).

Luff and Heath (1998) identify three types of *spatial mobility*: *Micro-mobility* refers to how small artifacts can be mobilized and manipulated by hand. *Local mobility* involves real-time interaction between people and technology at the same location. Finally, *remote mobility* supports both synchronous and asynchronous collaboration among individuals who move around in distant physical locations.

Several taxonomies of remote mobility are discussed in the literature. Kristiansen and Ljungberg (1999) distinguish between *traveling*, *visiting* and *wandering*. *Traveling* is movement between different locations in a vehicle. *Visiting* is a prolonged period spent in one location before moving back to the original location or on to another one. *Wandering* is moving about—usually on foot—in the local area. Esbjörnsson (2001) adds *mobile work* proper, i.e., work that by nature entails moving about. Since m-commerce is also relevant in nonworking situations, we may extend this category to *mobility proper*, i.e., activities where mobility is an essential aspect of the activity itself, such as going on a hike or scenic tour.

Bergquist (1999) proposes a typology that distinguishes between the *nomad*, the *pilgrim* and the *tourist*, and which can be seen as an extension of Kritiansen and Ljungberg's *traveling*. *The nomad* is constantly on the move, well off and unpredictable. Movements are not necessarily irrational but inherently not able to be planned and to a large extent improvised. The nomad thus cannot trust existing infrastructures. *The pilgrim* also moves between destinations, but movement is essentially planned, and the pilgrim can utilize existing infrastructure to a certain extent. Finally, *the tourist* moves in a predictable and planned way. Whereas the nomad and pilgrim travel in relation to their work, the tourist travels for recreation and entertainment.

People also "move" between different ways of organizing their time, i.e., a kind of *temporal mobility*. Hall (1976) distinguishes between *monochronicity* and *polychronicity*. In the former, people seek to structure their time sequentially doing only one thing at a time, if possible according to a plan. In the latter, people accept—and maybe prefer—doing several things simultaneously, placing less importance on planned order. The new technologies seem to be increasing monochronicity in some situations and polychronicity in others. On the one hand, increased monochronicity appears in the interfaces of many contemporary enterprise systems, which often require business process steps to be carried out in strict sequence with little flexibility for individual variations of temporal order and few possibilities for carrying out several processes in parallel. Because mobile devices have small screen sizes and memories, monochronicity is strengthened, because it is less convenient to operate on several windows that run different applications in parallel. On the other hand, polychronicity is increased in many mobile settings. A larger proportion of work is today done by what is termed *symbolic analysts* (Thomson and Warhurst, 1998), whose main work resources and work products are symbol structures and who do their work in (often many concurrent) knowledge-intensive projects, often in dynamically networked organizations. Symbolic analysts, such as consultants, reporters and researchers, will typically have many tasks going on concurrently and will be interested in a lot of different information there and then, much of which cannot be anticipated fully beforehand. There might be a need for learning on the fly, but also for capturing interesting situations to feed back to the organization supporting knowledge management and learning. We predict that many of these workers would like to be supported also in their more complex personal processes involving, e.g., governmental agencies, banks, and insurance companies using the same system, if possible, as the distinction of work and private life is blurred for many. This is similar to how many people nowadays use PDAs, where the calendar is used for capturing both private and professional tasks and appointments.

In sum, mobile technologies inherently tend towards providing monochronic services, while at the same time they tend to place their users in contexts with polychronic demands.

DIFFERENCES BETWEEN M-COMMERCE SYSTEMS AND TRADITIONAL INFORMATION SYSTEMS

From an application point of view, m-commerce systems differ from more traditional information systems in several ways (Siau, 2001; Hirsch, 2001). These differences are both technical and user-oriented, and the latter ones will be focused on in this chapter. We will discuss differences at several levels:

- user level,
- group level,
- organizational level, and
- inter-organizational level.

Whereas none of the differences are revolutionary in themselves, m-commerce systems increase their importance because (1) each difference is *amplified by* the m-commerce setting and (2) the difference can be *combined* in ways that are not yet understood.

User level

Because the new mobile devices integrate functions that were previously offered by physically separate tools, they probably signal the arrival of a plethora of new types of applications. New mobile devices (as of 2002) already integrate hand-held or palmtop PCs with mobile phone, Internet connectivity and browsing, fax, cameras, GPS, and office functionality. M-commerce systems on these devices address both traditional ICT users and new groups of users, meaning that user-interfaces should feature prominently and early in the development process and that user-interfaces should be extremely simple. Existing usability approaches provide suitable support for developing stationary, in-office ICT but do not cover the necessary span of contexts where mobile devices are used. Usability testing has traditionally been done in usability labs that mimic an office environment. Field studies and user observations have also been aimed at use in some form of office or work location. The new mobile technology allows the users to utilize business systems in such locations as on a bus, in a meeting or at home. Developers often cannot presume that users are acquainted with computers at all, and input and output facilities may be severely restricted (no keyboard, small screen-size, etc.) or based on new modalities (speech-recognition and synthesis, etc.). This means that *personalization* of m-commerce systems becomes increasingly important, where user-interface details such as commands and screen layout are tailored to personal preferences and hardware using information about the current context and context trace (Ralhff, 2001) of the user. Personalization refers to both m-commerce systems that automatically *adapt* themselves to the preferences of the user and systems that can be explicitly *tailored* by users through a specific user-interface. The separation between content and medium found in m-commerce systems poses new important challenges to developers. New systems should thus provide maximal personalization from minimal assumptions about physical devices.

The focus of traditional usability research is limited, because ICT-artifacts are seen only as means to complete user tasks, with focus on how users can complete their tasks faster or according to some measure of user performance (Jordan, 1997). This tradition broadly ignores the affective part of the user experience obtained. However, some products do not require a formal task to be performed, but rather provide a notion of "having fun" or obtaining an "optimal experience" (Brandtzæg and Følstad, 2001). The central element in an optimal experience is that the *activity* is a goal in itself (Csikszentmihalyi, 1990). As leisure, social, and work activities increasingly involve interacting with or communicating via ICT, personal, intrinsically motivated tasks start to merge with easier to define and extrinsically motivated tasks.

Group level

Just as systems at the individual level should be tailored to personal preferences, functions at the work level should fit the work processes, processes that typically involve other persons as well. At the group level, awareness of the status of knowledge resources is increasingly important in a mobile setting. Awareness here also includes the status of technical knowledge resources. Given that knowledge resources include both individuals and technology that can be mobile, one should look into *interactive* systems to improve group performance. Peter Wegner's interaction framework (Wegner, 1997; Wegner, 1998) was triggered by the realization that machines involving users in problem solving can solve a larger class of problems than algorithmic computing systems in isolation. The main characteristic of an *interaction machine* is that it can pose questions to human actors (users) during computation. The problem solving process is no longer just a user providing input to the machine that processes the request and provides an answer (output), it is a multistep conversation between the user and the machine, each being able to take the initiative.

A major research question in this area is how to specify and utilize interaction machines within a mobile infrastructure to make them more usable than traditional, algorithmic machines.

Traditional usability work typically concerns single users using one single technology to solve well-defined tasks in an undisturbed environment. Experiences from Naturalistic Decision Making have identified that solving tasks in this manner requires different strategies, other sources of support, and are accompanied by other problems than the solving of "real-world tasks," which are typically ill-defined and take place in environments characterized by time stress and multiple players (Zsambok, 1997). The identification of contextual variables of significant importance to the usability of systems in real-world use is important in the development and evolution of m-commerce systems.

Organizational level

M-commerce systems are often radical, and therefore reward increased focus on idea generation in the early development phases. This also means that there are not always existing services or situations in which to anchor problem analysis efforts, e.g., using As-is analysis as a starting point for To-be design (Rolland and Prakash, 2000). Mobile technology still develops rapidly, meaning that idea generation should not be limited too much by what is currently possible. It also means that systems must be designed for change. This applies to at least two levels:

- *User interface design*: m-commerce development methods should provide ample room for creativity and evaluation of alternatives at the very early stages, probably as part of requirements elicitation, in particular since there is still no clear convergence towards a dominant user-interface standard.
- *Overall system architectures:* m-commerce systems should be developed based from analyses of stable components and grouped into back-end systems that reflect responsibility and competence areas within the provider organizations. Importantly, system architectures must *not* be derived from unstable and technology-dependent usage scenarios. Appropriate, methods for *architecture evolution* as new services are provided and new devices supported should also be in place.

M-commerce calls for highly distributed solutions that comprise new user-interface systems on the client side, new and existing back-end systems, as well as new bridging systems (which port information between other systems.) The new technologies therefore highlight the need for principled, long-term IS-architecture management in the organization and for integrating architecture management with software development methodologies. Often there is a need to interface to existing enterprise systems and *enterprise information architectures* to enable the new workflow.

Inter-organizational level

Since mobile devices will typically be used for a mix of personal and professional tasks (often for many different organizations in parallel), new security and privacy aspects arise. M-commerce systems pose additional challenges to information systems security, most importantly by rendering traditional firewall thinking unusable.

The emerging technologies provide many different ways to offer the same or similar services to customers. For example, broadcast news in the future will be available through plain-old television (POTS), interactive television (iTV), Internet TV, 3G mobile phone, and numerous other information appliances (the door of the glove compartment in your car, your living-room wall, etc.) This stresses the importance of developing and evolving system architectures that are independent of usage scenarios.

MODEL-BASED DEVELOPMENT AND EVOLUTION OF USABLE M-COMMERCE SYSTEMS

Developing and evolving m-commerce systems with high usability for different mobile devices is not trivial. The complex m-commerce systems, the technical infrastructure, the frequent context changes, the usability issues involved and the development and evolution process itself demands an approach that allows for a high degree of structure combined with flexibility and even creativity. In addition, the m-commerce systems themselves are only one small, integrated part of larger enterprise systems. There is therefore a great need to integrate usability-engineering methods into a model-based development approach.

A user-centered iterative development process is a joint effort between different experts and the end-users. The iterative process is a way to ensure alignment between the information gathered from various activities with the users, previous knowledge and the proposed solutions from the developers. The different methods used in a user-centered development process could vary much depending on when, who and where the techniques are used. This spans from the use of informal models (e.g., rich pictures) (Monk and Howard, 1998) early in the process to a rigid user evaluation at the end to verify that the requirements are met (Maguire, 1998). All these activities need to report and link their results into the technical system development and architecture evolution.

In general, a model-driven approach to information systems development has been found to provide the following advantages (Krogstie and Sølvberg, 2000):

- Explicit representation of goals, organizations and roles, people and skills, processes and systems.

- An efficient vehicle for communication and analysis.
- Basis for design and implementation, either through traditional code generation, generating calls to prespecified components (whose characteristics need to be modeled), or as documentation.
- Readily available documentation as a basis for extensions and personalization.

One striking aspect in connection to contemporary information systems development and evolution is that there is an increasing demand for shorter development time for new products and services (Pries-Heie and Baskerville, 2001). This is specifically evident for m-commerce systems, where the convergence of different platforms continuously creates opportunities for new functionality. Some would argue that this highly dynamic situation would make model-based approaches impractical. To the contrary, we claim that for new technologies to develop rapidly, idea generation should not be limited by currently available technologies and systems must be developed for change.

There is little accumulated experience on how to develop software for the new technologies. As a consequence, lightweight design techniques and early prototyping are natural choices for practical m-commerce systems development at the moment. In addition, research is needed on how to accumulate experience from early development projects and package this knowledge into comprehensive, integrated and model-supported development methodologies. There is a need to develop new user-centered approaches that are "quick and clean" (Wichansky, 2000) rather than "quick and dirty" to address this.

Looking in more detail on the levels in our taxonomy from section 3, we can identify the following areas for potentially increased utility of techniques earlier developed as part of model-driven development:

User level

The mobile context of the user is changing all the time and m-commerce systems should therefore be able to adapt to new contexts when relevant for the user. It is for instance very different to access a system from a desktop and through a PDA. One way to make this more efficient is to apply the navigation in simple (process) models to give the necessary context for efficient access and use of enterprise-wide applications from anywhere using any kind of information appliance. In general, the relevant context parameters should be explicitly modeled to be able to give an overview of, support analysis of, and simulate the multitude of possible ways to adapt to the context and the use of context traces. The explicit modeling of goals and interrelationships between goals and activities becomes in connection to this more important to be able to take into account all aspects of user experience.

As mentioned above, people will to an increasing degree need to access what is logically the same system from a multitude of platforms (PDAs, PCs, mobile phones, TVs, etc.). Recently, work within user interface modeling has focused increasingly on modeling mobile user interfaces (Eisenstein, 2000; Eisenstein, 2001; Muller, 2001; Pribeanu, 2001; Schneider, 2001). This is often done to facilitate some level of common models for the mobile user interfaces and more traditional ones. A central element in this is the development of model-based approached that are powerful enough to be used as a basis for the development of user-interfaces on the multitude of platforms needed, but still general enough to represent the commonalties in a single place. One approach is to define user-interface patterns with general usability principles as powerful building blocks. A main challenge for model approaches for developing multi-interfaces is to have a set of concepts that are, on the one hand, abstract and general enough to express specifications across a number of quite different platforms and, on the other hand, powerful and expressive enough to support mapping to different platforms. Thus, there is a need to combine generalization and specialization. A model-based technique that is abstract enough to be able to describe user interfaces with significant differences may run the risk of being banal. By this we mean that the model is not able to describe sufficient number of aspects of the user interfaces in a way that renders it possible to transform the models to concrete user interfaces without adding so much additional information to the mapping process for each platform that the interfaces might as well have been developed from scratch on each platform.

Group level

Process support for users of m-commerce systems in a group setting extends the needs found in traditional systems. Experiences with the use of PDAs have shown us that many people will wish to be supported in carrying through many processes at one time (both private and business) by the same device and will wish support for both well-defined and emerging processes at the same time. Process support technologies are typically based on process models, which need to be available in some form for people to change them to support their emerging goals. Thus, *active models* should be supported (Jørgensen and Carlsen, 1999; Jørgensen, 2001). The outset for this thinking is that models can be useful tools in a usage situation, even if the models are changing and are partly incomplete and inconsistent. The user is included as an interpreter and changer of the models, based on underlying interaction machines. Emergent workflow systems (Jørgensen and Carlsen, 1999) represent a different approach to static and adaptive workflow systems with respect to their use of models. They target very different kinds of processes: Unique, knowledge-intensive processes where the structure emerges. It can be argued that this is not specific for mobile information systems utilizing GPRS and UMTS technology. On other hand, this area will be even more pronounced in such systems since future mobile devices (and a multitude of services across the network) will be always available (and thus more likely to be used in an emergent or ad hoc fashion). Instead of the specification of algorithmic machines, this brings forward the need to specify interaction frameworks.

Organizational level

Siau (2001) highlights the development of m-commerce business models as an important application-oriented research area. Within e-commerce, many new business models have appeared. The mobile environment in which m-commerce applications reside will require further adaptations of these models. In order for m-commerce to succeed, it is vital to ensure that all the related applications and services can be accessed with ease and little cost. Thus, in addition to externalizing the business models in a computing independent way, it is important to integrate these models with the internal enterprise models and enterprise architecture to be able to pinpoint the links to, e.g., internal systems for efficient billing of the m-commerce services. For organizational use, it is important to model the roles and responsibilities, the role-structures, and the tasks that are relevant and allowed for users that fill the different roles. It is also important to be able to use the models for these aspects across different enterprise systems and different mobile-client applications. To enhance social mobility, organizations and industries need to develop "social ontologies" that define the significance of social roles and their associated behavior and context. These ontologies should be available as explicit models to be useful for the development and evolution of m-commerce systems. Role distinctions are also relevant on an inter-organizational level (below).

Another aspect is that there are currently (and will be for some time) a multitude of competing technologies that provide the underlying infrastructure for m-commerce systems. A central element when addressing this is the development of model-based approaches that are powerful enough to be used as a basis for development and evolution of systems that run on a large number of mobile infrastructures but are still general enough to represent the commonalties at one place only. The Object Management Group's (OMG) current major initiative on *Model-Driven Architectures* (MDA (Poole, 2001)) specifies both platform-independent and platform-specific modeling notations, including refinement techniques between these notations, and thereby highlights current industrial focus on such an approach. In connection to this, it is interesting to note how meta-modeling techniques and domain-specific modeling (DSM) have found a special application for the design of mobile phone software (Kelly, 2001). m-commerce systems can be argued as a particularly appropriate area for domain-specific modeling:

- The software (on the client side) is partly embedded and must therefore be more reliable than traditional software. This can be supported by restricting choices through adding modeling rules and code generation.
- You often need to develop and evolve many very similar variants of the same application.
- There are several standards to adhere to, and the technology and standards change rapidly. For example, one wants to define GSM only once, use this definition in a range of products, and, when necessary, plug in a UMTS or a US analog component in its place. A single developer cannot know all the standards

that might be useful; there is a need to insulate developers from the plethora of technologies and standards.

Looking at the architecture for existing general m-commerce solutions (e.g., Celesta, 2001), we notice that their generic architecture is geared towards the modeling of data, business processes and tasks, events and behavior, rules, user interfaces and general context information.

Inter-organizational level

An important aspect on the inter-organizational level is the dependability of the systems made. Laprie defines dependability as the "ability to deliver service that can justifiably be trusted" and identifies six dependability attributes: availability, reliability, safety, confidentiality, integrity and maintainability. Model-based development will in general be able to support dependability analyses, i.e., use of methods, techniques and tools for improving and estimating dependability, e.g., risk analyses, probabilistic safety assessment, testing, formal walkthrough, simulation, animation, exhaustive exploration and formal verification. Many of the dependability areas will be even more complex in m-commerce systems than in traditional business systems. As an example, consider new issues arising in connection to security and privacy. The same device will often be used as a personal tool across the user's commitment to many different organizations and projects, and one must assure that data does not "leak" across different domains. Another area of concern is the users (lack of) control over the context traces they leave behind when they use location-based or other context-based general services.

This section has discussed modeling-based approaches to development and evolution of usable m-commerce systems. Although this chapter has focused more on development issues than on evolution, the need for rapid and continuous development of new releases of different variants of m-commerce systems brings forward a need for a higher degree of reuse and integration of models of different nature. A major research question in connection to these areas is to what extent existing modeling techniques, e.g., based on the UML, can be applied, when these techniques should be extended and when they need to be replaced all together. Early experiences with this kind of problem indicate that the development of some technical areas within m-commerce system can benefit from building on and extending the existing techniques within UML. For example, advanced dependability and risk-analysis is currently done extending UML (den Braber, 2002). On the other hand, Kelly (2001) reported that at least for certain types of software development on mobile platforms, the current version of UML has been found insufficient. For the business and process-oriented aspects we likewise see the potential uses of traditional enterprise and workflow modeling approaches in connection to m-commerce systems. A main problem, also found in more traditional development, is the bridging of enterprise and design modeling, an area that is not properly addressed either in theory or practice. Although the proposals for the next version of UML attempts to make the language better also for enterprise modeling, there is no sign on a that a good solution for bridging enterprise and system modeling will be presented in UML2.0. Another unsolved area is how to integrate the user-interface models with other parts of the requirements and design models; for instance, the class model, process model and goal model. On both the process and the user-interface side, the challenges can be attacked with extensions of existing approaches to modeling, although research is needed to investigate both which techniques should be extended and how they could be best adapted to the new problem areas.

CONCLUSION

The large-scale application of m-commerce is in its infancy and, unsurprisingly, limited work has so far been done on m-commerce systems, on usable m-commerce systems and on model-based development and evolution of usable m-commerce systems. On the other hand, the upcoming, 3G (UMTS) infrastructure provides higher bandwidth and constant connection to the network from virtually everywhere, and the number of m-commerce applications is therefore predicted to explode. The m-commerce community should therefore be at the forefront of the development. Obviously we are not starting this work from scratch; it is possible to build on existing work within the usability, user interface and modeling fields, specifically on techniques for modeling of functional and nonfunctional requirements, process modeling, usability requirements, model-driven architectures, requirements specifications of Web applications, domain-specific modeling and dependability analysis.

ACKNOWLEDGEMENTS

We would like to thank our colleagues Erik Gøsta Nilsson, Jan-Håvard Skjetne and Asbjørn Følstad at SINTEF Telecom and Informatics and Konrad Morgan, Anders Mørch and Bjørnar Tessem at the University of Bergen for input and discussions related to this chapter. We would also like to thank the anonymous reviewers for helping us to improve the focus of the chapter.

REFERENCES

den Braber, F., Dimitrakos, T., Gran, B. A., Stølen, K., Aagedal, J. Ø. (2002): Model-based risk management using UML and UP to appear in *Proceedings of IRMA International Conference.*

Brandtzæg, P. B., & Følstad, A. (2001). How to understand fun: Using demands, decision latitude and social support to understand fun in human factor design. In *Proceedings of the International Conference on Affective Human Factors Design,* Singapore.

Celesta (2001). Universal mBusiness Platform [online]. Available at http://www.celesta.com/pdf/products/mBusiness_Platform.pdf.

Csikszentmihalyi, M. (1990). *Flow: The Psychology of Optimal Experience.* New York: HarperPerennial.

Eisenstein, J., Vanderdonckt, J., & Puerta, A. (2000). Adapting to mobile contexts with user-interface modeling. In *Proceedings of IEEE Workshop on Mobile Computing Systems and Applications,* WCSMA 2000.

Eisenstein, J., Vanderdonckt, J., & Puerta, A. (2001). Applying model-based techniques to the development of UIs for mobile computers. In *Proceedings of ACM Conference on Intelligent User Interfaces* IUI 2001.

Esbjörnsson, M. (2001). Work in motion: Interpretation of defects along the roads. In S. Bjørnestad, R. Moe, A. Mørch, & A. Opdahl (eds.), *Proceedings of* IRIS24, University of Bergen, Norway.

Hall, E. T. (1976). *Beyond Culture.* Anchor Books, Doubleday.

Hirsch, R., Coratella, A., Felder, M., & Rodriguez, E. (2001). A framework for analyzing mobile transaction models. *Journal of Database Management, 12* (3).

Jordan, P. W. (1997). The four pleasures—taking human factors beyond usability. In *Proceedings of the 13th Triennial Congress of the International Ergonomics Association,* Helsinki, Finland.

Jørgensen, H. D., & Carlsen, S. (1999). Emergent workflow: Integrated planning and performance of process instances. In *Proceedings Workflow Management* '99, Münster, Germany.

Jørgensen, H. D. (2001). Interaction as a framework for flexible workflow modelling. In *Proceedings of GROUP* 2001, Boulder, Colorado, October 2001.

Kakihara, M., & Sørensen, C. (2001). Mobility reconsidered: Topological aspects of interaction. In S. Bjørnestad, R. Moe, A. Mørch, & A. Opdahl, (eds.), *Proceedings of* IRIS24, University of Bergen, Norway.

Kelly, S. & Tolvanen, J. P. (2001). Visual Domain-Specific Modelling: Benefits and Experiences of Using Metacase Tools, Metacase Consulting.

Kristiansen, S., & Ljungberg, F. (2000). Mobility—From stationary to mobile work. In K. Braa, C. Sørensen, & B. Dahlbom (eds.), *Planet Internet, Studentlitteratur, Lund.*

Krogstie, J., & Sølvberg, A. (2000). Information systems engineering—Conceptual modeling in a quality perspective. Information Systems Groups, NTNU, Trondheim, Norway.

Krogstie, J. (2001). Requirement engineering for mobile information systems. In *Proceedings of* REFSQ 2001, Interlaken, Switzerland.

Luff, P., & Heath, C. (1998). Mobility in collaboration. In *Proceedings of the* CSCW'98, Seattle, USA, pp. 305-314.

Maguire, M. C. (1998). User-centred requirements handbook (Report D5.3): HUSAT Research Institute.

Monk, A., & Howard, S. (1998). Methods & tools: The rich picture, A tool for reasoning about work context. *Interactions, 5* (2), 21-30.

Muller, A., Forbig, P., & Cap, C. (2001). Model based user interface design using markup concepts. In *Proceedings of the Eighth Workshop on the Design, Specification and Verification of Interactive Systems.*

Poole, J. (2001). Model-driven architecture: Vision, standards, and emerging technologies. ECOOP 2001, Workshop on metamodelling and adaptive object models.

Pribeanu, C., Limbourg, Q., & Vanderdonckt, J. (2001). Task modelling for context-sensitive user interfaces. In *Proceedings of the Eighth Workshop on the Design, Specification and Verification of Interactive Systems.*

Pries-Heie, H., & Baskerville, R. (2001). eMethodology. In *Proceedings of the IFIP TC 8 Conference on Developing a Dynamic, Integrative, Multidisciplinary Research Agenda in E-Commerce/E-Business*, Salzburg, June 22-23.

Rahlff, R., Rolfsen, R. K., & Herstad, J. (2001). Using personal traces in context space: Towards context trace technology. *Personal and Ubiquitous Computing, Special issue on situated interaction and context-aware computing, 5* (1).

Rolland, C., & Prakash, C. (2000). Bridging the gap between organisational needs and ERP functionality. *RE Journal, 5* (3), 180–193.

Schneider, K., & Cordy, J. (2001). Abstract user interfaces: A model and a notation to support plasticity in interactive systems. In *Proceedings of the Eighth Workshop on the Design, Specification and Verification of Interactive Systems.*

Siau, K., Lim , Ee-P., & Shen, Z. (2001). Mobile commerce: Promises, challenges, and research agenda. *Journal of Database Management, 12* (3).

Thompson, P., & Warhurst, C. (1998). *Workplaces of the Future.* Macmillan Business.

Wegner, P. (1997). Why interaction is more powerful than algorithms. *Communications of the ACM, 40* (5).

Wegner, P., & Goldin, D. (1999). Interaction as a framework for modeling. In *Conceptual Modeling: Current Issues and Future Directions,* LNCS 1565. Springer.

Weiser, M. (1993). Some computer science issues in ubiquitous computing. *Communications of the ACM, 36* (7), 75-84.

Wichansky, A. M. (2000). Usability testing in 2000 and beyond. *Ergonomics, 43* (7), 998-1006.

Zsambok, C. E. (1997). Naturalistic decision making: Where are we now? In C. E. Zsambok, G. Klein (ed.), *Naturalistic Decision Making* (pp. 3-16). Mahwah, NJ: Lawrence Erlbaum Associates.

Chapter X

Managing the Interactions Between Handheld Devices, Mobile Applications, and Users

Maristella Agosti and Nicola Ferro
University of Padua, Italy

ABSTRACT

In this work we present the problem of managing the interactions between handheld devices, mobile applications and users that are the three main entities involved in providing information access in the mobile scenario.
After introducing a general framework, which gives the reader a comprehensive view of the features of a handheld device and presents some techniques to overcome the constraints imposed by handheld devices during the interaction with users and mobile applications, we focus our attention on the problem of searching for information using a search engine through a handheld device. We present the motivations for studying this kind of search engines and a discussion on their features and design choices, explaining the relevant aspects that need to be addressed during the design and the development of these mobile applications.

INTRODUCTION

The goal of this chapter is to introduce the reader the range of issues that need to be addressed for managing the interactions between handheld devices, mobile applications and users. Specific attention is given to the interactions between handheld devices and mobile applications that use search engines for information seeking and discovery, because search engines are often used in mobile applications that provide information and services.

There are three main types of entities involved in the interactions:

- *handheld devices*: Examples of handheld devices[1] include *Personal Digital Assistants* (PDAs) and mobile phones;
- *mobile applications*: Among the mobile applications which can possibly be considered for implementation, this chapter will focus on Web search engines. Search engine services are particularly useful to end users who seek information for work or pleasure using handheld devices;
- *users*: There are different categories of mobile users looking for information resources to solve some emergency needs or to resolve their *Anomalous States of Knowledge* (ASK) (Belkin, Oddy, & Brooks, 1982).

The relationships between the three types of entities are illustrated in Figure 1, where it is shown that one has to consider human/computer interactions when presenting content on the handheld devices and user requirements when designing search capabilities for mobile applications. Another aspect that needs to be considered is related to the interaction and communication between the handheld devices and the mobile applications, as the search capabilities of mobile applications are affected by the content presentation characteristics and the limitations of handheld devices.

Figure 1: Main entities involved in the mobile computing scenario

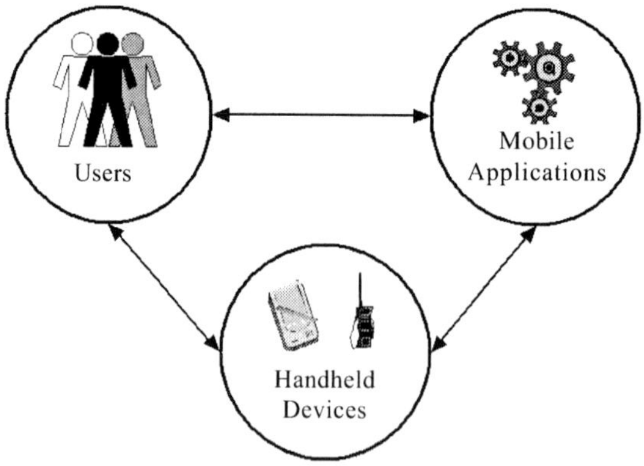

This chapter begins with a section that motivates the study on managing the interactions between entities in mobile scenarios. It concentrates on the interactions between handheld devices and mobile applications. Several problems need to be addressed in the use of handheld devices to present information and to adapt information content available on the Web to the handheld devices. Many examples are given on how these problems can be addressed. A general framework is introduced, giving the reader a comprehensive view of the features of a handheld device, and the differences between the handheld device and a desktop or portable computer including the constraints imposed by these differences. Some approaches will then be introduced to overcome the constraints imposed by handheld devices.

The chapter then presents the problems concerning the development of a search engine for handheld devices. Indeed, it could be thought that the problems of retrieving information end when the user has a good browser, but that would be a misleading conclusion. It is necessary to provide the user with a tool that allows him to find and reach relevant sites that otherwise would be difficult to know about. It will be shown how the considerations concerning surfing have to be changed and how everything is more complex, due to the features of handheld devices. In this way the reader will be made aware of the interaction problems with search engines when using handheld devices and the architecture required to address these problems.

MOTIVATIONS OF MANAGING THE INTERACTION BETWEEN HANDHELD DEVICES, MOBILE APPLICATIONS AND USERS

In *Mobile Commerce* (m-commerce), users will have to use handheld devices such as PDAs and mobile phones to access different types of services. These kinds of devices have special features that distinguish them from a desktop or portable computer.

Typically, through the handheld devices, users can access the services provided by the ordinary electronic commerce (e-commerce) sites. However, these e-commerce services designed for desktop computers usually do not suit handheld devices well. Due to their unique features, the handheld devices could become a bottleneck in accessing e-commerce services—not allowing the user to fully appreciate the information and services offered by the e-commerce sites.

The key approach to overcome the limitations of handheld devices is *content adjustment* that refers to the information presentation techniques and techniques required to manage user's interactions considering the features of handheld devices.

Over the years, many enterprises have invested in applications to manage their data, and now these are the legacy applications. Legacy applications were built in an environment where users interacted with the applications through dumb terminals. Now enterprises need to extend their reach to customers and business partners using pervasive devices and the Internet, but the potential risks and costs prohibit companies from porting these existing applications to any new technology, as observed by Britton et al. (2001). Using content adjustment, the legacy applications ported to the Internet environment could still be accessed by a variety of handheld devices, allowing more effective *Business to Business* (B2B) and *Business to Consumer* (B2C) interactions.

Finally, *Third Generation Mobile System* (3G) introduces the "any time, anywhere, any device" paradigm; and, in order to respect the "any device" constraint, the content needs to be independent of the handheld devices, and so it has to be adjusted to the various devices on which it will be displayed.

DESIGN ALTERNATIVES FOR OVERCOMING THE CONSTRAINTS OF HANDHELD DEVICES

We now analyze the handheld device features that constrain the design of a mobile application and which need to be addressed by content adjustment:

- screen size;
- input method;
- wireless link;
- small memory;
- slow *Central Processing Unit* (CPU);
- energy consumption.

Screen Size

The screen of a handheld device is a very limited and valuable resource. The small screen reduces the amount of information that can be displayed at any one time and can either make a Web page, that has been designed for a desktop computer, difficult to read, or it can force the user to do a great deal of scrolling.

Most Web pages are designed to be displayed on the screen of a desktop computer, with colour and with a resolution of at least 640×480 pixels, while other pages may be designed with higher resolutions in mind. For a PDA, the most common resolution is 160×160 pixels. This means that only 1/12 of the Web page can be displayed on the PDA. This may lead to a good deal of scrolling, both vertically and horizontally. If the PDA allows a long horizontal line to be wrapped into multiple lines, it may save some horizontal scrolling but will significantly compromise the readability of the page.

Nowadays, many Web browsers for handheld devices display Web pages without considering their screen sizes. Thus, pages do not fit properly onto these small screens, and Web surfing becomes very tedious, reducing a site's effectiveness.

In order to address these problems, various approaches can be adopted, as classified in Bickmore and Schilit (2002):

- device-specific authoring;
- multiple-device authoring;
- client-side navigation;
- automatic reauthoring.

Device specific authoring involves authoring a set of Web pages for the display device of a particular handheld device, be it a mobile phone or a PDA. The users will only have access to a selected set of services. All pages for these services are designed up-front for the display of some particular handheld device(s).

This is the approach taken in *Handheld Device Markup Language* (HDML) by *Phone.com, Wireless Application Protocol* (WAP) and *Compact HTML* (C-HTML) by *Nippon Telephone & Telegraph* (NTT) DoCoMo.

Screen size directly affects the amount of information displayed; and page layout needs to be planned accurately, so that information can be easily accessed by the user.

As suggested in Metter & Colomb, (2000), the page designer should:

- maintain user orientation during navigation by means of a title showing at what point of the site the user is at;
- verify that long text lines do not exceed screen dimension, compromising readability; and
- rearrange complex tables, made up of many columns, into lists so that they do not exceed screen limits.

Furthermore, scrolling should be reduced by:

- placing surfing aids, such as menu bars, in a fixed place near the top of a page;
- placing key information at the top of a page; and
- reducing the amount of content in a page.

Finally, small-screen users seem to choose and prefer direct access strategies to information, as illustrated in Jones, Marsden, Mohd-Nasir, Boone, and Buchanan (1999); Jones, Marsden, Mohd-Nasir, and Buchanan (1999); Jones, Mohd-Nasir, and Buchanan (1999); and Buchanan and Jones (2000), so the designer should:

- provide a search mechanism within a website; and
- organise the information so that the navigation is focused; for example, suggest to the users a list of goals they might want to achieve and present a framework designed for facilitating such access.

Multiple-device authoring identifies a range of target devices and defines the mapping from a single source document to a set of rendered documents that apply to the different target devices.

For example, StrecthText (Cooper & Shuebotham,[2] 2002) allows the page designer to tag portions of the document with a "level of abstraction measure." When the user receives the document, he can select the desired level of abstraction and the document is presented with the corresponding degree of detail.

Another example is *Cascading Style Sheets* (CSS) (Lie & Bos, 1999) of *HyperText Markup Language* (HTML), which allow the page designer to define, within a single style sheet, a set of display attributes for different structural portions of a document. For example, all top-level headings have to be in bold or the anchor text of a link has to be red. In general, a set of style sheets can be associated with a document, and the same style sheet can be shared by different documents. When the user retrieves a document, he also retrieves the set of style sheets associated with that document. For each style sheet in the set, there is a weight that measures the degree of satisfaction of applying that style sheet to the document. The CSS cascade mechanism assigns a weight to each style rule. When several rules apply, the one with the greatest weight takes precedence (W3C, 2002). It is relevant to note that in October 2001 the *World Wide Web Consortium* (W3C) produced the CSS Mobile Profile 1.0 (W3C, 2002).

Client-side navigation allows the user to navigate interactively within a single page, modifying a portion of it that is displayed at a time.

A trivial example is the use of scroll bars. Another example is PAD++ (Bederson & Hollan, 1994), in which the user can zoom in and out over portions the document. Active outlining (Hsu, Johnston, & McCarthy, 2002) is also an example of client-side navigation, because the user can dynamically expand or collapse sections of the document under their section headings.

Automatic re-authoring requires the development of a software capable of processing an arbitrary Web document designed for a desktop computer and adjusting it through a series of transformations to the screen of an handheld device. This process can be made on the client, on the server or on a proxy server, which has the single task of providing conversion services.

There are many possible reauthoring techniques, which can be categorised along two dimensions: *syntactic* vs. *semantic* and *transformation* vs. *elision*, as classified by Bickmore & Schilit (2002).

Syntactic techniques operate on the structure of the page, while semantic techniques require a certain degree of understanding of the content of the page. Transformation techniques modify the content or the presentation of a page, while elision techniques remove some content, leaving the rest unchanged.

So we can have the following cases:

- *syntactic elision*: portions of text or objects within a document are removed, considering the structure of the document or the kind of object. Examples are active outlining, displaying only the first paragraph of each phrase and removing images;

- *syntactic transformation*: portions of text or objects within a document are reorganized, considering the structure of the document or the kind of object. Examples are transforming a complex table into a list and image reduction;
- *semantic elision*: portions of text or objects within a document are removed, upon an understanding of their meaning. An example is the removing of duplicate links; and
- *semantic transformation*: portions of text or objects within a document are reorganized, upon an understanding of their meaning. Examples are text summarisation and reorganisation of all links within a page into a list.

A full example of automatic reauthoring is Power Browser (Buyukkokten, Garcia-Molina, Paepcke, & Winograd, 2000; Buyukkokten, Garcia-Molina, & Paepcke, 2000; Buyukkokten, Garcia-Molina, & Paepcke, 2001b; Kaljuvee, Buyukkokten, Garcia-Molina, & Paepcke, 2001; Buyukkokten, Garcia-Molina, & Paepcke, 2001a; Buyukkokten, Kalijuvee, Garcia-Molina, Paepcke, & Winograd, 2002), which utilises semantic and syntactic techniques to improve the user's interaction on a PDA.

Power Browser is a proxy server that transforms Web content before delivering it to a handheld device.

During navigation, a set of link descriptions is shown, heuristically generated by anchor text, *Uniform Resource Locator* (URL) structure and ALT tag. This structure, organised in a tree, not only includes links on a single page but also a hierarchical structure of links on linked pages, so the user can directly retrieve a page from a link description which is visible on the screen.

Once the desired page is found, the user can view some of its content. Images are not displayed, and the text contained in the ALT tag is displayed instead of images; images can be shown on user demand, after a refinement step of adjustment of the image to the low-screen resolution, where the step is performed by the proxy server.

A summary is presented for each page, obtained by partitioning an original Web page into fragments, for example paragraphs, lists or ALT tag describing images, and selecting the most important fragment as a summary.

Power Browser also tries to address the problem of filling in forms on a PDA screen.

Forms constitute a problem on a PDA and other handheld devices because input controls and associated textual explanations occupy too much space on the screen; therefore, it is difficult for users to gain an overview of the form's content and of what is required of them. For example, if the user has to fill in a form registering for a service, he could interpret the name field as a name and surname field and fill in those two data instead of the name only. When he proceeds by scrolling the bar he finds that there is another field asking for his surname. At this point the user has to go back to the previous field and correct the wrong information submitted. The problem of filling in forms needs to be carefully addressed in an m-commerce site, because asking users for information is quite a frequent operation.

Instead of showing all the input controls at once, Power Browser initially shows only minimal textual prompts for the input controls. When the user is ready to fill in information, a pen gesture on a textual prompt causes the associated input control to be displayed, while all other input controls remain hidden. In general, all the text other than labels is ignored, so each visible string is a label for an input control. Buttons are expressly identified and links are handled as usual.

A different solution is WebViews (Freire, Kumar, & Lieuwen, 2001), which allows end-users to create and maintain simplified views of Web content with a *Video Cassette Recorder-* (VCR) style interface.

In the following, "WebViews" means the name of the software, "Web view" is a view of the Web obtained using WebViews, and "Web views" is a set of views of the Web obtained using WebViews.

A user can create a Web view from a desktop computer by simply browsing the desired page and selecting the content of interest within the page, while the WebViews recorder registers all intermediate steps and actions undertaken, such as interacting with *Common Gateway Interface* (CGI) scripts and so on. Therefore, a Web view is much more than a simple bookmark because it also stores the information needed to interact with scripts. The Web view created is robust to some changes in page layout so, if the site does not change too much, the user can avoid recreating the Web view. When the user accesses the Internet through a handheld device, he can recall the created Web view from the Web views' server which accesses the previously selected content on the desired Web page, clips it and returns an *eXtensible HyperText Markup Language* (XHTML) response to the client. The client could also be an intermediate proxy, which performs content adjustment and translates the XHTML content into various formats, e.g., *Wireless Markup Language* (WML).

WebViews facilitates and shortens the subsequent phase of automatic reauthoring, since it provides the reauthoring module of a simplified version of the Web page; this contains only the content desired by the users and avoids all the intermediate steps.

To address the problem of filling in forms on a handheld device screen, WebViews allows the user during the recording phase to specify which field values are to be stored in the Web view specification itself and which ones are to be requested by the user every time the Web view is executed.

The *International Business Machines* (IBM) Websphere Transcoding Publisher (Britton et al., 2001; IBM, 2002) is a commercial product which can be considered as being halfway between a multiple authoring and an automatic reauthoring device. It uses, in fact, users' profiles and both *eXtensible Markup Language* (XML) and *eXtensible Stylesheet Language* (XSL) style sheets for content adjustment, just as in multiple device authoring. Another important characteristic is that it transforms HTML into WML *cards* and *decks*; to perform such a transformation, it uses syntactic techniques such as image reduction and semantic techniques, such as rearranging text into WML cards and decks.

Strengths and weaknesses of those different approaches

"Device specific authoring" typically gives the best results, but it limits users to a small subset of Web pages. Furthermore, managing and keeping a site updated is an onerous task, requiring hours of manual work.

"Multiple device authoring" requires less effort for a single document than device-specific authoring, but it requires a considerable amount of manual design, and it limits the users to a subset of Web pages.

"Client-side navigation" is a promising approach if a set of good techniques can be developed. In fact, the PAD++ "magnifying glass" approach can be very awkward if the documents are large; the active outlining approach can have a limited applicability, since many Web pages are not organised into sections and subsections.

"Automatic reauthoring" is thus the ideal approach to provide a broad access to the Web through a wide range of handheld devices, where it is possible to produce legible and aesthetically pleasing documents that can be easily surfed without loss of information.

An approach such as that taken in WebViews can help the automatic reauthoring process, but it still requires a certain degree of user interaction to prepare the Web view, and that same Web view needs to be recreated if many changes are made to the original page. Moreover, the process of creating a Web view is managed on a desktop computer, and that requires the user to plan which pages he will access from the handheld device.

Input Method

During normal Web surfing, the user can utilise both the keyboard and mouse to reach every object on the screen and to provide input to forms easily.

The input in a PDA is done through a pen, which makes both gestures and character input possible. An advantageous aspect of this input method is that the engine for character recognition could be shared with the engine for gesture recognition, as illustrated in Moran, Cheyer, Julia, Martin, & Park (1997).

Characters are written by the user in a special area of the screen and are then recognised via *Optical Character Recognition* (OCR). With this approach text input is subject to mistakes, and the user has to write characters with some care to ensure the system recognises them; this process increases the time required to write a word. An alternative is to draw a virtual keyboard on the PDA's screen so that the user can choose the characters with the pen and compose a word. Even this approach is time consuming and can lead to the user making mistakes.

The use of a pen allows the user to reach any object placed on the screen, simplifying the design of objects layout, but not input difficulties. Furthermore, gestures made with a pen are generally more varied than those that can be done with a mouse. For example, a top-down gesture could mean vertical page scrolling.

Input method on a mobile phone is a great constraint, as there is the numeric keypad and a scrolling key. The WAP Forum requires at least vertical scrolling, so no horizontal or diagonal scrolling is guaranteed. A consequence is that the process of surfing a page can be difficult and frustrating, looking for links to other pages or consulting long lists. Moreover, the user has to press the keys many times to form words. In order to address the page-surfing problems, page layout should be carefully planned—making frequently used functions easily reachable. Metter & Colomb (2000) suggest helping the user by gathering all the links within a page in a permanently available list, eventually combining these links with keys on the numeric keypad.

Word completion techniques should be employed to manage text input problems, both for PDAs and mobile phones. We can distinguish between general purpose techniques, meaning techniques developed generally to overcome the constraints of the device, and specialised techniques, meaning techniques tailored to the user and the application.

General purpose techniques try to complete words according to a predefined dictionary, differentiating between one word and another as characters are entered.

A technique developed for handheld devices generally is *Predictive cOmposition Based On eXample* (POBox) (Masui, 1999), which is organised into two steps: filtering step and selection step. During the filtering step, as the user enters characters, the system dynamically uses those characters to look for candidate words in a dictionary. During the selection step, the user can select the desired word from among candidate words. Then the next input words are predicted from the context and passed on to the next filtering step.

Many solutions were studied to facilitate text input through the numeric keypad of mobile phones, proposing a way to introduce words by pressing keys only once, such as in Tegic (Tegic, 2002) *Text on 9 keys* (T9), which is a software adopted on many mobile phones and which is available for many languages. Every time there is an ambiguity between words, T9 shows the user a list (by means of an internal database) with the possible terms in order of utilisation frequency, so that the user can choose the correct one. Furthermore, the internal database can be updated with new words added by the user, when he introduces words unknown to the system. GSMBox (2002) reports a study made by Nokia according to which text input with T9 is twice as fast as conventional methods, and the Nokia study confirmed the user's desired word is the first choice 95% of the time.

Specialised techniques suggest words to users, choosing them from a well-defined set personalised for the user, his needs, and features of the applications with which the user is interacting.

For example, Power Browser offers support for keyword entry during a local search session within a site: for every site that the user is visiting, word completion and measures of keyword selectivity are provided. As users enter successive keywords, the system shows how many pages match the query, and users can submit queries only when the keywords are selective enough.

The approach adopted by WebViews facilitates text input, because it avoids all the intermediate steps, thus requiring less input for surfing the Web; input forms can be partially or fully filled in during the Web view creation phase, decreasing the problem of inserting words with a handheld device.

So, if in general we can expect the presence of general purpose techniques on the device, peculiarities of the user or of the application can require the developing of more specialised techniques.

A very attractive alternative for solving text input problems is speech recognition. It is certainly very useful but still not widely adopted, due to the problems of implementing voice-independent recognition systems, where little or no information is provided about the domain to which the text to be recognised pertains.

Wireless Link, Small Memory, Slow CPU and Energy Consumption

A handheld device uses a wireless link to transfer data to and from a server. The bandwidth available nowadays is very low, and as a result the link between the client and the server is very slow and does not allow large data transfer. This problem will be partially overcome in the future by 3G networks. An example of this trend is the *General Packet Radio Service* (GPRS), which is an evolution of the *Global System for Mobile Communication* (GSM) and a technology emerging today that achieves performances that are roughly the same as those of a standard analog modem, in a favourable case scenario. Furthermore, wireless networks suffer from high latency.

The WebViews approach addresses low bandwidth problems, since it minimises the amount of data transferred between the handheld device and the server, as well as high latency problems because it avoids repeated access to the Internet for fetching intermediate pages needed to reach the desired page.

A handheld device has a small amount of memory, and it is impossible to manage complex pages, such as a page containing a large image or a high amount of data. The CPU is quite slow and computational intensive operations should be avoided, as otherwise the user has to wait far too long. Moreover, battery energy needs to be saved in order to guarantee the autonomy of the device, and this is a constraint for the operations that can be done on the client, since the CPU operations are expensive in terms of energy consumption.

The problems regarding computational load distribution affect the choice of which component has to provide the various functions to help the user—either the server or the handheld device: certainly the server has the power required to manage the various tasks needed to facilitate surfing, but response time from the server to the handheld device needs to be considered, since it can compromise performances. So there is a trade-off, as some computational load has to be moved to the client to guarantee the system some efficiency. This is at least until the slowdown caused by the processing action on the client is below the time required to do the same processing on the server and to send data to the handheld device.

Ojanen & Veijalainen (2000) give an example of these considerations and the method according to which various parameters have to be evaluated. Ojanen & Veijalainen (2000) study WML code compression before it is sent to a mobile phone, so that the amount of bytes to be transmitted decreases and the transmission time is reduced. It is explained how to determine whether this choice is convenient or if the overheads introduced by compression and decompression of the code outweigh the gain obtained from reduced transmission time.

MOBILE SEARCH ENGINES

In the previous section, the adjustment of contents to handheld devices was analysed, which constitutes a part of the interaction process for the user seeking information. In this section, we analyse the other part of the interaction process, i.e., that of having a *Search Engine* (SE) available for use and access from handheld devices.

This type of SE can be called a *Mobile Search Engine* (MSE), stressing that these SEs need to be designed and implemented for direct access by any type of mobile device.

The first part of this section clarifies the motivations for studying MSEs. The second part presents a general discussion on MSE features and related problems. The final part presents the design of Odysseus, an ongoing MSE project initiated by the authors of this chapter.

Odysseus is the name of our project for designing and implementing an MSE. At present Odysseus is only in the early stages of design, so it is used here as an example for discussing relevant aspects that need to be addressed in designing mobile applications to retrieve information for a user not using a desktop computer but a mobile device. We will explain the concepts, architecture and solutions for Odysseus, which are all under careful evaluation.

Motivations for Studying MSEs

Search engines are often used to discover and find information on the Web. To search the Web using handheld devices, a specific type of search engine will be required.

The user could utilise a search engine on handheld devices for the following reasons:

- to choose between services offered, including those of m-commerce, and to find one which best satisfies his needs; or
- to retrieve information of every-day interest for work and pleasure.

With regard to the first point, an analysis of the situation leads to the consideration that the amount of applications offered and the range of resources available require a tool that allows the retrieval of services and information in a quick, simple and easy manner. If there are sites offering online trading, m-commerce, entertainment and so forth, it is necessary to have an application which finds what is required, since the burden of knowing and remembering various sites cannot be passed on to the user.

The *Universal Mobile Telecommunication System* (UMTS) Forum (UMTS Forum, 2002) also considers the mobile portal as the way of providing access to 3G services and, within the mobile portal, search engine is an important component.

With regards to the second point, the user, in addition to coping with prepacked proposals such as those that are managed by the organisation he works with, may desire to use the power of the Internet as a source of resources and, therefore, to search for documents and information on specific topics of interest for work and pleasure, as with traditional search engines.

At this point someone could wonder how the inquiring and retrieving can take place on a handheld device. The following scenarios can be shown:

- *directly legible document*: if the size of the document is not too big, it can be displayed directly on the handheld device's screen, after a content adjustment phase;
- *not directly legible document*: if the size of the document does not allow for its online reading, or, if the kind of file does not permit its browsing (such as a binary file), the search result can be sent to a workplace capable of managing the resource; in this case, the actual use of the document would be possible later. While the user is moving, the search can serve to find useful documents and some other tool can arrange in advance a source of interest at destination, which can be ready when the user arrives at destination, so that it can be used immediately—improving efficiency;
- *printable document*: if the user has a portable printer or can connect to a printer or a fax, he can immediately print the search result and use it, even if the search result is not easily readable on a screen. This option conforms with the concept of "any device," and it is already in line with the Bluetooth (Bluetooth, 2002) standard for wireless interconnection of heterogeneous devices.

The study of searching by means of handheld devices should not be limited to SEs but should be extended to *Information Retrieval Systems* (IRS)s generally and *Digital Libraries* (DL)s. Indeed, mobility could improve the utilisation of a DL: for example, a person can perform a search on a DL while moving between the physical shelves of a library, as observed in Marshall, Golovchinsky, & Price (2001).

Furthermore, as suggested in Marshall et al. (2001), searching and reading a DL can be done together with other activities, such as working with colleagues, alternating searching with writing and organising materials. During teamwork, one person can add annotations to documents of a DL and another person can search the DL for documents with a specific annotation. Merging content and wireless communication can develop ubiquitous access to DLs, improving well-established collaborative practices and exploiting physical and digital resources.

Content Adjustment for MSE

In the previous section the motivations that can lead to the use of search engines from mobile devices were shown. Those considerations suggest that a Web search engine cannot be used directly from a handheld device, but its use should be adjusted to the peculiar features of the types of those devices. Thus, this section addresses the different aspects that have to be taken into account when the content, which constitutes the result of a query to a search engine, needs to be adjusted to a handheld device. The presentation follows the same sequence of presentation of device features that has been given previously, so it constitutes a parallel analysis and study of the previous one.

Screen size. The output of a query to a traditional search engine is shown, in multiple pages, as a list of links with a small paragraph of text, taken from the original Web page, describing each Web document of the result. Considering the small screen of a handheld device, search results cannot be displayed in a traditional way, because the information displayed could be difficult to read and the user would be compelled to scroll a long document to look for interesting links.

Therefore, it is necessary to identify and design a more compact way of displaying search results, different from the simple traditional ranked list. This new way of presenting results should allow the user to gain an overview of the search results and to reach the most relevant links easily.

Input method. Another problem is the input method: if during a simple navigation this problem could be less urgent, the user's interaction is greater in the use of a search engine, since the user inserts the query, the search engine returns a set of results, and, eventually, this cycle can be repeated as many times as is necessary to improve the search.

Therefore, the user needs to be facilitated with a general feature of word completion and a more focused suggestion of relevant keywords, tailored to the user's areas of interest.

Wireless link. The low bandwidth and high latency problems also need to be borne in mind, as a generic query could produce a long list of results and the query refinement process leads to frequent communications between the handheld device and the server.

To address these problems, a measure of the selectivity of a query should be estimated so that the user can actually submit the query when he is sure of not receiving too many results; this reduces the number of steps otherwise necessary to reduce the size of the set of results to improve the precision of the query.

Caching can help to face these problems—both local caching on the handheld device, as it allows the return to previous results, and server caching, which allows pages' prefetching while the user is reading the results.

Customizing the search engine. As illustrated generally for 3G services (UMTS Forum, 2002), a necessary feature for a search engine for handheld devices is the possibility of customising the application according to the user's needs. Today, Internet services are already offered as portals, highly configurable, and the user of handheld devices can expect to have similar services.

Customisation is important since most users mainly use a specific search engine, because it satisfies their needs adequately in terms of topics' coverage; also, they know its features and query language better. Therefore, the interface of the preferred search engine needs to be shown to a specific user, and it needs to be tailored to the specific device.

Customisation, as better knowledge of a user's needs, is also advantageous for the application, since it allows anticipation of the user's requests, offering a better service, and gets to know a user's vocabulary and topics of interest, which information is important for word completion and keyword suggestion.

Method of implementation. At present the change to 3G services is made as smoothly as possible, and this is witnessed from the introduction of 2.5G networks, as, for example, for GPRS networks. From this point of view, the implementation of a search engine to be used from handheld devices should be planned so that it will not require the redesign of existing systems; rather, it should be added to them as a new layer, permitting the start of the path towards 3G.

The current proposal, at protocol level, to bring the Web to mobile phones is WAP, which was developed by a consortium of manufacturers with the aim of making a standard.

However, WAP is designed for current mobile phones, but it is improving to ensure the support of those that are 3G. Once 3G mobile phones are on the market, other solutions such as XML (Leavitt, 2000) could be used.

The guidelines, which will be used in the design of an MSE, should not be bound too much to current standards; they should rather resolve the problems related to the kind of device and application offered, than exploit today's technology features, so that the design of an MSE is easily portable in the event of a change of standards and technology.

On the other hand, it should be shown how the design choices and solutions adopted could be fulfilled in an actual application. From this point of view, developing a prototype with WAP could be advisable; it would allow accurately documenting the phases of the implementation process, so that these could be used as a paradigm.

A relevant application for consideration is "Pirate," the recent proposal for a search engine for the Palm made by IBM. Pirate is the *Palm Information Retrieval Application for Text sEarch* (Pirate) (Aridor, Carmel, Maarek, Soffer, & Lempel, 2001; Aridor, Carmel, Maarek, Soffer, & Lempel, 2002), which allows users to predefine a topic of interest and then capture a very small and representative set of Web pages for the topic, storing it in a persistent repository called *Knowledge Agent Bases* (KAB). The process of creating a KAB can be called from a desktop computer or from a PDA. The knowledge acquisition is then performed on a desktop computer, and the resulting KAB is downloaded to the PDA.

The KAB is created starting from a set of four to six queries supplied by the user for defining the topic and/or from a set of seed URLs which the user considers relevant to the topic. A *Knowledge Agent* (KA) then browses the Web looking for pages relevant to the topic.

Stored in the KAB are a topic-specific lexicon, a small number (roughly 100) of core pages whose full text is stored in the KAB, a larger number (roughly thousands) of Web pages pointed from the core pages, for which only the URL and anchor text are stored, and an index for searching the KAB.

Pirate offers both word and query completion by using a domain-specific vocabulary from which it suggests, as keys are pressed, the most frequent words in the vocabulary consistent with the input prefix. Furthermore, query completion is performed by suggesting terms related to the query terms already introduced using a global semantic word network.

This solution has the advantages of allowing users to perform a search task with few or no data exchanges with the server, addressing low bandwidth and high latency problems, and completing words and queries in a very focused way, facing text input problems.

On the other hand, Pirate requires users to arrange in advance the KAB, not offering support and help for an online search of topics not stored in the KAB. Moreover, Pirate does not implement advanced features for addressing limited screen and scrolling problems.

Odysseus: An Example of an MSE

In order to understand all the problems related to the development of search engines for use on handheld devices, it seems important to underline the most relevant key features for their design. In Ferro (2001) a proposal called Odysseus was introduced. The most relevant characteristics of that proposal are reported by Ferro (2001) and are used here as a useful example, permitting the presentation of all the different aspects that have to be addressed when designing and developing a mobile search engine.

Odysseus is organised according to the client-server paradigm: the client sends a user's query to the middleware server, which forwards it to the selected search engine, collects the search results, and organises them for their display them on the client. The general architecture of Odysseus is shown in Figure 2.

Figure 2: Odysseus architecture

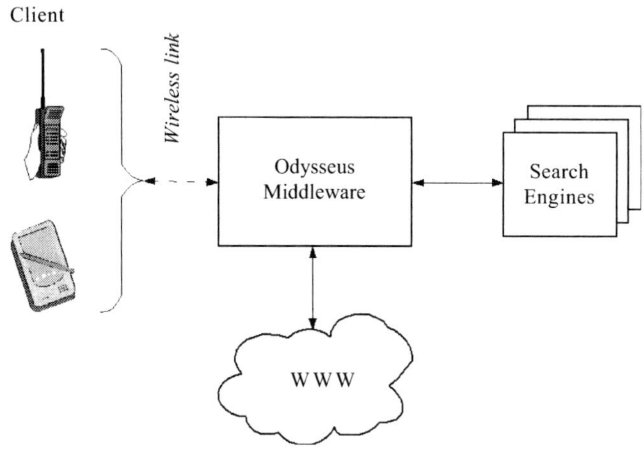

The middleware server provides, in general, two kinds of connection: a wireless connection towards mobile clients, and a wired connection with the Internet. The wired connection can be further specialised, at conceptual level, into a general connection to the Internet, which allows the middleware server to fetch required Web pages, and a focused connection, which is the interface toward search engines.

The aim of this architecture is to make the features of a Web search engine usable for handheld devices, without redesigning the technology already in existence. This is done by introducing an intermediate layer, which takes care of the mediation between the mobile clients and the Internet, and by adjusting Web content to the particular features of a handheld device. Odysseus architecture fits into the roadmap suggested by the UMTS Forum (UMTS Forum, 2002) to reach the mobile portal.

Functionalities of Odysseus middleware server

The middleware server executes the following tasks: it waits for a user's query and forwards it to a search engine (selected from a list of available search engines).

Once the list of search results has been obtained, Odysseus estimates whether there are too many search results to be successfully displayed on the handheld device. If that is the case, Odysseus tries to suggest other words to the user and invites him to refine the query to make it more selective.

The suggested terms together with the selectivity assessments for each term are uploaded from Odysseus to the client, so that the whole process of query refinement is done on the client, minimizing the data exchange between the client and Odysseus to address high latency problems.

Furthermore, Odysseus suggests terms to the user in a focused way—choosing them according to the user's needs, as explained later, integrating a general word completion feature, which can already be offered by the device, and assisting the user in the difficult phase of text input.

When the number of search results is small enough, the middleware server extracts a concise and significant description for each result and indexes the pages referred by search results. In order to perform this task, the middleware server accesses the Web repeatedly.

There are two kinds of visualisation to satisfy the requirements of adjustment to a device's features: a textual visualisation and a graphical visualisation, which give the user an overview of obtained results. These visualisations will be explained in more detail later.

The functioning of the middleware server is illustrated in the flow diagram of Figure 3.

By observing the tasks performed by the middleware server and considering that it communicates directly with handheld devices through the wireless link, you can notice that some features of Odysseus are common to a browser for handheld devices. Firstly, the middleware server has to manage the user interaction. Secondly, Odysseus has to show search results to the user. Then, when the user chooses a result, Odysseus has to present the desired page to the user, and this means rendering the page on the device's screen. Finally, Odysseus represents for the client the access point to the *World Wide Web (WWW),* and so it should also offer browsing functionalities. Thus, it could be a good choice to merge the functionalities of Odysseus with those of a browser for handheld devices.

We will now explain in more detail the techniques for managing the interaction between a search engine and a handheld device, considering that suggestions about content adjustment for Web browsing have been already given in the section on design alternatives for overcoming the constraints on handheld devices.

Odysseus middleware server architecture

Within the middleware server the modules (needed to take into account the three main entities described at the beginning of the chapter) have to be planned. Indeed users, devices, applications and their relationships should be borne in mind during the design and development of an application accessible through handheld devices.

The architecture of the middleware server is illustrated in Figure 4.

Four main functional modules are illustrated in Figure 4:

- *User Modelling Engine*: this module models the various kinds of users, taking into consideration the users' preferences, their topics of interest, and the way in which they prefer to receive answers from the system, e.g., a summarisation of information or a list of keywords;
- *Device Modelling Engine*: this module models the various kinds of handheld device the system can interact with, bearing in mind basic functionalities offered by the device and special features that can be exploited;

Figure 3: Tasks executed by Odysseus middleware server

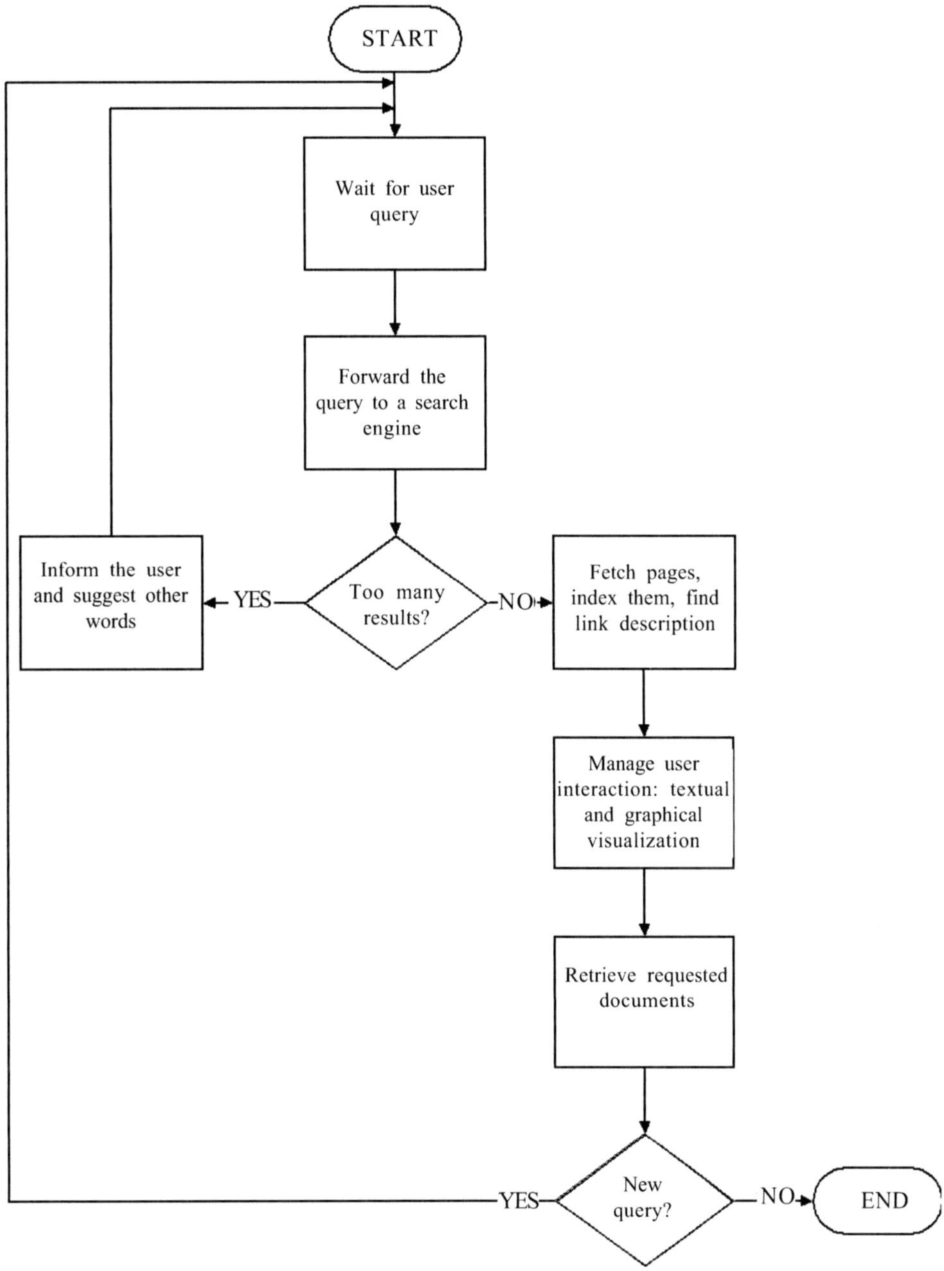

- *Application Modelling Engine*: this module models some aspects regarding the kind of application and special application needs (for example, an SE is characterized by a user's interaction greater than by browsing—in an m-commerce application filling in forms is a more substantial problem than browsing);
- *Engines Manager*: this module is a supervisor and coordinates the interaction of the other modules.

With this architecture, only one user model is maintained for each user, and so the user can access the application through various devices, keeping his work environment unchanged when he passes from one device to another, such as in Virtual Home Environment (VHE) of UMTS Forum (UMTS Forum, 2002).

On the other hand, only one device model is maintained for each device, and so different users, using the same device, have the same device model. This approach conforms with the *Composite Capability/Preference Profiles* (CC/PP) initiative (W3C, 2002).

We now see in more detail the function of each component within the middleware server.

Figure 4: Architecture of Odysseus and its modules

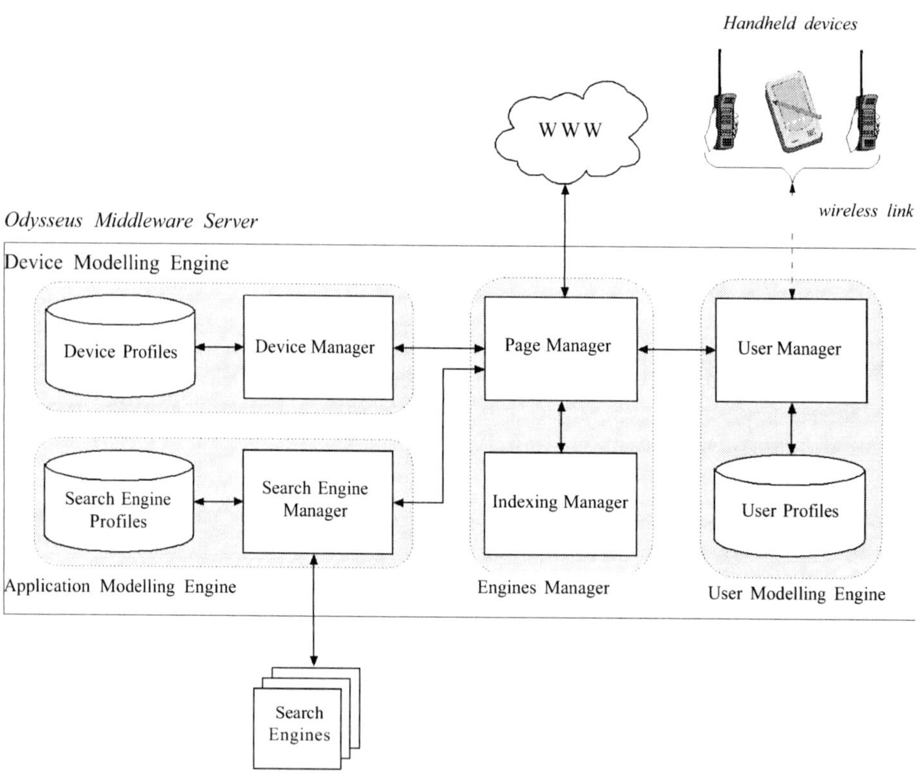

User Manager controls all the communication to and from the wireless client. It represents for the client the access point to the Internet. Furthermore, by using the *User Profiles* database, it recognises each user and his preferences, and stores personalised information for each user; this could be topics of interest, previous searches along with query terms, links judged to be relevant and visited pages, keywords within these pages, the preferred search engine—thus allowing the customisation of the service.

User knowledge is a progressive and incremental process, because information about a user's behaviour is gathered as he utilises the system, enters queries and browses the Web.

Device Manager keeps information about various handheld devices in the *Device Profiles* database, information such as screen resolution, available characters dimension, input characteristics and data protocol used by the handheld device, e.g., WAP, HDML and so forth.

The information on a handheld device's characteristics is used in the phase of content adjustment to tailor the behaviour of Odysseus to the specific device utilised by the user.

Search Engine Manager co-ordinates interactions with the various search engines supported by Odysseus: it forwards the user's query to the desired search engine, it retrieves the results proposed by the search engine, parses them, and translates them in an internal format (based on XML), so that the other components of the middleware server can perform their computation without considering the particular search engine used.

Furthermore, the *Search Engine Manager* can communicate with search engines not only through *HyperText Transfer Protocol* (HTTP), but also through CGI script, if the search engine provides advanced functionalities.

The *Search Engine Manager* keeps information about a search engine in the *Search Engine Profiles* database. It stores the name and the URL of the search engines, the parameters regarding the interface of the search engine, such as the relative position of objects, an object's type and function, information about special features implemented by the search engine, for example finding all the pages pointing to a desired URL, the search engine's available operators, modality of communication, e.g., HTTP, CGI, and a brief guide to the search engine's functionalities for helping the user.

Page Manager performs complex tasks and coordinates the components of the middleware server. It performs the following tasks:

- *management of number of results*: using a device dependent threshold, the *Page Manager* decides if there are too many search results and, in which case, it warns the user and suggests more terms.

- *query expansion*: the query expansion technique can be used when a user has entered his query, but that query would produce too many results; in an automatic way, more terms can be added to the original query to reduce the result set to be presented to the user; to reach this objective, the *Page Manager* asks the *User Manager* for terms related to the query by using the user profile and it extracts the more significant terms from the set of retrieved documents;
- *results description*: the result page itself is used for extracting a description sentence. Eventually, even the pages pointing to the desired page can be utilised: within these pages, the paragraph containing the link to the desired page is analysed and used to extract the description. An approach of this kind is adopted in InCommonSense (Amitay, 2000). To extract the summary sentence, the *Page Manager* uses the *Indexing Manager*;
- *page rendering and user interaction*: to render the page on the client, the *Page Manager* asks the *Device Manager* for information about the device's physical features and the language to use for describing the page, e.g., WML, HDML and so on.

Indexing Manager receives as input an HTML page and splits it into sentences of simple text, using structural elements such as paragraph breaks, lists, tables, titles and sections, punctuation.

Then each sentence is indexed, extracting tokens, filtering the stop words and performing the stemming, and is weighted according to the tf $\times$ idf (*term frequency, inverse document frequency*) weight:

$$w_{ij} = \mathrm{tf}_{ij} \cdot \log_2\left(\frac{N}{n}\right)$$

where:

- w_{ij} is the weight of the word T_j in the sentence S_i;
- tf_{ij} if the frequency of the word T_j in the sentence S_i;
- N is the number of sentences in the collection; and
- n is the number of sentences where the word T_j appears at least once.

Then a similarity score is computed between each sentence and the query. The sentence with the highest score is chosen as a description for a result.

Results visualisation

Odysseus offers two kinds of visualisation: a textual visualisation, similar to the ranked list, and a graphical visualisation, which lets the user choose from among three different types of representation of the results set.

Textual visualisation displays the description of a link and, if there is enough space on the screen, the URL as well. The URL is an important element, because from it the user can ascertain which site it is, if he already knows the site and if the site is interesting; thus, the URL should be displayed. If screen size allows it, a link description and a URL can be displayed at the same time while, if there is no space, the link description is displayed first and, upon user demand, the URL too.

Depending on the device used, one or two links can be displayed at the same time, and to see those that follow, the user can utilise a scroll key. Once the user reaches the desired link, he presses a confirmation key and the corresponding page is displayed.

The link order in textual visualisation is that returned by the search engine.

Graphical visualisation operates by offering the user the choice between various methods of visualisation and allowing him to move from one to another; once an object of interest is selected, Odysseus shows the link description, as used in textual visualisation, and it is possible to choose to see the Web page or to restart the exploration process.

In graphical visualisation, search results are represented as points in a bidimensional space. The aim is to give the user an overview of the search results, so that he can understand the relationships between the query and the results, and the relationships between the results themselves, without requiring the user to read each result to assess its relevance. In this way the information retrieval process is simplified and sped up.

A problem with graphical visualisation is input: how can the user move among various objects displayed? This is a problem especially for a mobile phone, because on a PDA, any object can be selected by the pen. For a mobile phone, a movement can be used which allows the probably more significant document to be visited first. For example, a clockwise or anticlockwise spiral movement, according to whether the user presses the up or down key, begins from the result most similar to the query and then moves on to those less similar.

When the desired object is reached, pressing a confirmation key permits the display of pertinent information.

Assuming a vector space model, every graphical visualisation tries to move from the *n*-dimensional space of terms of the documents to a two-dimensional space; what is changed from one graphical visualisation to another is the way in which point position is calculated within two-dimensional space.

Odysseus offers three graphical visualisations:

- The first visualisation tries to discover the links between the results, showing the inter-document similarities and which documents regard the same topic, through clustering techniques. If two results are similar, the corresponding points will be close while, if the points are far apart, it means that page content is very different. This approach is taken in Lighthouse (Leuski & Allan, 2000b; Leuski & Allan, 2002; Leuski & Allan, 2000a; Leuski & Allan, 1998).

A circle can be used to represent a judged or estimated relevant document, while a square can represent a judged or estimated nonrelevant document.
A simple corollary of Cluster Hypothesis is that if a relevant document is found, other relevant documents should be in the neighbourhood of this relevant document. As a consequence, with this visualisation, circles of relevant documents move toward circles of other relevant documents. Therefore, finding interesting information should be as easy as inspecting a circle near the circle of a document of known relevance. Similarly, squares of non-relevant documents tend to be brought together.
An example of this visualisation is shown in Figure 5.

- The second visualisation tries to highlight the relationships between the query and the results, in terms of relevance and similarity to the query and, in the case of multiple queries, it shows results common to the queries and the relationship of these results with different queries. This approach is taken in Hyperspace (Beale, McNab, & Witten, 1997).
  A circle can be used to represent a document, while a rhombus can represent a query.
  Independent and completely separate queries make a series of "dandelion heads," i.e., sets of unconnected circles, each one centred on the query that generated it. More interesting modalities are given by correlated queries, because if the same document belongs to a set of search results of two or more queries, it is connected to various rhombi.
  An example of this visualisation is shown in Figure 6.
- The third visualisation lets the user analyse the relationships both between the query and the result and between the results and the terms of the query. Thus, it is possible to know that a document was judged relevant rather by the presence of some terms than the one of others.

Figure 5: Visualization of interdocument similarities and document clustering

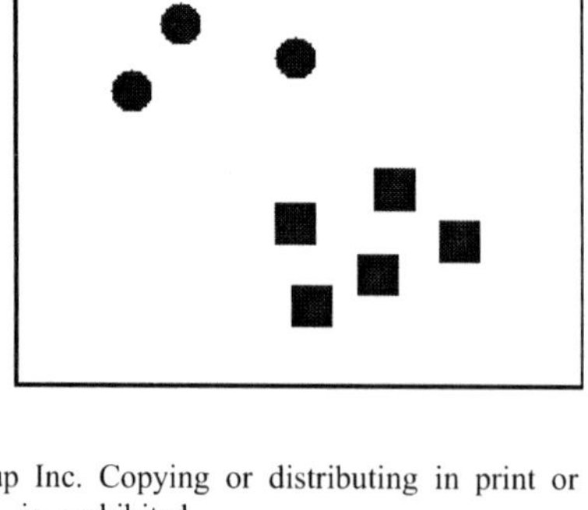

Figure 6: Visualization of relationships between results and query

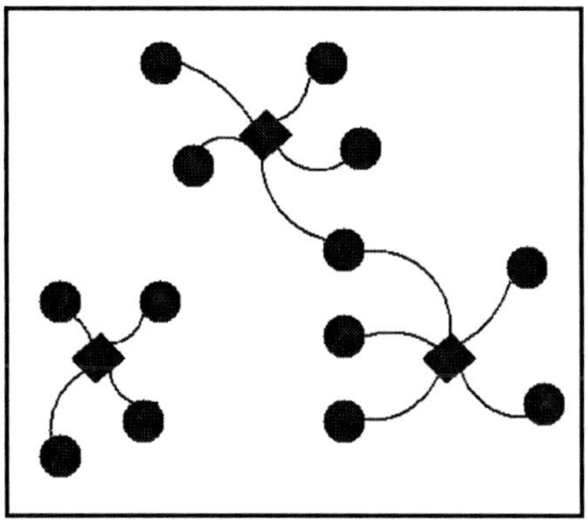

A three-terms query could be represented as a pyramid with a triangular basis, which is reminiscent of a Native American Indian tepee. Each face of the tepee is utilised to represent one of the three terms of the query. A pendulum (representing found documents) is used to show which terms are more relevant. The pendulum is attracted by the tepee's face, which represents the most relevant term, and its length is a measure of the overall relevance of the document. This approach is taken in Tepee (Grewal, Jackson, Burden, & Wallis, 1999; Grewal, Jackson, Wallis, & Burden, 2002).

The intersection between the pendulum and pyramid's basis can be determined, so that a bidimensional graph can be obtained, as in Figure 7. A circle can be used to represent a document, while a rhombus can represent the query.

The extension to queries with more terms is done simply by using a pyramid with polygonal basis, with as many faces as there are query terms.

Figure 7: Three-terms query and three results

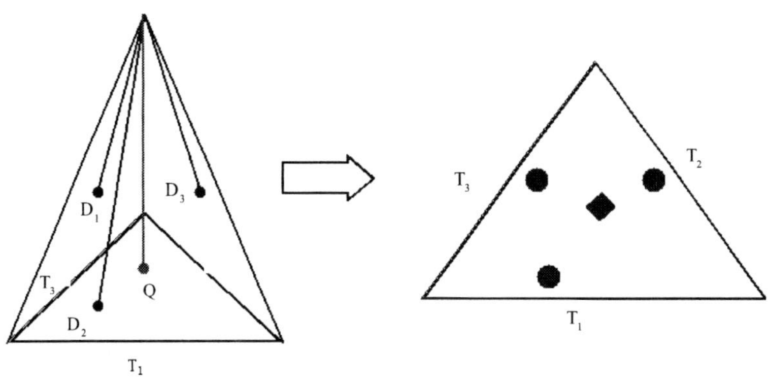

CONCLUSIONS

Generally, users perceive small screen size and input difficulties make Web surfing on a handheld device a poorer experience than that on a desktop computer. To facilitate better access to the mobile applications, the latter should know if a user request originates from a desktop computer or a handheld device; in this latter case, the content should be adjusted so that it is easy to use, making user interaction more effortless.

On the other hand, content adjustment should also preserve the functional features of the mobile applications as much as possible; the user should be just as at ease using an application on a handheld device as on a desktop computer.

The responsiveness of mobile application influences user satisfaction. The interactions between mobile applications and users should be designed to reduce the processing and data transfer time. At present, a problem experienced by WAP is the long time required for the initial connection set-up with GSM (in several seconds). To reduce processing time, the mobile applications should not overload the handheld devices with computationally intensive operations.

The message emerging from these considerations is clear: the content needs to be adjusted to the handheld device, but the required processing should neither force the user to wait too long nor should it consume too much energy, compromising the user's mobility.

The aim is to encourage users to leave fixed Internet connections for wireless ones, without distressing them with different features of devices and discouraging them with great difficulties in Web fruition.

We have described the problems of managing the interaction between handheld devices, mobile applications and users, particularly when designing search engines for handheld devices. The project of extending the use of pervasive devices leads to the concept of content adjustment, which is needed for various reasons:

- there are different types of handheld devices, and the "anytime, anywhere, any device" paradigm of 3G systems underlines that the same content needs to made available and usable for different devices;
- handheld devices are very different from desktop or portable computers, and the content developed for these computers cannot be moved to handheld devices as it is;
- legacy applications should be used on handheld devices to extend an enterprise's business opportunities.

ACKNOWLEDGEMENTS

We are especially grateful to the editors of the book and to the anonymous referees for their most valuable comments on the earlier version of this chapter.

During the initial development of this work, Nicola Ferro was attending a training course in the stimulating environment of IBM Technology Deployment of Segrate, Milan, Italy. He wishes to thank his manager, Antonio Caizzi, and his colleagues for the time spent with him and the stimulating discussions.

REFERENCES[2]

Amitay, E. (2000). InCommonSense—Rethinking Web search results. In *Proceedings of the IEEE International Conference on Multimedia and Expo* (ICME 2000) (pp. 1705-1708).

Aridor, Y., Carmel, D., Maarek, Y. S., Soffer, A., & Lempel, R. (2001). Knowledge encapsulation for focused search from pervasive devices. In *Proceedings of the Tenth International World Wide Web Conference* (pp. 754-764).

Aridor, Y., Carmel, D., Maarek, Y. S., Soffer, A., & Lempel, R. (2002). Knowledge encapsulation for focused search from pervasive devices. *ACM Transactions on Information Systems (TOIS), 20* (1), 25-46.

Beale, R., McNab, R. J., & Witten, I. H. (1997). Visualising sequences of queries: A new tool for information retrieval. In *Proceedings of the IEEE Conference on Information Visualization* (pp. 57-62).

Bederson, B. B., & Hollan, J. D. (1994). Pad++: A zooming graphical interface for exploring alternate interface physics. In *Proceedings of the ACM symposium on User Interface Software and Technology* (pp. 17-26).

Belkin, N., Oddy, R., & Brooks, H. (1982). ASK for information retrieval. *Journal of Documentation, 38*, 61-71 (part 1); 145-164 (part 2).

Bickmore, T. W., & Schilit, B. N. (2002). Digestor: Device-independent access to the World Wide Web. *Proceedings of the Sixth International World Wide Web Conference* [online]. Available at: http://www.scope.gmd.de/info/www6/technical/paper177/paper177.html.

Bluetooth (2002). The official Bluetooth website [online]. Available at http://www.bluetooth.com/.

Britton, K. H., Case, R., Citron, A., Floyd, R., Li, Y., Seekamp, C., Topol, B., & Tracey, K. (2001). Transcoding: Extending e-business to new environments. *IBM Systems Journal, 40* (1), 153-178.

Buchanan, G., & Jones, M. (2000). Search interfaces for handheld Web browser. In *Poster Proceedings of the Ninth International Word Wide Web Conference* (pp. 86-87).

Buyukkokten, O., Garcia-Molina, H., & Paepcke, A. (2000). Focused Web searching with PDAs. In *Proceedings of the Ninth International World Wide Web Conference* (pp. 213-230).

Buyukkokten, O., Garcia-Molina, H., & Paepcke, A. (2001a). Accordion summarization for end-game browsing on PDAs and cellular phones. In *Proceedings*

of the Conference on Human Factors and Computing Systems (SIGCHI 2001) (pp. 213-220).
Buyukkokten, O., Garcia-Molina, H., & Paepcke, A. (2001b). Seeing the whole in parts: Text summarization for Web browsing on handheld devices. In *Proceedings of the Tenth International World Wide Web Conference* (pp. 652-662).
Buyukkokten, O., Garcia-Molina, H., Paepcke, A., & Winograd, T. (2000). Power browser: Efficient Web browsing for PDAs. In *Proceedings of the Conference on Human Factors and Computing Systems* (CHI 2000) (pp. 430-437).
Buyukkokten, O., Kalijuvee, O., Garcia-Molina, H., Paepcke, A., & Winograd, T. (2002). Efficient Web browsing on handheld devices using page and form summarization. *ACM Transactions on Information Systems (TOIS), 20* (1), 82-115.
Cooper, I., & Shufflebotham, R. (2002). Pda Web browsers: Implementation issues [online]. Available at: http://www.cs.ukc.ac.uk/pubs/1995/57/index.html.
Ferro, N. (2001). Online information access through handheld devices. Laurea's thesis in telecommunications engineering. Department of Electronics and Computer Science. University of Padua, Italy (In Italian).
Freire, J., Kumar, B., & Lieuwen, D. (2001). WebViews: Accessing personalized Web content and services. In *Proceedings of the Tenth International World Wide Web Conference* (pp. 576-586).
Grewal, R. S., Jackson, M., Burden, P., & Wallis, J. (1999). A novel interface for representing search-engine results. In *Proceedings of the IEE Colloquium Lost in the Web—Navigation on the Internet* (ref. no. 1999/169) (pp. 7/1 - 7/10).
Grewal, R. S., Jackson, M., Wallis, J., & Burden, P. (2002). Using visualisation to interpret search engine results [online]. Available at: http://seed.scit.wlv.ac.uk/papers/activeweb99.html.
GSMBox. (2002). Studies confirm T9 text input is fastest method for word entry on mobile phones [online]. Available at: http://uk.gsmbox.com/news/mobile_news/all/2091.gsmbox.
Hsu, J., Johnston, W., & McCarthy, J. (2002). Active outlining for HTML documents: An X-Mosaic implementation. Proceedings of the Second International World Wide Web Conference [online]. Available at: http://archive.ncsa.uiuc.edu/SDG/IT94/Proceedings/HCI/hsu/hsu.html.
IBM. (2002). Websphere transcoding publisher [online]. Available at http://www.ibm.com/software/webservers/transcoding/.
Jones, M., Marsden, G., Mohd-Nasir, N., Boone, K., & Buchanan, G. (1999). Improving Web interaction in small screen displays. In *Proceedings of the Eighth International World Wide Web Conference* (pp. 51-59).
Jones, M., Marsden, G., Mohd-Nasir, N., & Buchanan, G. (1999). A site-based outliner for small-screen Web access. In *Poster Proceedings of the Eighth International World Wide Web Conference* (pp. 156-157).

Jones, M., Mohd-Nasir, N., & Buchanan, G. (1999). An evaluation of WebTwig—a site outliner for handheld Web access. In H. W. Gellerson (ed.), *Lecture Notes in Computer Science*, LNCS 1707 (pp. 343-345). Springer-Verlag.

Kaljuvee, O., Buyukkokten, O., Garcia-Molina, H., & Paepcke, A. (2001). Efficient Web form entry on PDAs. In *Proceedings of the Tenth International World Wide Web Conference* (pp. 663-672).

Leavitt, N. (2000). Will WAP deliver the wireless Internet? *Computer, 33* (5), 16-20.

Leuski, A. (2002). Evaluating a visual presentation of retrieved documents [online]. Available at http://cobar.cs.umass.edu/pubfiles/ir-159.ps.

Leuski, A., & Allan, J. (1998). Evaluating a visual navigation system for a digital library. In *Proceedings of the Second European Conference on Research and Advanced Technology for Digital Libraries* (ECDL '98) (pp. 535-554).

Leuski, A., & Allan, J. (2000a). Improving interactive retrieval by combining ranked list and clustering. In *Proceedings of the RIAO 2000 Conference* (pp. 665-681).

Leuski, A., & Allan, J. (2000b). Lighthouse: Showing the way to relevant information. In *Proceedings of the IEEE Symposium on Information Visualization 2000* (InfoVis 2000) (pp. 125-129).

Leuski, A., & Allan, J. (2002). Details of Lighthouse [online]. Available at http://cobar.cs.umass.edu/pubfiles/ir-212.ps.

Lie, H. W., & Bos, B. (1999). *Cascading Style Sheets: Designing for the Web* (2nd ed.). Addison-Wesley.

Marshall, C. C., Golovchinsky, G., & Price, M. N. (2001). Digital libraries and mobility. *Communications of the ACM, 44*, 55-56.

Masui, T. (1999). POBox: An efficient text input method for handheld and ubiquitous computers. In *Proceedings of the International Symposium on Handheld and Ubiquitous Computing* (HUC 99) (pp. 288-300).

Metter, M., & Colomb, R. (2000). WAP enabling existing HTML applications. In *Proceedings of the First Australasian User Interface Conference* (AUIC 2000) (pp. 49-57).

Moran, D. B., Cheyer, A. J., Julia, L. E., Martin, D. L., & Park, S. (1997). Multimodal user interfaces in the open agent architecture. In *Proceedings of the International Conference on Intelligent User Interfaces* (pp. 61-68).

Ojanen, E., & Veijalainen, J. (2000). Compressibility of WML and WML script byte code: Initial results [wireless mark-up language]. In *Proceedings of the Tenth International Workshop on Research Issues in Data Engineering* (RIDE 2000) (pp. 55-62).

Tegic. (2002). T9 [online]. Available at http://www.t9.com, http://www.tegic.com.

UMTS Forum (2002). Enabling UMTS/third generation services and applications [online]. Available at http://www.umts-forum.org/reports/report11.pdf.

UMTS Forum (2002). Shaping the mobile multimedia future [online]. Available at: http://www.umts-forum.org/reports/report10.pdf.

UMTS Forum (2002). The UMTS third generation market—structuring the service revenues opportunities [online]. Available at http://www.umts-forum.org/reports/report9.pdf.

W3C (2002). CSS Mobile Profile 1.0—W3C Candidate Recommendation, 24 October 2001 [online]. Available at http://www.w3.org/TR/css-mobile.

W3C (2002). Cascading style sheets, level 2—CSS2 specification [online]. Available at http://www.w3.org/TR/REC-CSS2.

W3C (2002). Composite Capability/Preference Profiles (CC/PP): A user-side framework for content negotiation [online]. Available at: http://www.w3.org/TR/NOTE-CCPP.

ENDNOTES

1 A handheld device is also known as a mobile device. In this chapter, we use both terms interchangeably.

2 The date on the availability of documents on the Web is 2002 for all documents, because the presence of all those digital documents has been checked at the given URL during the late months of the year 2002.

Chapter XI

Mobile Commerce and Usability

Susy S. Chan and Xiaowen Fang
DePaul University, USA

ABSTRACT

This chapter analyzes the critical issues confronting usability for mobile commerce (m-commerce) applications. Limited bandwidth and multiple form factors pose constraints for user interface design in terms of the amount and format of content presentation, navigation, and site structure. Mobile tasks performed on handheld devices—such as wireless PDAs, Pocket PCs and WAP phones—challenge developers to adopt new methods and design guidelines that take into account contextual variations in a mobile environment. At this early stage of mobile commerce, careful mapping of e-business strategies, mobile tasks, and technology characteristics will be critical for wireless interface design. Future research in these areas is needed to improve the usability of mobile commerce.

INTRODUCTION

Usability refers to how well an application is designed for users to perform desired tasks easily and effectively. Usability issues involve user interface design, methods for development, testing, and deployment. The current wireless technology poses many constraints for effective interface design. These constraints include limited connectivity and bandwidth, diverse yet simplistic devices, the dominance of proprietary tools and languages, and the absence of common standards for application development. Many of these problems are similar to early Web developments (Ramsay & Nielsen, 2000). However, users of m-commerce are likely to have already experienced e-commerce technology (Anckar & D'Incau, 2002) and therefore may have heightened expectation for data and services.

The convergence of mobile Internet and wireless communications has not yet resulted in expected growth in mobile commerce. Many factors may influence the adoption of m-commerce (Zhu, Nah, & Zhao, 2002). Consumer adoption of m-commerce has been slow, even in countries that have broadly adopted wireless technology (Anckar & D'Incau, 2002). The enterprise and business use of wireless technology holds greater promise for business growth and increased competitiveness, but it demands the transformation of business processes and infrastructure (Kalakota & Robinson, 2001). Poor usability of mobile Internet sites and wireless applications for commerce activities stands out as a major obstacle for the slow adoption of mobile solutions. Such difficulty discourages users from accessing mobile Internet sites (Chan, Fang, Brzezinski, Zhou, Xu, & Lam, 2002) or choosing m-commerce as a distribution channel (Shim, Bekkering, & Hall, 2002). Even with the latest 3G phones in Japan, consumers still find the small screen display and small buttons on these devices difficult to use (Belson, 2002).

Researchers suggest that interface developers need to consider the interaction among the interface design of user tasks, form factors, and purposes of applications (e.g., Chan et al., 2002). Application developers should also consider the interaction between the context of the environment and the device (Johson, 1998). Some researchers question whether the existing user interface design guidelines for e-commerce are still applicable for mobile application development, or if new ones should be created (Tarasewich, Nickerson, & Warkentin, 2002). A comprehensive methodological and comparative framework for evaluating the usability of m-commerce applications is also necessary (Siau, Lim, & Shen, 2001).

The main objective of this chapter is to provide a critical analysis of usability issues confronting the interface design, development, deployment and adoption of m-commerce applications. We focus primarily on Internet-based solutions in North America that use WAP (Wireless Application Protocol) phones, wireless PDAs (Personal Digital Assistants), Pocket PCs, and wireless two-way pagers. In addition, we examine the unique characteristics of mobile technology that affect user interface design, discuss the incorporation of user interface in the development of m-commerce applications, and suggest topics for future research and development.

This chapter consists of four sections. We begin by reviewing definitions of m-commerce and usability research for both e-commerce and m-commerce. Based on this review, we identify five sets of issues—technology, user goals and tasks, content preparation, application development, and the relationship between mobile and wired e-commerce. We then examine emerging trends, make recommendations, and conclude with a new research agenda.

RELATED RESEARCH

Defining Mobile Commerce

Mobile commerce broadly refers to the use of wireless technology, particularly handheld mobile devices and the mobile Internet to facilitate transactions, information searches, and user task performance in consumer, business-to-business, and intra-enterprise communications (Chan & Fang, 2001). These activities are typically grouped under the term "m-business." Such a broad definition is consistent with several proposed m-commerce frameworks. One such framework presents 12 classes of m-commerce applications, ranging from retail, auction, mobile office, and entertainment to mobile inventory emphasizing the potential of mobile B2B and intra-enterprise applications (Varshney & Vetter, 2001). Another framework groups m-commerce into goods, services, content for consumer e-commerce, and activities among trading partners (Kannan, Chang, & Whinston, 2001). M-commerce applications can also be categorized as: communication and interaction, value added service, information and data access, remote control and decision support, transactions, and entertainment (Lehner & Watson, 2001). While successful m-commerce business models in North America are rare, until recently there has been a belief in the potential for wireless technology to enhance a broad range of commerce processes and activities, particularly in the B2B and intra-enterprise arena.

There are two visions for the potential and opportunities of m-commerce (Waters, 2000). One perspective argues that the mobile, wireless channel should be viewed as an extension of the current e-commerce channel or part of a company's multichannel strategies for reaching customers, employees, and partners. The second and more radical view suggests that mobile commerce can create new markets and new business models. In recent years, the slow and selective adoption of wireless technology has indicated that the first perspective is more realistic, especially considering the current limitations of wireless technology.

Several recent studies provide empirical support for the first perspective. Consumers have shown relatively low willingness to use m-commerce, but adopters of e-commerce are more likely to embrace this new technology (Anckar & D'Incau, 2002). Another study shows that major e-commerce sites implement their mobile Internet sites as an extension of wired e-commerce to support existing customers (Chan et al., 2002). The perceived difficulty of use can affect the consumer's choice of m-commerce as a distribution channel (Sim et al., 2002). These findings suggest that in a multichannel environment, m-commerce supplements e-commerce rather than becoming a substitute for e-commerce.

Enterprise and business applications of m-commerce technologies seem to hold greater promise because it is easier for companies to standardize and customize the applications and the devices to enhance current work processes. Except for the retail industry sector, most industries view m-commerce as being vital for growth and efficiency strategies, but not necessarily for generating new revenue (Ernst & Young, 2001). Most mobile applications implemented in the United States are for business purposes—extending or enhancing existing work processes and business models geographically (Jarvenpaa, 2001). More comprehensive integration of the wireless platform in an enterprise requires significant structural transformation (Kalakota & Robinson, 2001). It will be challenging for companies to undertake the significant process redesigns and breakthrough strategies necessary for transitioning into a multidevice, multichannel computing environment. Successful transformation can increase an organization's capability for real-time interaction with its customers, employees, and suppliers.

An essential goal of m-commerce is to search for mobile values for individual users (Keen & Mackintosh, 2001). Anckar and D'Incau (2002) present a framework that differentiates between the value offered by wireless Internet technology (wireless value) and the value arising from the mobile use of the technology (mobile value). Wireless values are best represented by convenience, cost savings, and cell phones. For example, users can attain the wireless value by using the wireless PDA to perform tasks available through e-commerce. Services that deliver strong mobile values make m-commerce a dominant channel. These services meet the following five needs:

- Time-critical needs and arrangements,
- Spontaneous needs and decisions, such as auctions, email, and news,
- Entertainment needs,
- Efficiency needs and ambitions, and
- Mobility related needs.

Based on their survey of consumers in Finland, Anckar and D'Incaur (2002) indicate that, at present, consumers are most interested in services with high mobile values that meet time spontaneous and time critical needs. Furthermore, e-commerce users are more likely to adopt m-commerce services. These findings have implications for companies in North America for cross-channel coordination and highlight the relationship between e-commerce and m-commerce.

Usability Research for E-Commerce

Usability gauges the quality of a user's experience in interfacing with a product or system, be it a website, a software application, mobile technology, or any user-operated device. Generic usability principles (Nielsen, 2001) have guided the development of e-commerce applications:

- Ease of learning: How fast can a novice user learn the interface sufficiently well to perform basic tasks?
- Efficiency of use: Once an experienced user has learned to use the system, how fast can he or she accomplish tasks?
- Memorability: Can a past user remember enough to use the system more effectively the next time?
- Error frequency and severity: How often do users make errors, how serious are these errors, and how easy is it to recover from an error?
- Subjective satisfaction: How much does the user *like* using the system?

Usability has gained increasing attention in e-commerce website engineering. Industry consultants have incorporated these principles into a set of guidelines for e-commerce website design. Guidelines developed by the Nielsen/Norman Group (Nielsen, Farrell, Snyder, & Molich, 2000a, 2000b, 2000c, 2000d) focus on the design of category pages, the checkout and registration process, product pages, and user's trust. Several unpublished research papers (Rehman, 2000; Hurst & Gellady, 2000; Hurst & Terry, 2000) also examine customer experience in the e-commerce process and suggest a broad set of design guidelines:

- *Homepage.* Web pages should be clean and not cluttered with text and graphics. Horizontal scrolling should be avoided.
- *Navigation.* Text on the links or buttons should be descriptive and self-explanatory. Links to another product-related website should be direct.
- *Categorization.* Products should be categorized meaningfully with no more than three levels in depth.
- *Product information.* Accurate, consistent, and detailed descriptions of products should be provided along with full pictures. Inventory information and related charges should be presented up front. The size of products should be shown in a measurable and comparable way.
- *Shopping cart.* There should be a link directing the customer back to the page he/she left in order to resume shopping.
- *Checkout and registration.* The vender should only ask for necessary and meaningful information, such as name and address, not asking marketing questions. Customers should be allowed to browse the site without logging in.
- *Customer service.* Customers should be provided with a 1-800 telephone number on every page of the site.

In contrast to these broad design guidelines, formal usability studies tend to focus on specific tasks or specific user behaviors in the e-commerce context. Henderson and his colleagues point out that enjoyment in using an electronic system and peer norms can significantly contribute to the online shopping experience (Henderson, Rickwood, & Roberts, 1998). Kim and Moon (1998) indicate that manipulation of the visual design factors of the customer interface induces a target emotion, such as trustworthiness. The use of a combination of navigation features (neighborhood, top, and index) can generate the optimal link structure and increase the degree of shopping pleasure and convenience (Kim & Yoo, 2000). These studies did not validate any design guidelines and focused on very narrow aspects of e-commerce sites, which typically are feature and information rich.

Usability Research for M-Commerce

Usability research on wireless applications has usually focused on addressing the design constraints imposed by bandwidth limitations and small displays of handheld devices. Several studies address the effective input and output of mobile devices, such as the gesture recognition for PDAs (Sears & Arora, 2001), the effects of keyboard for mobile devices (Zha & Sears, 2001), and PDA Web browsing through eye movement analysis (Bautsch-Vitense, Marmet, & Jacko, 2001).

Research on the applications on mobile devices addresses various content presentation issues. Jones and his colleagues point out that direct access methods are more effective for retrieval tasks with small displays (Jones, Marsden, Mohd-Nasir, Boone, & Buchanan, 1999). Novice WAP phone users perform better when using links instead of action screens for navigation among cards and when using lists of links instead of selection screens for single-choice lists (Chittaro & Cin, 2001). HTML-based Web content can be converted automatically and online to WML by using a conversion proxy server following certain guidelines (Kaasinen, Anltonen, Kolari, Melakoski, & Laakko, 2000). However, these studies did not test handheld devices based on appropriate tasks and did not validate any framework or guidelines.

Ramsay and Nielsen (2000) point out that many WAP usability problems echo issues identified during the early stage of website development for desktop computers. These problems can be alleviated by applying good user interface design. Such design guidelines for WAP applications may include: (1) short links, (2) backward navigation on every card, (3) minimal level of menu hierarchy, and (4) headlines for each card (Colafigi, Inverardi, & Matricciani, 2001). Similar design guidelines were validated in a separate usability study of WAP phones (Buchanan, Farrant, Jones, Thimbleby, Marden, & Pazzaini, 2001), which include: (1) direct, simple access to focused valuable content, (2) simple hierarchies, (3) reducing the amount of vertical scrolling, and (4) reducing the number of keystrokes. These studies focused solely on WAP phones.

Researchers also conducted usability studies on other platforms, such as PDAs and Pocket PCs. In a study on the methods of text summarization for Web browsing on handheld devices, Buyukkokten, Garcia-Molina, and Paepcke (2001) found that keyword/summary was the best method. Sugimoto (1999) studied single hand keys input schemes for pocket computers.

Diverse form factors offer different functionalities and have different interface requirements. Chan et al. (2002) systematically evaluated 10 wireless websites—ranging from travel, financial services, retail, news, and Internet portals—across three form factors: wireless Palm, WAP phones, and Pocket PCs. They found that user tasks for the wireless sites were designed with steps similar to the wired e-commerce sites and primarily for experienced users. Many usability problems, such as long downloads and broken connections, information overload, and excessive horizontal and vertical scrolling, are common to all three form factors. They point out that interface design flaws are platform independent, but the more limitations imposed on the form factors, the more acute design problems become. They recommend eight guidelines:

- Avoid scrolling,
- Use a flat hierarchy,
- Design a navigation system consistent with a regular Web browser,
- Design a back button,
- Provide a history list,
- Provide an indication of signal strength,
- Reduce user's memory load, and
- Limit the search scope to improve search efficiency.

Several recent studies have investigated usability issues of mobile applications from different perspectives. Findings from these studies suggest usability for m-commerce goes beyond content presentation and form factors. Usability also pertains to mobile users' use contexts (Kim, Kim, Lee, Chae, & Choi, 2002) and socio-technical systems (Palen & Salzman, 2002), and their access of people and information anytime, anywhere in the work environment (Perry, O'Hare, Sellen, Brown, & Harper, 2001). In addition, ways to provide feedback about changes in the state of systems connectivity can improve usability of wireless applications (Ebling, John, & Satyanarayanan, 2002). These studies offer new ways of examining factors affecting usability for m-commerce applications.

Context factors have special impact on wireless usability. A study conducted by Kim and his colleagues (2002) reveals that three use context factors—hand (one or two hands), leg (walking or stopping), and co-location (alone or with others)—result in different usability problems. In this study, users reported more problems of site structure when they accessed wireless sites with one hand instead of two hands. When accessing wireless sites while moving rather than stopping, users experienced more difficulty with site representation. Content presentation also posed more usability problems for those who were stopping or alone. These findings suggest that the design of user interface has to consider various use contexts.

Based upon the results of a study of 19 novice wireless phone users who were closely tracked for the first 6 weeks after service acquisition, Palen and Salzman (2002) describe the wireless telephony system as having four sociotechnical components: hardware, software, "netware," and "bizware." They indicate that each of these four components has to be designed as user-friendly as possible. This research suggests a systems-level usability approach.

Perry and his colleagues (Perry et al., 2001) propose different facets of accessing remote people and information anytime, anywhere for mobile workers. They identify four key factors in mobile work: the role of planning, working in "dead time," accessing remote technological and informational resources, and monitoring the activities of remote colleagues.

Mobile users access information from different sources and often experience problems caused by a wide range of network connectivity. Ebling et al. (2002) conducted a study addressing the importance of translucence in mobile computing systems. Their findings suggest that with the presence of the illusion of connectivity provided by translucent cashing even when network performance is poor or nonexistent, novice users performed almost as well as experienced users.

KEY ISSUES

Based on the above review, we identify five issues concerning user interface for M- commerce in the following areas: technology, user goals and tasks, content preparation, application development, and the relationship between M- and e-commerce. These issues correspond to the four sociotechnical components in the wireless telephony system: hardware, software, netware, and bizware proposed by Palen and Salzman (2002).

Technology Issues

Limitations of bandwidth and form factors pose special constraints on the interface design of m-commerce applications.

Limitation of Bandwidth

First of all, the data rate for mobile communication is much lower than regular modems (Sepenzis, Pfau, & Lum, 2000). Most mobile communication standards only support data rates that are less than 28.8 kbps. Only a few provide higher data rates in select locations. Secondly, the connection to the wireless service base station is unstable because signal strength changes from place to place, especially on the move. This limitation has several implications for the interface design of mobile applications and devices.

- The amount of information exchanged between the device and the base station should be limited. Lengthy text messages and graphics are not suitable for mobile applications. A large amount of information takes longer to download, and the download process can be interrupted by broken connections. This constraint will therefore affect the functionalities provided by a wireless website.

- Indication of the download progress is necessary. Due to the lower data rates, the download process on a mobile device is significantly longer than on a regular desktop. Chan et al. (2002) observed that it took on average more than 20 minutes to download a regular Web page on Pocket PC devices because most wireless sites did not tailor their content to fit Pocket PCs. For customers used to speedy downloading on the Internet, 20 minutes would be intolerable. Furthermore, the Pocket PC reviewed in that study did not provide a good indication of the download progress. Most participants felt frustrated and frequently quit the process. Therefore, indicating the download progress can help users to have a better sense of control.
- Translucent cashing can be helpful. Ebiling et al. (2002) indicate that with translucent cashing in the presence of disconnected or weakly connected operations, novice users could perform almost as well as experienced users. They also found that it is important to make users aware of the state of network connectivity. Providing feedback about connectivity may reduce user frustration and enhance perceived usefulness for mobile users.
- Friendly recovery from broken connections is essential. If the connection between the mobile device and its base station is broken during the process of receiving or sending information, the device or the application must be able to resume the process once the connection returns. Otherwise, users may not be able to complete a single task due to frequent interruptions. It is also important to explicitly provide users with information about the recovery.

Form Factors

Mobile commerce services are accessible through multiple platforms. These platforms use different operating systems and offer different functionalities. Interface design for different platforms varies. Some popular platforms are:

- *Wireless PDA devices using Palm OS.* These devices provide computing and personal information management systems. Devices using Palm OS have an LCD display that comes in monochrome and color. The screen accepts pen and finger input, as well as handwriting on an electronically sensitive pad. Internet-enabled Palm OS devices have extensive support from third-party developers for applications (e.g., timers, spreadsheets, databases, games, even small Web browsers and email applications.) The small screen size and input device are two primary constraints in the interface design with Palm OS handheld devices. Due to the small size of the screen, Web clipping application pages running on wireless Palms do not support several common Web page features: named typefaces, style sheets, image maps, frames, nested tables, scripts/applets, and cookies. The graphics presented on Palms should be no larger than a full page: 63KB, 153 pixels wide, with a depth of 1 or 2 bits. Palm handheld devices also display some query form elements differently.

- *Pocket PCs running Microsoft Windows CE/Pocket PC OS.* This type of devices represents a stripped-down counterpart to Windows 95/98. This operating system has much the same look and feel of its desktop counterpart but does not have all the same features. In general, Pocket PCs have more memory and functions, slightly larger screen, and a higher resolution than Palm OS devices do. The input device of Pocket PCs is very similar to Palms. However, the screen is still too small compared to regular desktops. Furthermore, Pocket PCs do not support some common Web page features such as frames.
- *WAP phone.* This type of mobile phones can access Internet sites and services via WAP technology. With a much smaller screen than Palms and Pocket PCs, WAP phones use the phone keypad as the primary input device. Fewer functions are supported by WAP phones as compared to Palms and Pocket PCs.
- *Two-way pagers.* This type of device, such as Research in Motion (RIM), can both send and receive information from the base station. They also allow users to send and receive emails and browse the Web. The small screens on these devices provide only a keyhole view of the Web, and the slow data-transfer speeds make even a simple search a major chore.

These four platforms support different functions. Even within the same platform, the design of each type of handheld device varies. Consequently, the interface for different devices changes dramatically. A developer needs to consider the unique characteristics of the form factor when developing an m-commerce application.

User Goals and Tasks

In a fixed environment, such as an office or home with a desktop or a laptop computer, users can fully focus on their tasks or spend hours exploring a regular website. Rich information is far more important than time. M-commerce assumes that users access the Internet or wireless applications either on the move or while stationary but away from office or home. Because mobile users can spare only limited time and cognitive resources for performing a task, services that emphasize mobile values, meeting time-critical and spontaneous needs, such as checking flight schedules, checking stock prices, and submitting bids for auctions, are more useful for m-commerce users.

Three issues emerge in the design of tasks for m-commerce applications. First, when developing m-commerce applications, one should determine what tasks are essential to mobile users and what tasks are suitable for wireless applications. Because of the user's mobility, the interaction between the user and the mobile device is usually very short. It is important to design tasks in such a way that users can perform them in a timely manner. Currently, most wireless websites only allow users to perform simple tasks such as checking flight status and searching for a movie. Table 1 shows the projected market shares of different activities on mobile devices, suggesting that data transfer, short messaging, location services, playing games, shopping, doing e-business, and checking daily news services are the most suitable tasks for mobile commerce (Reinhardt, 2001). While popular in Europe and Asia, short messaging services (SMS) have not yet been widely adopted by consumers and enterprises in North America as a viable form for mobile communication. Table 1 also shows that voice calls will continue to be the primary activity performed using mobile devices. When designing tasks to be performed on wireless devices, a user's perceived usefulness plays an important role (Perry et al., 2001). As suggested by Perry and his colleagues, mobile workers demonstrate the following four access modes: planning, working in "dead time," accessing remote technological and informational resources, and monitoring the activities of remote colleagues. This finding has implications for selecting and designing tasks to support collaboration and communications in the enterprise environment

Second, how to handle differences in interface design for expert and novice users should be considered. Many wireless websites currently assume that users have experience with regular websites and are familiar with the task flow (Chan et al., 2002). The wireless site is basically a simplified version of its regular counterpart. This strategy makes sense in that it allows existing customers to transfer knowledge about a website to the wireless site. However, this strategy fails to accommodate

Table 1: Projected market shares of different activities on mobile devices

Activities	Market Share in 2005(%)
Data transfer	22
Short messaging	9
Location services	5
Games and gambling	2
Ads	~1
E-business	~1
Shopping	~1
Daily news service	0.5
Voice calls	~60

the needs of novice users who have never visited the regular website. For example, many wireless sites ask users to log in on the main screen without instructing them how to create a new account. Better interfaces should support both experienced and novice users.

Third, the best use of location-based technology to support user task performance for m-commerce presents additional challenges. Mobile networks can pinpoint a caller's location and supply that information to providers of geographically targeted services. According to Table 1, the projected market for location service will be 5% in 2003. The question is when to use that information. Based on the location theory, Mannecke and Strader (2001) suggest that location-based technology should be employed for applications for geographical differentiation and for low-involvement purchases that do not require extensive information search. Further research is needed in the proper use of location-based technology to enhance mobile tasks.

Content Preparation

Designing a wireless application is fundamentally different from designing a website because of the limited screen space and bandwidth. It is challenging to prepare content for a wireless application or to convert a regular website to a wireless website. Most of the design guidelines for e-commerce (Hurst & Gellady, 2000; Hurst & Terry, 2000; Niesen et al., 2000a, 2000b, 2000c, 2000d) support the development of rich product information sets and a complete process of shopping cart, checkout, registration, and order tracking. In contrast, a wireless website has to simplify the content presentation. The depth of a wireless website's structure, navigation, and format will require different design guidelines.

Amount of Information

A regular website usually contains very rich information to help users make decisions. This paradigm works fine for regular websites with large screens and high bandwidth. However, it is not feasible to download and present a large amount of information on handheld devices. Therefore, information in a regular website must be filtered when the website is to be presented on handheld devices. Research is needed to determine the amount of information suitable for handheld devices. In general, users should have sufficient, if not rich, information to accomplish the goals for the application.

Navigation

Navigation systems vary from one form factor to another because the design of handheld devices differs. Some devices such as RIM two-way pagers provide function keys on the device for users to move back and forth. Some other devices such as Pocket PCs require users to use the toolbar to navigate a site. There is no consensus yet on what functions or features should be provided by the application or built into the device itself.

Depth of Site Structure

Since mobile users can only spend very limited time browsing a wireless application, the organization of information is critical in order to improve efficiency. Information in most wireless applications is organized into a hierarchy. Numerous studies have addressed the hierarchy design issue of computer menu systems (e.g., Jacko, Salvendy, & Koubek, 1995; Jacko & Salvendy, 1996). These studies generally assume the response time of each step is constant and users have some basic training before using the system. However, wireless applications require much more time and effort to connect to the server and to download a page. Therefore, a flatter structure with fewer steps would allow users to review more options in the same step and to locate the desired information in less time. However, a theoretical framework of menu design for wireless applications is still needed in order to better justify a particular interface design.

Graphics or Text

Once the content of a wireless application is determined, the next question is how to present the information. In general, graphics as metaphors are better at catching a user's attention. Reading text is time consuming and also requires more human cognitive resources. Due to the constrained screen size and low resolution of handheld devices, text is probably a better choice in most cases, although small-size graphics can be displayed on most of handheld devices. However, better technology may improve the screen quality of handheld devices for displaying more complicated graphics. When determining the format of information to present, it is important to consider the form factor that may potentially pose additional constraints to the format.

Development Environment

Methods for mobile application development and usability testing should also be re-examined. User interface design and testing have traditionally been conducted in a "fixed" context of use, i.e., "a single domain, with the users always using the same computer, to undertake tasks alone, or in collaboration with others" (Johnson, 1998). Mobile computing alters this assumption. Traditional means of user interviews or usability testing in a laboratory environment cannot reveal insights into users' activities and mobility in their real life. Context becomes a critical consideration in gathering information about user requirements. The method of contextual inquiry (CI) can augment user interface design by exploring the versatility of usage patterns and usage context (Väänänen-Vainio-Mattila & Ruuska, 1998). This method employs an ethnographic approach, i.e., observing user activities in a realistic context. While contextual inquiry may help developers to secure a realistic understanding about contextual factors affecting user behavior in motion, conducting nonobtrusive observation and inquiries remains a challenge.

In addition, developers for mobile applications need to consider a broad range of contextual variations in infrastructure, applications, systems, and location (Rodden, Chervest, & Davies, 1998). Variations in the infrastructure context, resulting from interactions between the mobile device used and the supporting wireless infrastructure, can affect mobile user interface. It is important to ensure that the application corresponds to the state of the supporting infrastructure and the interaction style. Application context, on the other hand, refers to the relationship between the mobile device and the goals and tasks performed by the user. User interfaces need to be designed accordingly. Furthermore, advanced mobile application systems are distributed across multiple channels, devices, and infrastructures. Usability needs to be addressed at the systems level. It is necessary to consider the overall functionality of the application and to develop an architecture that provides appropriate access to different levels of functionality for different tasks. As discussed earlier, contextual variations due to the interactions between location and tasks also warrant considerations in interface design.

Functionalities of a complex wireless application have to be distributed across multiple wireless infrastructures, platforms, and form factors. Mapping platforms and form factors to user tasks and data needs becomes a critical decision in

Table 2: Wireless applications of United Airlines

	Alpha-numeric Pager	Web Phone	Wireless PDA
Book a Flight		X	X
My Itinerary		X	X
Flight Availability		X	X
Flight Status		X	X
Flight Paging	X	X	X
Mileage Plus Summary		X	X
Mileage Plus Upgrade Status			X
Mileage Plus Award Travel Availability			X
Contact United		X	X

application development. For example, United Airlines offers a range of wireless applications in pagers, WAP phones, and wireless PDAs. As shown in Table 2, their wireless applications vary by form factors, operating systems, and user needs for information (United Airlines, 2001). The time dimension of the task is inherent in this mapping. Information searches that are time sensitive and require low involvement, such as checking flight status, can be accessed via wireless PDAs and WAP phones. For tasks that involve more data exchange, such as Mileage Plus Upgrade Status and Mileage Plus Award Travel Availability, wireless PDAs offer a better platform for showing detail because of its larger screen size and greater input capability. More complicated trip planning searches require time for browsing and decision making; these tasks would be better suited to the desktop platform.

A major architectural issue for context-aware applications concerns the way in which contextual issues cut across the whole system design. However, structures apparent at the user interface often do not match those necessary for efficient implementation and sound software engineering (Rodden et al., 1998). Form factors may also pose conflicts with the development of context-aware applications. For example, the toolkits and the architecture for supporting Windows CE were designed for fixed environments. The requirements for mobile user interfaces may not be met by simply modifying some of the functionalities. More research is needed to examine new approaches for mobile application development that consider user interface issues in the context of the whole system's architectural and functional complexity.

Relationship between M-Commerce and E-Commerce

The wireless channel for e-commerce has raised many new questions. How does this new channel enhance e-commerce? What commercial activities or tasks can be performed on the mobile handheld devices? How can wireless user interfaces be coordinated between the two channels? Kannan et al. (2001) postulate that because wireless technology renders m-commerce "transaction aware" and "location aware," consumers would be likely to increase impulse purchases, especially in low-value, low-involvement product categories, such as books and CDs. Appropriate user interfaces for facilitating push-advertising and dynamic pricing based on location should be explored. As the wireless technology holds promise as a new marketing channel, a better understanding of customer profiles and preferences can make interface design and content presentation more efficient.

Many regular websites have extended the wireless channel to their current customers and subscribers to leverage relationships with existing customers (Chan & Fang, 2001). For example, Amazon only offers the one-click order option for purchasing from its wireless site. This feature does not allow users to review order details before submitting the order, and, once the order is submitted, it is difficult to navigate to the right screen on the handheld device to cancel the order. Therefore, only the experienced users who have built trust in Amazon and this special interface would find it efficient to order its product on the move. Novice users would be intimidated. In the case of accessing e-Bay by a wireless PDA device, users often encounter a large number of results from a product search. This high volume of data transfer often results in connection errors or user frustration. Only seasoned e-Bay users on the move would benefit from using a handheld device to monitor the bid in progress. Using these two examples, a careful mapping about what tasks would be most suitable for the wireless channel should be performed to support the mobile users of these m-commerce websites.

Given the current state of technology and the poor usability of most wireless sites, it is difficult to expand m-commerce as an independent channel. It is more appropriate to view m-commerce sites as an extension of existing e-commerce sites. Many researchers believe that the wireless channel has the potential to strengthen relationships with customers. Four factors make the mobile Internet an ideal channel to implement customer relationship management (CRM). They are its ability to (1) personalize content and services, (2) track consumers or users across media and over time, (3) provide content and service at the point of need, and (4) provide content of highly engaging characteristics (Kannan et al., 2001). The challenge is the coordination of user interfaces and contents across multiple channels so the experienced users and repeat customers can handle multiple media and platforms with satisfaction.

FUTURE TRENDS

This section discusses emerging trends in three areas: technology, application development, and business models.

Technology Trends

Four technology trends will affect the user interface design for mobile commerce in North America. First, multiple standards for wireless communication will not be resolved quickly. Second, the high cost of third-generation (3G) technology will delay the availability of broadband technology for complex functionality and content distribution for mobile applications. Investments made by mobile operators and infrastructure providers in Europe already proved to be premature (Gartner Group, 2002). Third, instead of the convergence of functionalities into a universal mobile handheld device, there will more likely be a variety of communication devices operating in harmony to support users in their everyday lives (Ruuska-Kalliokulju, Schneider-Hufschmidt, Väänänen-Vainio-Mattila, & Von Niman, 2001). Fourth, input and output format will expand to incorporate voice and other formats (Ruuska-Kalliokulju et al., 2001) and expandable keyboards, which are already being marketed by several product vendors. The introduction of the voice-based interfaces may complement the text-based interface and remedies some of the information input/display problems tied to the handheld devices. These trends point to the opportunities to conceptualize wireless user interface beyond text-based interaction.

New challenges are to design better interfaces for inter-device communication in order to simplify tasks for mobile users.

Development Trends

Alternative methods for interface design and evaluation will be necessary to support m-commerce applications development. First, requirement analysis has to focus on the context of mobile users' behaviors and tasks. Contextual inquiry and other methods may be developed to facilitate the understanding of interaction between mobility and usability. Second, usability testing has to be conducted with an understanding of contextual variables besides user behavior. Third, mapping form factors, user tasks, data needs and content across multiple channels and platforms is important in order to synchronize content and coordinate functionality in a distributed system. These trends require a fresh look at the methodology in use and determine new ways of incorporating user interface design and usability testing for distributed wireless application development. Other than contextual inquiry, traffic logs and more research-based design guidelines are important for improving user interface design for mobility. Furthermore, the complexity in contextual variations (Rodden et al., 1998) points to the need for a more comprehensive development framework in which user interface can be considered. For example, Roth (2001) has proposed mobility patterns as a design framework to allow designers to address multiple problems through reusing design elements, such as user interface, mobile data and mobile service, as building blocks. Since usability research in m-commerce is new, the need to understand commerce workflow and consumer behaviors will further complicate the development framework.

M-Commerce Business Models

Wireless technology holds promise for business enterprises but is still in the early stages of development. Companies typically take a "wait and see" posture until technology issues are resolved and a clearer picture of the potential return on investment emerges. A survey by Cap Gemini Ernst & Young (2001) of the 90 largest companies in Sweden indicates that the retail industry appeared most unsure about the future of mobile solutions, while 80% of other industries viewed m-commerce as being central to growth and efficiency strategies. This study has implications for the North American companies. The result is indicative of several trends. For intra-enterprise and business-to-business uses, wireless technology provides location-aware and mobility-aware solutions for mobile workers. The range of possibilities is broad. The deployment can be controlled more easily. Content distribution can be integrated with the enterprise systems. Variation in infrastructures and mobile devices can also be limited. Context-based applications, interfaces, functionality, and even devices can be customized according to the mobile tasks and user groups. For example, United Parcel Service (UPS) recently consolidated its multiple operating systems and devices to Windows CE and three devices (Nelson, 2001). This approach makes application development, deployment, and integration easier to manage. User interface design for these tasks and contexts can fall back to more familiar methodology. Distribution of content and data between the wired and wireless, fixed and mobile environment can be better synchronized.

In contrast, it is far more challenging to manage the design, development, and deployment of wireless applications for customers. The technology's capability for personalization seems to be the strongest argument for establishing a wireless channel. Mobile CRM is likely to be one of the first areas to embrace wireless solutions. A careful mapping of tasks, data, form factors, and the CRM process will become essential for user interface design. Location technology and personalization of services and content are critical for content presentation, navigation, and search. Differences between novice and experienced users will also be important, as well as approaches for development and usability testing.

CONCLUSIONS

In this chapter, we have identified the constraints, challenges and opportunities for usability issues affecting the adoption of m-commerce. At this early stage of mobile commerce, it is likely that the wireless technology and mobile Internet will continue to undergo rapid advancement. Until the technology matures and bandwidth improves, wireless applications will gear toward those requiring limited bandwidth, short exchange of data and text, and simple functionality. Two areas of wireless applications, CRM and enterprise efficiency, may reap better benefits.

Within this context, user interface design for m-commerce applications may take two directions. The consumer e-commerce websites will need to focus on the selection of tasks that are most suitable for the wireless channel, especially for experienced users. These are services and applications that demonstrate mobile values. Such mapping process requires a good understanding of the CRM strategy, user's preference, and constraints imposed by a mobile environment. For enterprise adoption, consolidating the wireless platforms and form factors will facilitate interface design. In either case, research to improve usability for mobile commerce is essential.

Research is urgently needed in the following areas:

1. Bandwidth limitation: How should wireless applications and handheld devices be designed to remedy the bandwidth problem?
2. Form factors: How should wireless applications be designed for different form factors?
3. Tasks: What tasks are most suitable for m-commerce applications? How will these tasks supplement the e-commerce process?
4. Content preparation: How should the navigation system be designed for wireless applications? What functions are best provided by devices and what functions are best provided by applications? How should a wireless site be

structured to facilitate information retrieval?

5. Development methodology: How should wireless interface design be incorporated into the development of distributed, multichannel systems? What alternative methods for interface design and usability testing should be considered for m-commerce applications?
6. M-commerce applications: How should the wireless technology be used to facilitate CRM? What criteria should be considered to guide the mapping of e-business strategies, tasks, and technology choices for wireless applications?

REFERENCES

Anckar, B., & D'Incau, D. (2002). Value creation in mobile commerce: Findings from a consumer survey. *Journal of Information Technology Theory & Application, 4* (1), 43-64.

Belson, K. (2002). Japan is slow to accept the latest phones. *The New York Times*, April 22, C4.

Bautsch-Vitense, H., Marmet, G., & Jacko, J. (2001). Investigating PDA Web browsing through eye movement analysis. In M. Smith, G. Salvendy, D. Harris, and R. Koubek (eds.), *Proceedings of the 9th International Conference on Human-Computer Interaction.* Mahwah, NJ: LEA.

Buyukkokten, O., Garcia-Molina, H., & Paepcke, A. (2001). Seeing the whole in parts: Text summarization for Web browsing on handheld devices. *Proceedings of the Tenth International World Wide Web Conference.* New York, NY: ACM.

Buchanan, G., Farrant, S., Jones, M., Thimbleby, H., Marsden, G., & Pazzani, M. (2001). Improving mobile Internet usability. *Proceedings of the Tenth International World Wide Web Conference*, 673-680. New York, NY: ACM.

Chan, S., & Fang, X. (2001). Usability issues for mobile commerce. *Proceedings of the Seventh Americas Conference on Information Systems*, 439-442.

Chan, S., Fang, X., Brzezinski, J., Zhou, Y., Xu, S., & Lam, J. (2002). Usability for mobile commerce across multiple form factors. *Journal of Electronic Commerce Research, 3* (3), 187-199.

Chittaro, L., & Cin, P. D. (2001). Evaluating interface design choices on WAP phones: Single-choice list selection and navigation among cards. In M. D. Dunlop & S. A. Brewster (ed.), *Proceedings of Mobile HCI 2001: Third International Workshop on Human Computer Interaction with Mobile Devices.*

Colafigli, C., Inverard, P., and Martriccian, R. (2001). InfoParco: An experience in designing an information system accessible through WEB and WAP interfaces. *Proceedings of the 34th Hawaii International Conference on System Science,* Los Alamitos, CA: IEEE Computer Society Press.

Ebling, M., John, B., & Satyanarayanan, M. (2002). The importance of translucence in mobile computing systems. *ACM Transactions on Computer-Human Interaction, 9* (1), 42-67.

Ernst & Young (2001). Global online retailing: An Ernst & Young special report. Cap Gemini Ernst & Young.

Gartner Group (2002). Conference on wireless access, mobile business solutions, March 11-13, 2002, Chicago.

Henderson, R., Rickwood, D., & Roberts, P. (1998). Beta test of an electronic supermarket. *Interacting with Computers, 10,* 385-399.

Hurst, M., & Gellady, E. (2000). White paper one: Building a customer experience to develop brand, increase loyalty and grow revenues. Unpublished report.

Hurst, M., & Terry, P. (2000). The dotcom survival guide. Unpublished report.

Jacko, J. A., Salvendy, G., & Koubek, R. J. (1995). Modelling of menu design in computerized work. *Interacting with Computers, 7* (3), 304-330.

Jacko, J. A., & Salvendy, G. (1996). Hierarchical menu design: Breadth, depth and task complexity. *Perceptual & Motor Skills, 82* (3), 1187-1201.

Jarvenpaa, S. (2001). Developing mobile commerce capabilities. Presentation to the SIM Advanced Practices Council, May 7-8.

Johnson, P. (1998). Usability and mobility: Interactions on the move. In C. Johnson (ed.), *Proceedings of the First Workshop on Human Computer Interaction with Mobile Devices.*

Jones, M., Marsden, G., Mohd-Nasir, N., Boone, K., & Buchanan, G. (1999). Improving Web interaction on small displays. *Computer Networks: The International Journal of Distributed Informatique,* (31), 1129-1137.

Jones, M., Marsden, G., Mohd-Nasir, N., Boone, K., & Buchanan, G. (1999). Improving Web interaction on small displays. *Computer Networks: The International Journal of Distributed Informatique,* (31), 1129-1137.

Kaasinen, E., Aaltonen, M., Kolari, J., Melakoski, S., & Laakko, T. (2000). Two approaches to bringing Internet services to WAP devices. *Computer Networks: The International Journal of Distributed Informatique,* (33), 231-246.

Kalakota, R. & Robinson, M. (2001). *M-Business: The Race to Mobility.* New York: McGraw-Hill.

Kannan, P., Chang, A., & Whinston, A. (2001). Wireless commerce: Marketing issues and possibilities. *Proceedings of the 34th Hawaii International Conference on System Sciences.* Los Alamitos, CA: IEEE Comp Soc.

Keen, P. & Mackintosh, R. (2001). *The Freedom Economy.* Berkeley, CA: Osborne/McGraw-Hill.

Kim, J. & Moon, J. (2000). Designing towards emotional usability in customer interfaces – Trustworthiness of cyber-banking system interfaces. *Interacting with Computers,* (10), 1-29.

Kim, K., Kim, J., Lee, Y., Chae, M., & Choi, Y. (2002). An empirical study of the use contexts and usability problems in mobile Internet. *Proceedings of the*

35th Annual Hawaii International Conference on System Sciences. Los Alamitos, CA: IEEE Computer Society Press.

Kim, J., & Yoo, B. (2000). Toward the optimal link structure of the cyber shopping mall. *International Journal of Human-Computer Studies,* (52), 531-551.

Lehner, F., & Watson, R. (2001). From e-commerce to m-commerce: Research directions. Unpublished paper.

Mennecke, B., and Strader, T. (2001). Where in the world on the Web does location matter? A framework for location based services in m-commerce. *Proceedings of the Seventh Americas Conference on Information Systems,* 450-455.

Nelson, M. (2001). For UPS, wireless technology has become a tradition. Informationweek.com, June 11, 2001. Available at http://www.informationweek.com/841/ups_side.htm.

Nielsen, J. (2000). *Designing Web Usability: The Practice of Simplicity.* Indianapolis, IA: New Riders.

Nielsen, J. (2001). What is usability? Available at http://www.zdnet.com/devhead/stories/articles/0,4413,2137671,00.html.

Nielsen, J., Farrell, S., Snyder, C., & Molich, R. (2000a). E-commerce user experience: Category pages. Unpublished report. Nielsen/Norman Group.

Nielsen, J., Farrell, S., Snyder, C., & Molich, R. (2000b). E-commerce user experience: Checkout & registration. Unpublished report. Nielsen/Norman Group.

Nielsen, J., Farrell, S., Snyder, C., & Molich, R. (2000c). E-commerce user experience: Product pages. Unpublished report. Nielsen/Norman Group.

Nielsen, J., Farrell, S., Snyder, C., & Molich, R. (2000d). E-commerce user experience: Trust. Unpublished report. Nielsen/Norman Group.

Palen, L. & Salzman, M. (2002). Beyond the handset: Designing for wireless communications usability. *ACM Transactions on Computer-Human Interaction, 9* (2), 125-151.

Perry, M., O'hara, K., Sellen, A., Brown, B., & Harper, R. (2001). Dealing with mobility: Understanding access anytime, anywhere. *ACM Transactions on Computer-Human Interaction, 8* (4), 323-347.

Ramsay, M. & Nielsen, J. (2000). WAP usability: Déjà vu, 1994 all over again. Unpublished report. Nielsen/Norman Group.

Reinhardt, A. (2001). Wireless Web woes. Available at http://www.businessweek.com/magazine/content/01_23/b3735602.htm.

Rehman, A. (2000). Holiday 2000 e-commerce. Unpublished report. New York, NY: Creative Good.

Rodden, T., Chervest, K., & Davies, N. (1998). Exploiting context in HCI design for mobile systems. In C. Johnson (ed.), *Proceedings of the First Workshop on Human Computer Interaction with Mobile Devices.*

Roth, J. (2001). Patterns of mobile interaction. In M. D. Dunlop & S. A. Brewster (eds.), *Proceedings of Mobile HCI 2001: Third International Workshop on Human Computer Interaction with Mobile Devices.*

Ruuska-Kalliokulju, S., Schneider-Hufschmidt, M., Väänänen-Vainio-Mattila, K., & Von Niman, B. (2001). Shaping the future of mobile devices: Results of the CHI2000 workshop on future mobile device user interfaces. In M. D. Dunlop & S. A. Brewster (eds.), *Proceedings of Mobile HCI 2001: Third International Workshop on Human Computer Interaction with Mobile Devices.*

Scholtz, J. (1998). WebMetrics: A methodology for producing usable websites. *Proceedings of the Human Factors Society* 1998 *Annual Meeting*, 1612. Santa Monica, CA: HFES.

Sears, A. & Arora, R. (2001). An evaluation of gesture recognition for PDAs. In M. Smith, G. Salvendy, D. Harris, and R. Koubek, (eds.), *Proceedings of the 9th International Conference on Human-Computer Interaction.* Mahwah, NJ: LEA.

Schenkman, B. & Joensson, F. (2000). Aesthetics and preferences of Web pages. *Behaviour & Information Technology, 19* (5), 367-377.

Sepenzis, T., Pfau, D., & Lum, E. (2000). Wireless Internet: Overview and outlook. Unpublished report. CIBC World Markets, Inc.

Sharma, S. & Deng, X. (2002). An empirical investigation of factors affecting the acceptance of personal digital assistants by individuals. *Proceedings of the Eighth Americas Conference on Information Systems,* 1829-1834.

Shim, J. P., Bekkering, E., & Hall, L. (2002). Empirical findings on perceived value of mobile commerce as a distributed channel. *Proceedings of the Eighth Americas Conference on Information Systems,* 1835-1837.

Sugimoto, M. (1999). Application of Single Hand Keys Input Scheme to Pocket Computer. *Fujitus Science Technology Journal, 35* (2), 181-190.

Siau, K., Lim, E., & Shen, Z. (2001). Mobile commerce: Promises, challenges, and research agenda. *Journal of Database Management, 12* (3), 4-13.

Tarasewich, P., Nickerson, R., & Warkentin, M. (2002). Issues in mobile e-commerce. *Communications of the Association for Information Systems, 8*, 41-64.

United Airlines, Inc. (2002). Wireless applications. Available at http://www.ual.com/site/primary/0,10017,1972,00.html.

Väänänen-Vainio-Mattila, K., & Ruuska, S. (1998). User needs for mobile communication devices: Requirements gathering and analysis through contextual inquiry. In C. Johnson (ed.), *Proceedings of the First Workshop on Human Computer Interaction with Mobile Devices.*

Varshney, U. & Vetter, R. (2001). A framework for the emerging mobile commerce applications. *Proceedings of the 34th Hawaii International Conference on System Sciences.* Los Alamitos, CA: IEEE Comp Society.

Waters, R. (2000). Rival views emerge of wireless Internet. *Financial Times FT-IT Review,* March 1, 2000, 1.

Wells, J. & Fuerst, W. (2000). Domain-oriented interface metaphors: Designing Web interfaces for effective customer interaction. *Proceedings of the*

Hawaii International Conference on System Sciences 2000, 155. Los Alamitos, CA: IEEE Comp Society.

Zha, Y. & Sears, A. (2001). Data entry for mobile devices using soft keyboards: Understanding the effect of keyboard size. In M. Smith, G. Salvendy, D. Harris, & R. Koubek (eds), *Proceedings of the 9th International Conference on Human-Computer Interaction.* Mahwah, NJ: LEA.

Zhu, W., Nah, F., & Zhao, F. (2002). Factors influencing adoption of mobile computing. *Proceedings of the 2002 Information Resources Management Association Conference: Issues & Trends of Information Technology Management in Contemporary Organizations*, 536-539.

Chapter XII

Using Continuous Voice Activation Applications in Telemedicine to Transform Mobile Commerce

James A. Rodger
Indiana University of Pennsylvania, USA

ABSTRACT

This chapter is designed to relate the rationale used by the Department of Defense (DoD), for the military to adapt the principles of Mobile and Voice Commerce to meet increasing global crises and to find ways to more effectively manage manpower and time. A mobile Telemedicine package has been developed by the Department of Defense to collect and transmit near-real-time, far-forward medical data and to assess how this Web-based capability enhances management of the battlespace. Telemedicine has been successful in resolving uncertain organizational and technological military deficiencies and in improving medical communications and information management. The deployable, mobile teams are the centerpieces of this telemedicine package. These teams have the capability of inserting essential networking and communications capabilities into austere theaters and establishing an immediate means for enhancing health protection, collaborative planning, situational awareness, and strategic decision making through Web-based internet applications. In order to supplement this mobile commerce aspect of telemedicine, U.S. Navy ships have been utilized to integrate voice commerce interactive technologies to improve medical readiness and mobility. An experimental group was tasked to investigate reporting methods in health and environmental surveillance inspections to develop criteria for designing a lightweight, wearable computing device with voice interactive capability.

This chapter is also designed to relate the rationale used by the Department of Defense and the Test and Evaluation (T&E) Integrated Product Team, in order to determine the military utility of the Joint Medical Operations—Telemedicine Advanced Concept Technology Demonstration (JMO-T ACTD) and continuous voice activation applications. Voice interactive computing devices are used to enhance problem solving, mobility and effectiveness in the battlespace. It improves efficiency through automated user prompts, enhanced data analysis, presentation, and dissemination tools in support of preventive medicine. The device is capable of storing, processing, and forwarding data to a server. The prototype devices have enabled quick, efficient, and accurate environmental surveillance. In addition to reducing the time needed to complete inspections, the device supported local reporting requirements and enhanced command-level intelligence.

This chapter further focuses on developing a holistic model of implementing a strategy for mobile telemedicine. The model synthesizes current thinking on transformation into a holistic model and also explains the integrative influence of vision on the other four model components: environment, people, methodology, and IT perspective. The model was tested by Testing and Evaluating (T&E) the JMO-T ACTD. JMO-T ACTD has developed a very successful training program and is very aware of the importance of planned change. Top military officials, such as the Commander in Chief (CINC), are actively involved in change and are committed to people development through learning. The model served an applied purpose by allowing insights into how well the military organization fit current theory. The model also fit a theoretical purpose by organizing a holistic, comprehensive framework. Accordingly, we have organized and synthesized the literature into five interrelated components that act as a fundamental guide for research. The model also helped to identify a theoretical link and apply it to the internal operations of the military and its adaptation of mobile e-commerce principles to more effectively deliver telemedicine benefits to military personnel.

INTRODUCTION TO TELEMEDICINE, MOBILE AND VOICE COMMERCE ISSUES

Telemedicine is an approach of providing care for patients that are geographically separated from a doctor. Telemedicine allows a doctor and a patient to interact with each other using computer networks. Telemedicine, when used in military, has a potential to heal patients in the war zone where doctors may not be readily available. The U.S. national strategy for military preeminence is based on technological superiority. Through new discoveries in advanced science and technology, the goal of the Department of Defense under Joint Vision 2010 (JV 2010) is to develop the ability to directly and decisively influence events ashore and at sea—anytime, anywhere—to meet current and future challenges.

To successfully counter these challenges, the DoD must continue to move forward in its effort to incorporate telemedicine into its prime mission—to keep every service member healthy and on the job, anywhere in the world, to support combat operations, as well as humanitarian, peacekeeping, and disaster relief missions. Telemedicine supports the DoD's goal by electronically bringing the specialist to the primary provider who directly cares for service members in austere, remote, and isolated environments (Floro, Nelson, and Garshnek, 1998). Telemedicine also creates an opportunity to provide rapid, accurate diagnosis and therapeutic recommendations (Garshnek and Burkle, 1998). The end result is that telemedicine helps to maintain the health of service personnel and their ability to quickly return to duty, minimizing logistically burdensome, inconvenient, and expensive transportation to distant specialty care (Bangert, Doktor, and Warren, 1998).

For telemedicine methods to be successful, however, their operational effectiveness, suitability, and importance to the warfighters' mission must continuously be tested, evaluated, and proven (Oliver, Sheng, Paul and Chih, 1999). In 1997, the U.S. Army, in partnership with the Navy and Air Force, was tasked to develop exercises to explore the integration of advanced technologies with existing systems and architectures to meet the requirements established under JV2010.

These technologies are all aligned with the Joint Vision 2010 concepts of Dominant Maneuver, Precision Engagement, Focused Logistics and Full Dimensional Protection. The technology initiatives utilize dedicated, small mobile teams, with a sophisticated IT infrastructure, to provide telemedicine capabilities wherever they are needed in the medical battlespace (Mann, 1997). This IT Infrastructure includes novel Medical Equipment Sets (MES) with digital capture devices such as digital cameras, digital scopes, digital blood and urine laboratories, physiological monitors, advanced digital radiography, and digital ultrasound (Perednia and Allen, 1995). Other, associated items of equipment include novel software, such as the Pacific Virtual Health Care System. This package offers electronic medical record archiving capability that enables automated, standardized teleconsultation by forward medics to higher echelon physicians (Rodger and Pendharkar, 2000).

The Joint Medical Operations-Telemedicine Advanced Concept Technical Design (JMO-T ACTD) has charged itself with operating within the concept of Focused Logistics and Full Dimensional Protection. It is, therefore, pertinent to understand just how this ACTD can accomplish its missions/objectives and meet the operational concepts of JV2010. This operationalization is embodied in the following quote: "To protect the force, the Army will rely on a technically advanced, operationally simple network of multicomponent intelligence sources capable of detecting and locating forces, active and passive obstacles, in-flight aircraft, ballistic and cruise missiles and their launch sites, chemical and biological agents, electronic jamming sources and a host of still-developing threats."

One technology that is mentioned in the document that applies to this ACTD is the use of "advanced soldier technologies." It is necessary for this ACTD to fit within this concept and provide the warfighter with information that identifies, early on, those countermeasures that can be used to defeat medical threats (Dardelet, 1998). It is also important to recognize other action that may be used to defeat enemy deployment of weapons of mass destruction (WMD), especially biological agent dispersal. Focused Logistics makes only one mention of "telemedicine." "For the Army, Focused Logistics will be the fusion of logistics and information technologies, flexible and agile combat service support organizations, and new doctrinal support concepts to provide rapid crisis response to deliver precisely tailored logistics packages directly to each level of military operation." The document portrays medical support to Focused Logistics in the form of "Internet triage" and "telemedicine" in order to enhance the survivability of the joint force (Zajtchuk, 1995).

Achieving 21st century medical support capability demands significant advances in the military's ability to provide force health care and medical protection and to deploy medical communications and information management in tactical operations (Institute of Medicine, 1996). The broad mission of Telemedicine in the military is to assess advanced mobile applications that can potentially meet such demands (Paul, Pearson, and McDaniel, 1999).

U.S. military has adapted a suite of software, databases, and architecture standards to provide deployable medical information management (Tanriverdi and Venkatraman, 1998). The Theater Medical Core Services (TMCS) is a database that stores data locally and is capable of sending encrypted e-mail to several redundant database servers via store-and-forward (Rasberry, 1998). The database servers aggregate information and store it in databases for distribution. Web servers supply data to medical personnel as customized encrypted reports.

The Medical Workstation (MeWS) is a network-based workstation equipped with portable medical devices, clinical support capabilities, medical information support, and a graphical user interface. The MeWS will support multipatient monitoring, interface with the patient's clinical record, and provide access to a searchable database. It will also provide full Personal Information Carrier (PIC) read and write implementation. MeWS will collect, store, and forward medical device data and images. By utilizing a Global Positioning System (GPS), MeWS has the capability to enter the patient's geographical location. The various software components of the MeWS help to facilitate clinical data entry, acquisition and retrieval. MeWS enables the generation of medical facility status reports, the monitoring of disease surveillance, the updating of supplies, and tracking of evacuation requirements.

The Field Medical Surveillance System (FMSS) is an expert system that systematically detects and monitors epidemiological trends and profiles patient populations. FMSS integrates patient information into the Global Infectious Disease and Epidemiology Network (GIDEON) knowledge base. Demographic and symptomatic information is used to arrive at a presumptive diagnosis or classify the patient using discriminate analysis. FMSS is also capable of providing incidence and prevalence trends for infectious diseases.

The Libretto is a commercial-off-the-shelf (COTS) handheld computer, manufactured by Toshiba. It has the capability to automate field medic Personal Identification Card (PIC) card software by reading service member's demographic information from the PIC into the software on the Libretto. It can also write GPS medical encounter information to the PIC and store the information as a pre formatted message for transmission. Tactical medical communications require updating of the existing IT infrastructure. The previously mentioned novel hardware, software, and interfaces were implemented in order to enable this change and facilitate the transmission of medical unique information over the existing communications hardware and command, control, communication, computers, intelligence, surveillance, and reconnaissance (C4ISR) networks. However, telecommunications from the operational area of responsibility (AOR) to the medical sustaining base uses the existing Defense Information Systems Network (DISN) that has been set up by the military in conjunction with the existing C4ISR.

The technologies described above have been assembled into an exportable capability that is specifically tailored to meet the medical Information Management (IM) and Information Technology (IT) needs of the unit it is supporting. This assemblage of technologies is referred to as the Capability Package. The capability package technologies must work in concert with the unit's infrastructure, communications, tactical situation, and logistical constraints if the military is to realize its full potential in meeting today's global crises. For such technologies to be successful, however, their operational effectiveness, suitability, and importance to the Telemedicine mission must continuously be tested, evaluated, and proven. To perform this task, the military established a Test and Evaluation Integrated Product Team (T&E) to evaluate candidate mobile models and architectures. These technologies are examined in a rigorous test and evaluation (T&E) environment with extensive user participation as a means of assessing their mobile applications. The T&E-IPT have leveraged and optimized existing communications technologies to transmit medical data. Database technologies for mobile technologies are utilized for epidemiological and trend analyses utilizing data mining of these data warehouses.

The initial concept of operations (CONOPS) was to employ a tailored Joint Task Force (JTF) to accomplish missions in controlled environment demonstrations. The first series of demonstrations was used to test communication methodologies, functionality, and the field utility of collecting and sending patient data from the forward edge of the battlefield. By allowing data and commands to be entered into a computer without the need for typing, computer understanding of naturally spoken languages frees human hands for other tasks. Speech recognition by computers can also increase the rate of data entry, improve spelling accuracy, permit remote access to databases utilizing wireless technology, and ease access to computer systems by those who lack typing skills.

SPEECH RECOGNITION APPLICATIONS

Since 1987, the National Institute of Standards and Technology (NIST) has provided standards to evaluate new voice interactive technologies (Pallett, Garofolo, & Fiscus, 2000). In a 1998 broadcast news test, NIST provided participants with a test set consisting of two 1.5-hour subsets obtained from the LinguisticData Consortium. The task associated with this material was to implement automatic speech recognition technology by determining the lowest word error rate (Herb & Schmidt, 1994; Fiscus, 1997; Greenberg, Chang, & Hollenback, 2000; Pallett, 1999). Excellent performance was achieved at several sites, both domestic and abroad (Przybocki, 1999). For example, IBM-developed systems achieved the lowest overall word error rate of 13.5%. The application of statistical significance tests indicated that the differences in performance between systems designed by IBM, the French National Laboratories' Laboratoire d'Informatique pour la Mechanique et les Sciences de l'Ingenieur, and Cambridge University's Hidden Markov Model Toolkit software were not significant (Pallett, Garfolo, and Fiscus, 2000). Lai (2000) also reported that no significant differences existed in the comprehension of synthetic speech among five different speech-to-text engines used. Finally, speaker segmentation has been used to locate all boundaries between speakers in the audio signal. It enables speaker normalization and adaptation techniques to be used effectively to integrate speech recognition (Bikel, Miller, Schwartz, & Weischedel, 1997).

The seamless integration of voice recognition technologies creates a human machine interface that has been applied to consumer electronics, Internet appliances, telephones, automobiles, interactive toys, and industrial, medical, and home electronics and appliances (Soule, 2000). Applications of speech recognition technology are also being developed to improve access to higher education for persons with disabilities (Leitch & Bain, 2000). Although speech recognition systems have existed for two decades, widespread use of this technology is a recent phenomenon. As improvements have been made in accuracy, speed, portability, and operation in high-noise environments, the development of speech recognition applications by the private sector, federal agencies, and armed services has increased.

Some of the most successful applications have been telephone-based. Continuous speech recognition has been used to improve customer satisfaction and the quality of service on telephone systems (Charry, Pimentel, & Camargo, 2000; Goodliffe, 2000; Rolandi, 2000). Name-based dialing has become more ubiquitous, with phone control answer, hang-up, and call management (Gaddy, 2000a). These applications use intuitive human communication techniques to interact with electronic devices and systems (Shepard, 2000). BTexact Technologies, the Advanced Communications Technology Centre for British Telecommunications (Adastral

Park, Suffolk, England), uses the technology to provide automated directory assistance for 700 million calls each year at its U.K. bureau (Gorham & Graham, 2000). Studies in such call centers have utilized live customer trials to demonstrate the technical realization of full speech automation of directory inquiries (McCarty, 2000; Miller, 2000). Service performance, a study of customer behavior, and an analysis of service following call-back interviews suggest user satisfaction with the application of speech automation to this industry (Gorham & Graham, 2000).

Speech recognition technologies could expand e-commerce into v-commerce with the refinement of mobile interactive voice technologies (McGlashan, 2000; Gaddy, 2000b; Pearce, 2000). As an enabler of talking characters in the digital world, speech recognition promises many opportunities for rich media applications and communications with the Internet (Zapata, 2000). Amid growing interest in voice access to the Internet, a new Voice-extensible Markup Language (VoiceXML™, VoiceXML Forum) has surfaced as an interface for providing Web hosting services (Karam & Ramming, 2000). VoiceXML promises to speed the development and expand the markets of Web-based, speech recognition/synthesis services as well as spawning a new industry of "voice hosting." This model will allow developers to build new telephone-based services rapidly (Thompson & Hibel, 2000). The voice-hosting service provider will lease telephone lines to the client and voice-enable a specific URL, programmed in VoiceXML by the client. This model will make it possible to build speech and telephony services for a fraction of the time and cost of traditional methods (Larson, 2000).

Haynes (2000) deployed a conversational Interactive Voice Response system to demonstrate site-specific examples of how companies are leveraging their infrastructure investments, improving customer satisfaction, and receiving quick return on investments. Such applications demonstrate the use of speech recognition by business. The investigation of current customer needs and individual design options for accessing information utilizing speech recognition is key to gaining unique business advantages (Prizer, Thomas, & Suhm, 2000; Schalk, 2000).

A long-awaited application of speech recognition, the automatic transcription of free-form dictation from professionals such as doctors and lawyers, lags behind other commercial applications (Stromberg, 2000). Due to major developments in the Internet, speech recognition, bandwidth and wireless technology, this situation is changing (Bourgeois, 2000; Pan, 2000).

Internationalizing speech recognition applications has its own set of problems (Krause, 2000). One such problem is that over-the-phone speech applications are more difficult to translate to other languages than Web applications or traditional desktop graphic user interface applications (Head, 2000). Despite the problems, internationalizing speech applications brings with it many benefits. Internationalization of an application helps to reveal some of the peculiarities of a language, such as differences in dialects, while providing insight on the voice user interface design process (Scholz, 2000; Yan, 2000). Furthermore, speech comprehension can work effectively with different languages; studies have documented both English and

Mandarin word error rates of 19.3% (Fiscus, Fisher, Martin, Przybocki, & Pallett, 2000).

Speech technology has been applied to medical applications, particularly emergency medical care that depends on quick and accurate access of patient background information (Kundupoglu, 2000). The U.S. Defense Advance Research Projects Agency organized the Trauma Care Information Management System (TCIMS) Consortium to develop a prototype system for improving the timeliness, accuracy, and completeness of medical documentation. One outcome of TCIMS was the adoption of a speech-audio user interface for the TCIMS Consortium prototype (Holtzman, 2000).

The Federal Aviation Administration conducted a demonstration of how voice technology supports a facilities maintenance task. A voice-activated system proved to be less time consuming to use than the traditional paper manual approach, and study participants reported that the system was understandable, easy to control, and responsive to voice commands. Participants felt that the speech recognition system made the maintenance task easier to perform, was more efficient and effective than a paper manual, and would be better for handling large amounts of information (Mogford, Rosiles, Wagner, & Allendoerfer, 1997).

Speech recognition technology is expected to play an important role in supporting real-time interactive voice communication over distributed computer data networks. The Interactive Voice Exchange Application developed by the Naval Research Lab, Washington, DC, has been able to maintain a low data rate throughput requirement while permitting the use of voice communication over existing computer networks without causing a significant impact on other data communications, such as e-mail and file transfer (Macker & Adamson, 1996).

Pilots must have good head/eye coordination when they shift their gaze between cockpit instruments and the outside environment. The Naval Aerospace Medical Research Lab, Pensacola, FL, has investigated using speech recognition to support the measurement of these shifts and the type of coordination required to make them (Molina, 1991). Boeing Company, Seattle, WA, has investigated ways to free pilots from certain manual tasks and sharpen their focus on the flight environment. The latest solution includes the use of a rugged, lightweight, continuous-speech device that permits the operation of selected cockpit controls by voice commands alone. This technology is being applied in the noisy cockpit of the Joint Strike Fighter (Bokulich, 2000).

EXISTING PROBLEMS—LIMITATIONS OF SPEECH RECOGNITION TECHNOLOGY

Even though applications of speech recognition technology have been developed with increased frequency, the field is still in its infancy, and many limitations have yet to be resolved. For example, the success of speech recognition by desktop computers depends on the integration of speech technologies with the underlying processor and operating system and the complexity and availability of tools required to deploy a system. This limitation has had an impact on application development (Markowitz, 2000; Woo, 2000).

Use of speech recognition technology in high-noise environments remains a challenge. For speech recognition systems to function properly, clean speech signals are required, with high signal-to-noise ratio and wide frequency response (Albers, 2000; Erten, Paoletti, & Salam, 2000; Sones, 2000; Wickstrom, 2000). The microphone system is critical in providing the required speech signal, and, therefore, has a direct effect on the accuracy of the speech recognition system (Andrea, 2000; Wenger, 2000). However, providing a clean speech signal can be difficult in high-noise environments. Interference, changes in the user's voice, and additive noise—such as car engine noise, background chatter, and white noise—can reduce the accuracy of speech recognition systems. In military environments, additive noise and voice changes are common. For example, in military aviation, the stress resulting from low-level flying can cause a speaker's voice to change, reducing recognition accuracy (Christ, 1984).

The control of the speech recognition interface poses its own unique problems (Gunn, 2000; Taylor, 2000). The inability of people to remember verbal commands is even more of a hindrance than their inability to remember keyboard commands (Newman, 2000). The limited quality of machine speech output also affects the speech recognition interface. As human-machine interaction becomes increasingly commonplace, applications that require unlimited vocabulary speech output are demanding text-to-speech systems that produce more human-sounding speech (Hertz, Younes, & Hoskins, 2000).

The accuracy of modeling has also limited the effectiveness of speech recognition. Modeling accuracy can be improved, however, by combining feature streams with neural nets and Gaussian mixtures (Ellis, 2000). The application of knowledge-based speech analysis has also shown promise (Komissarchik & Komissarchik, 2000). Pallett, Garofolo, and Fiscus (1999) pointed out that potential problems associated with the search and retrieval of relevant information from databases have been addressed by the Spoken Document Retrieval community. Furthermore, standards for the probability of false alarms and miss probabilities are set forth and investigated by the Topic Detection and Tracking program (Doddington, 1999). Decision Error Trade-off plots are used to demonstrate the trade-off between the miss probabilities and false alarm probabilities for a topic (Kubala, 1999). Security issues and speech verification are major voids in speech recognition technology (Gagnon, 2000). Technology for the archiving of speech is also undeveloped. It is well recognized that speech is not presently valued as an archival information source because it is impossible to locate information in large audio archives (Kubala, Colbath, Liu, Srivastava, and Makhoul, 2000).

STRENGTHS OF NVID

To ensure the health and safety of shipboard personnel, naval health professionals—including environmental health officers, industrial hygienists, independent duty corpsmen (IDCs), and preventive medicine technicians—perform clinical activities and preventive medicine surveillance on a daily basis. These inspections include, but are not limited to, water testing, heat stress, pest control, food sanitation, and habitability surveys. *Chief of Naval Operations Instruction 5100.19D,* the *Navy Occupational Safety and Health Program Manual for Forces Afloat*, provides the specific guidelines for maintaining a safe and healthy work environment aboard U.S. Navy ships. Inspections performed by medical personnel ensure that these guidelines are followed.

Typically, inspectors enter data and findings by hand onto paper forms and later transcribe these notes into a word processor to create a finished report. The process of manual note taking and entering data via keyboard into a computer database is time consuming, inefficient, and prone to error. To remedy these problems, the Naval Shipboard Information Program was developed, allowing data to be entered into portable laptop computers while a survey is conducted (Hermansen & Pugh, 1996). However, the cramped shipboard environment, the need for mobility by inspectors, and the inability to have both hands free to type during an inspection make the use of laptop computers during a walk-around survey difficult. Clearly, a hands-free, space-saving mode of data entry that would also enable examiners to access pertinent information during an inspection was desirable. The NVID project was developed to fill this need. The NVID project was developed to replace existing, inefficient, repetitive survey procedures with a fully automated, voice interactive system for voice-activated data input. In pursuit of this goal, the NVID team developed a lightweight, wearable, voice-interactive prototype capable of capturing, storing, processing, and forwarding data to a server for easy retrieval by users. The voice interactive data input and output capability of NVID reduces obstacles to accurate and efficient data access and reduces the time required to complete inspections. NVID's voice interactive technology allows a trainee to interact with a computerized system and still have hands and eyes free to manipulate materials and negotiate his or her environment (Ingram, 1991). Once entered, survey and medical encounter data can be used for local reporting requirements and command-level intelligence. Improved data acquisition and transmission capabilities allow connectivity with other systems. Existing printed and computerized surveys are voice activated and reside on the miniaturized computing device. NVID has been designed to allow voice prompting by the survey program, as well as voice-activated, free-text dictation. An enhanced microphone system permits improved signal detection in noisy shipboard environments. All of these capabilities contribute to the improved efficiency and accuracy of the data collection and retrieval process by shipboard personnel. Shipboard medical department personnel regularly conduct comprehensive surveys to ensure the health and safety of the ship's crew. Currently, surveillance data are collected and stored via manual data entry, a time-consuming process that involves typing handwritten survey findings into a word processor to produce a completed document.

This prototype system is a compact, mobile computing device that includes voice interactive technology, stylus screen input capability, and an indoor readable display that enables shipboard medical personnel to complete environmental survey checklists, view reference materials related to these checklists, manage tasks, and generate reports using the collected data. The system uses Microsoft Windows NT®, an operating environment that satisfies the requirement of the IT-21 Standard to which Navy ships must conform. The major software components include initialization of the NVID software application, application processing, database management, speech recognition, handwriting recognition, and speech-to-text capabilities. The power source for this portable unit accommodates both DC (battery) and AC (line) power options and includes the ability to recharge or swap batteries to extend the system's operational time. As the information and results were obtained the CONOPS was expanded to use additional activities. These activities are as follows:

- The deployment of mobile technologies and agents, called Theater Telemedicine Teams (TTTs), to medical treatment facilities (MTFs) to establish and conduct telemedicine operations; coordinate with signal and Command, Control, Communications, Computers, and Intelligence (C4I) assets to establish and maintain tactical medical networks; receive, verify, and log Command information provided from lower echelons.
- The use of advanced mobile information management models and technologies, such as software, databases, and architecture standards, that were adapted to provide deployable medical information management for advanced mobile applications.
- Two radio frequency (RF) networking technologies that were enhanced for user interface design in a battlefield setting.
- Modeling and simulation (M&S) capabilities provided through advanced mobile application software during training exercises.
- The validation of NVID project objectives and system descriptions by assessing the feasibility of voice interactive environmental tools and the NVID prototype's ease of use.
- All of these capabilities are being evaluated by the military. The goal of this evaluation is to first establish effective, interoperable mobile communications in the early stages of the exercises and to then implement more robust mobile database technology capabilities as the application matures.
- Mobile application.
- Types of mobile technologies that were identified and tested as potential candidates for enhancing Telemedicine capabilities

- Objectives of each mobile agent in the field
- Methods and applications of these mobile technologies
- Performance results of these mobile database technologies
- Recommendations, lessons learned, and feedback received from actual mobile users
- Overall findings and results of Telemedicine mobile field agents.

HOLISTIC MODEL

For this project, the author applied a holistic model to the DoD's mobile e-commerce re-engineering strategy. Strong evidence from prior case studies shows that holism offers a viable management model for successful transformation, or reengineering (Clark, et. al., 1997). Our model consists of five interdependent components—environment, people, methodology, information technology (IT) perspective, and vision (Paper, Rodger, and Pendharkar, 2000).

Environment

Basic environmental factors that lead to structural change include top management support, risk disposition, organizational learning, compensation, information sharing, and resources (Amabile, 1997; Lynn, 1998; O'toole, 1999). Innovation can come from any level of an organization, but environmental change originates at the top (Paper, 1999; Cooper, and Markus, 1995). When employees actually see top managers initiating process improvement changes, they perceive that their work is noticed and that it is important to the organization (Paper, and Dickinson, 1997; Paper, 1999).

It has been argued that the fear of failure must be limited and risk taking promoted for innovation to thrive (Nemeth, 1997). Many organizations make the mistake of trying to manage uncertainty with creative projects by establishing social control; however, it is the freedom to act that provokes the desire to act (Sternberg et al., 1997).

The ability to learn as an organization dictates whether and how fast it will improve (Harkness et al., 1996). Knowledge exchange between and among teams appears to give some organizations a distinctive competitive advantage (Lynn, 1998). Learning as part of the environment enables top management to disseminate its change message to the people who do the work (Gupta et al., 1999). Compensation has been attributed as a means of motivating employees to perform better (Pfeffer, 1998). Being rewarded for one's work sends the message to employees that their contributions to the organization are valued. It seems logical to conclude that people who are well compensated for risk taking, innovation, and creativity will continue that behavior (Paper, Rodger, and Pendharkar, 2000).

Information sharing enables people to better understand the business and what it requires to be successful (Paper, and Dickinson, 1997; Harkness, et al., 1996).

Restricting information, on the other hand, inhibits change. Resources can be viewed as a source for providing a variety of services to an organization's customers (Kangas, 1999). According to Barney (1991), an organization's resources can include all assets, capabilities, organizational processes, attributes, information, and knowledge that enable the organization to develop and implement strategies that improve its efficiency and effectiveness.

PEOPLE

Transformation success hinges on people and their knowledge, creativity, and openness to change (Cooper, and Markus, 1995). Real change will not occur without mechanisms in place to help people transform processes. Such mechanisms include training and education, challenging work, teamwork, and empowerment. "Education and training is the single most powerful tool in cultural transformation." (Wong, 1998) It raises people's awareness and understanding of the business and customer (Wong, 1998). Training helps develop creativity, problem solving, and decision-making skills in people previously isolated from critical roles in projects that potentially impact the entire enterprise. Business education is equally important in that people need to know how the business works in order to add value to business processes (Paper, 1999; Paper, and Dickinson, 1997). When work is challenging, people are more motivated, satisfied, and often more productive (Hackman et al., 1975). Challenge allows people to see the significance of and exercise responsibility for an entire piece of work (Cummings, and Oldham, 1997). Challenge stimulates creativity in people and gives them a sense of accomplishment (Amabile, 1997). People cannot reach their creative potential unless they are given the freedom to do so (Pfeffer, 1998). Management, therefore, needs to be sensitive to and aware of their role in creating a workplace that allows people freedom to act on their ideas.

Methodology

Methodology keeps people focused on the proper tasks and activities required at a specific step of a transformation project. It acts as a rallying point for cross-functional teams, facilitators, and managers as it informs them about where the project is and where it is going (Paper and Dickinson, 1997). It allows people to challenge existing assumptions, recognize resistance to change, and establish project buy-in (Kettinger et al., 1998). Of critical importance in the beginning stages is the buy-in and direction from top management, which is essential to identifying information technology opportunities, informing stakeholders, setting performance goals, and identifying BPR opportunities. Direction is important because large-scale re-engineering spans functional boundaries in which people from across the organization are involved (Paper, 1998).

Information Technology (IT) Perspective

The perspective of IT professionals toward change is critical because technology implementation is an organizational intervention (Markus, and Benjamin, 1996). As such, IT can either strengthen or weaken an organization's competitiveness (Kangas, 1999).

As introduced by Markus and Benjamin (1996), the three fundamental models of IT change agentry are traditional, facilitator, and advocate. Each model offers the dominant belief system or perspective of IT professionals toward the goals and means of work that shape what they do and how they do it. IT professionals with the traditional perspective believe that technology causes change. IT professionals with the facilitator perspective believe that people create change. IT professionals with the advocate perspective also believe that people create change. However, they believe that the advocate and the team are responsible for change and performance improvements. The facilitator perspective best characterizes the philosophy adopted at the DoD Telemedicine project.

Consistent with the change-agency theory, IT perspective is categorized rather than measured. IT perspective cannot really be measured because one has one belief system or another. The facilitator perspective views change as influenced by the people who do the work. Managers facilitate and guide the process. However, they do not control the process in any way. People control the processes, set their own goals, and are responsible for the consequences. However, managers share goal-setting tasks with the group, champion the effort, and are jointly responsible for the consequences. Mata et al.'s (1995) findings reinforce the facilitator model and suggest that two factors effectively contribute to an organization's competitive advantage: 1) Developing methods for strategy generation involving information resources management that emphasizes and enforces the learning of these skills across the entire organization, and 2) developing shared goals within the entire organization. This facilitator attitude toward common business processes and systems has been adopted by many organizations, including General Motors (Schneberger and Krajewski, 1999).

Transformation (Change) Vision

Vision offers a means of communicating the re-engineering philosophy to the entire organization and to push strategic objectives down through the process level and align the project with business goals. If the change vision is holistic, work is viewed as part of the whole system (Teng, et al., 1998). The underlying goal of a holistic change vision is to align employee goals with those of the organization and vice versa (Drucker, 1989). Change management, however, is very difficult because people tend to react negatively to it (Topchick, 1998). Hence, a top-down vision is imperative because it helps people understand the reasons for change. If people believe that change will benefit them or the organization, negativity is reduced. Top management has in its power the ability to influence how the organization perceives environment, people, IT, and methodology.

The vision can help open communication channels between IT and top management. One cannot be successful without frequent interactions between top management and IT change advocates (Markus, and Benjamin, 1996). Open communication can help inform top management of political obstacles, training issues, and budget problems before they stymie the project. It can also help top management disseminate information about the business and BPR progress across the organization. The more informed people are about the business, the better they feel about what they do. It is well known that organizations need information in order to compete (Ives and Jarvenpaa, 1993). The source for the following comments is the briefing Army Vision 2010. (Briefing is on the Web at URL www.army.mil/2010/introduction.htm.) This document and the efforts underway to achieve its objectives shape the Army's vision for the year 2010 and beyond. In the aggregate, the Army is seeking to "lighten up the heavy forces" and to "heavy up the capabilities of the light forces." From mission receipt through deployment, operations and transition to follow-on operations, Army elements will execute their responsibilities through a deliberate set of patterns of operation. These patterns are:

- ***Project the Force,***
- ***Protect the Force,***
- ***Shape the Battlespace,***
- ***Conduct Decisive Operations,***
- ***Sustain the Force, and***
- ***Gain Information Dominance.***

These patterns are all aligned with the Joint Vision 2010 concepts of Dominant Maneuver, Precision Engagement, Focused Logistics and Full Dimensional Protection. The technology initiatives utilize dedicated, small mobile teams, with a sophisticated IT infrastructure, to provide telemedicine capabilities wherever they are needed in the medical battlespace (Mann, 1997). This IT infrastructure includes novel Medical Equipment Sets (MES) with digital capture devices such as digital cameras, digital scopes, digital blood and urine laboratories, physiological monitors, advanced digital radiography, and digital ultrasound (Perednia and Allen, 1995). Other, associated items of equipment include novel software, such as the Pacific Virtual Health Care System. This package offers electronic medical record archiving capability that enables automated, standardized teleconsultation by forward medics to higher echelon physicians. This ACTD has charged itself with operating within the concept of Focused Logistics and Full Dimensional Protection. It is, therefore, pertinent to understand just how this ACTD can accomplish its missions/objectives and meet the operational concepts of JV2010. This operationalization is embodied in the following quote. "To protect the force, the Army will rely on a technically advanced, operationally simple network of multicomponent intelligence sources capable of detecting and locating forces, active and passive obstacles, in-flight aircraft, ballistic and cruise missiles and their launch sites, chemical and biological agents, electronic jamming sources, and a host of still-developing threats."

One technology that is mentioned in the document that applies to this ACTD is the use of "advanced soldier technologies." It is necessary for this ACTD to fit within this concept and provide the warfighter with information that identifies, early on, those countermeasures that can be used to defeat medical threats. It is also important to recognize other action that may be used to defeat enemy deployment of weapons of mass destruction (WMD), especially biological agent dispersal.

Focused Logistics makes only one mention of "telemedicine." "For the Army, Focused Logistics will be the fusion of logistics and information technologies, flexible and agile combat service support organizations, and new doctrinal support concepts to provide rapid crisis response to deliver precisely tailored logistics packages directly to each level of military operation." The document portrays medical support to Focused Logistics in the form of "Internet triage" and "telemedicine" in order to enhance the survivability of the joint force (Zajtchuk, 1995). This ACTD will best support this concept by demonstrating the ability to:

- capture the data,
- see the data,
- use the data,
- use decision tools to plan and prioritize,
- model and simulate, and
- utilize the GSSS strategy to accomplish the above.

That strategy is to develop the hardware, software, database, and network solutions that impact the computer-based patient record, medical threat identification, and command and control of medical units. This will be accomplished through management of information and information technologies, deployed throughout the battlespace. Most logisticians consider medical under their purview. Therefore, logistics organizations will be streamlined and "right-sized" to allow the delivery of service in a balance between "just in time" and "just in case = just enough." The operatives in the impact of Focused Logistics are "reduced footprint" and "tailoring on the fly" of units. This will provide for rapid crisis response, the tracking and shifting of assets while en route, and the delivery of tailored logistics packages and sustainment directly at the operational and tactical levels of operation. The JMO-T ACTD will tailor forces using novel modeling and simulation packages.

The most important facet of all of the JV2010 concepts is that the enablers and technologies will empower soldiers and not replace them. The enablers listed for Focused Logistics are germane to this ACTD as well. These are:

- Integrated Maneuver & Combat Service Support Systems Command and Control
- Total Asset Visibility

- Modular Organization
- Movement Tracking System
- Coupling computer recognition of the human voice with a natural language processing system that makes speech recognition by mobile computers possible
- Wireless Information Management Systems

Measurement of Issues and Findings

A series of measurements were conducted to test mobile communications methodologies and functionality. The field utility or application of mobile communications to field conditions consisted of collecting and transmitting near-real-time, far-forward medical data. It was examined and assessed as to how this improved capability enhanced medical management of the battlespace. This phase was also used to expand and improve the techniques for testing and evaluating the proposed mobile technologies and software enhancements.

The medical play in several of the demonstrations was robust enough to provide a rich opportunity to observe how these mobile technologies provided support to the user in an operational setting. These results were then used as a baseline for follow-on demonstrations and exercises.

Both the WaveLAN and JINC demonstrated their primary intended functions of mobile tactical networking capacity. The WaveLAN system provided superior bandwidth and full wireless local area network (LAN) capabilities, and the JINC provided tactical networking over low bandwidth military radio systems. Among the outcomes, it was found that mobile technologies could successfully replace wired LANs with wireless LANs and that mobile database technology software development and refinement should be continued.

The field exercises demonstrated the following capabilities:

- Theater Medical Core Services (TMCS) system—a mobile database application used to provide medical reports.
- Medical workstation (MeWS) —a mobile, functionally configured, network-based workstation designed to support the clinical and information support requirements of forward echelon providers ashore and afloat
- Toshiba Libretto end-user terminal (EUT)—a lightweight, handheld computer capable of reading, storing, and transmitting the soldiers' demographic information in the field.
- Desert Care II (DC II) Theater Clinical Encounter Application (TCEA) —a Web-based application that facilitates the user interface design, on the browser workstation, for mobile providers or medical technicians to record, view, and report patient encounter information in the field.
- Personal information carrier (PIC) —a small, portable storage device containing demographic and medical information pertaining to the soldier who is wearing or carrying the device.

- Theater Telemedicine Prototype Program (T2P2) —a Web-based delivery system of consultive care that gives healthcare providers from remote locations the ability to access the expertise of a regional facility for medical specialty consultations.
- Theater Telemedicine Team (TTT)—a mobile team composed of a leader with a clinical background, a visual systems operator, and an information systems operator who provide telemedicine capability to select, deployed MTFs.
- Aeromedical Evacuation (AE) Suitcase—a mobile system that provides critical voice and data communications to the AE mission of the U.S. Air Force (USAF) Air Mobility Command (AMC).

The tasks needed to achieve the objectives of the demonstration were carried out. These included the ability to collect and forward healthcare data in DC II and TMCS Lightweight Data Entry Tool (LDET), transmit it over existing communications (high frequency (HF) and International Maritime Satellite (INMARSAT)), extract it to a medical situational awareness system (TMCS), view those data in a Web environment on the TMCS server at Systems Center, San Diego (SSC SD), and conduct long-range clinical consultations. Although technical difficulties were experienced, the lessons learned from these exercises were evaluated, and solutions to these problems were incorporated into the next exercise. One good example of a lesson learned was the use of the wireless LAN to track patients within the MTF.

The exercises also indicated that essential data transport requirements of these mobile technologies can be met consistently, reliably, and cost effectively. Specific technologies were examined relative to each other for specific operational requirements of data throughput, transmission distance, time to setup, time to train, and actual costs to acquire, maintain and dispose. Among the significant achievements was the employment of the five-person mobile TTT, which successfully conducted clinical reachback capability.

Several parameters were not measured directly by the field exercise. These parameters can be determined through future exercises and battle laboratory testing and evaluation methods. For example, analysis still is not complete on the availability of mobile HF and very high frequency (VHF) radios, the overall reliability of the mobile laptops demonstrated, the software reliability of several of the communication modules, and the sustainability of several of the software database applications, hardware components, networks, and databases used in the exercise. As new data becomes available through future exercises and battle laboratory testing, a more complete picture of these advanced mobile applications of telemedicine will evolve.

Testing and evaluation of mobile Telemedicine applications have produced tangible evidence for the military utility of these technologies. Results from the field indicate that the essential data collection and dissemination requirements of these mobile technologies can be met consistently, reliably, and cost effectively. The mobile models and architectures demonstrate the potential to enhance data collection and dissemination of information through the use of quality database software and robust, mobile communications infrastructure. Through its efforts, these mobile agents have developed a consistent pattern of progression. From an initial state of uncoordinated, service-unique solutions to the building of an overall mobile framework, this architectural solution is being developed and refined by several different technological concepts. These concepts have been and will continue to be assessed for operational and technical feasibility. The results from these operational and technical assessments will ultimately lead to the development and insertion of an emerging architecture, which will encompass these advanced mobile applications. This first series of exercises was conducted to test communications methodologies, functionality, and the field utility of collecting and transmitting near-real-time, far-forward medical data and to assess how this improved capability enhanced medical management of the battlespace. This phase was also used to expand and improve the techniques for testing and evaluating the proposed technologies and software enhancements specified in the exercises.

Wireless WaveLAN

The WaveLAN system was developed and maintained from commercial off-the-shelf (COTS) wireless networking capabilities for the exercise. All JMO-T participation in this exercise was predicated on the work accomplished by the engineers to enhance the Lucent WaveLAN II system for military applications. In this regard, the WaveLAN represented an extension of a LAN via wireless means at data rates in excess of 2 million bits per second (Mbps).

Joint Internet Converter (JINC)

The JINC system is a tailored set of software and firmware that is geared toward providing lower bandwidth (i.e., 2.4–64 kilobytes per second (Kbps) data networking capabilities to existing military field radio systems). The basic concept behind JINC system development was to field a "programmable," mobile tactical networking system capable of exchanging digital data between ships, aircraft, combat vehicles, and individual soldiers in the field. The JINC system was enhanced from an existing COTS product to allow data connectivity between any two existing military radio systems without reliance on satellite communications (SATCOM). The intent behind this configuration was to avoid having the ACTD become involved in procuring and installing new generation radio systems. The JINC is composed of three elements operating together—the host computer, Team Care Automation System (TCAS) software, and a Micro-INC data controller device.

TCAS

The TCAS software installed on the JINC computer host provided automated network connectivity for distributed facilities, remote users, and individual units all interconnected using existing military communications media. TCAS software is based on object-oriented technology to enhance data exchange at low bandwidths. Fundamentally, TCAS software operates in two basic modes. The first mode emulates any specified data package as an "object" in an object-oriented database structure. Using a common database distributed throughout the entire JINC network, the software takes the "objects" and compresses them using a proprietary compression scheme and then transmits the "object" across the RF network. At the receiving node, the object is decompressed and translated back into its original protocol stack prior to delivery; thus, hosts on either end of a JINC-supported RF network see the expected data format in the form it was transmitted. Using this object compression scheme, JINC is able to deliver near full use of available low-bandwidth data links with very little administrative network overhead.

Micro-INC Data Controller

The Micro-INC (MINC) data controller provided the conversion from RS-232 serial data to a synchronous MIL-STD-1880-114 data stream. Each Micro-INC data controller can support up to two radio systems simultaneously. This data controller is normally located near the Single-Channel Ground and Airborne Radio System (SINCGARS) radio installation to reduce the length of the synchronous cable run. The controller requires no external or manual operation to function. All MINC functions are controlled by TCAS software.

Technologies Demonstrated

For this demonstration, a Mobile Medical Monitor (B) (M3B) computer system simulating a MeWS was connected to a SINCGARS via the TCAS. A Libretto system running TCAS was connected to a second Libretto via the WaveLAN Personal Computer Memory Card International Association (PCMCIA) wireless networking devices. Abbreviated discharge summary documents in Microsoft Word format were prepared on the M3B based on input from the various sensors attached to the M3B. This message was transmitted as a file attachment to a TCAS freetext email from the M3B to the first Libretto via SINCGARS. The Libretto then ported the data, via preset forwarding rules, from the SINCGARS net over the WaveLAN net to the second Libretto using the socket interface.

The computer systems selected for the exercise consisted of Libretto 110CT-NT computers, which were similar to the 100CT Libretto EUTs. The principal difference was that JMO-T Librettos required the Windows NT 4.0 operating system to support the TMCS system. The Librettos used in the exercise generally used Windows 95/98. In addition to the basic computer system, each JMO-T EUT was provided with a Quatech four-port serial expander PCMCIA card, which allowed the connection of the PIC reader along with the Garmin 12XL Global Positioning System (GPS) device. The second PCMCIA slot on the Libretto was occupied by the WaveLAN II 803.11 PCMCIA wireless network card.

During this exercise, far-forward Hospital Corpsman (HM) transmitted medical information from four far-forward first responder sites to the medical command onboard the USS Coronado. Data was entered via the Libretto 110CT-NT, which was equipped with a PIC Reader and TMCS LDET software. Three stationary sites were located at Area 41 in Camp Pendleton, California, and one mobile platform, a High-Mobility, Multipurpose Wheeled Vehicle (HMMWV), traveled to Yuma, Arizona. Because no specific medical exercise took place during the ELB phase, each user was given a set of preprogrammed PICs to scan into the system. The data were then periodically transmitted.

Initially, the Joint Medical Semi-Automated Forces (JMedSAF) simulation was to be used in conjunction with the scenario played out on the ground to give the staff onboard the USS Coronado a more robust "picture" of the battlespace; however, early in the exercise, it became apparent that bandwidth was at a premium on the network. The demonstration manager, therefore, elected to "shut down" the JMedSAF feed to the USS Coronado to keep essential data feeds open to the Enhanced Combat Operations Center (ECOC). As a result, very little data generated from the simulation runs eventually made its way to the TMCS database. Furthermore, the scenario of the "real" battlespace was disconnected from the "virtual" battlespace.

The NVID project was developed and tested to demonstrate proof of concept for replace existing, inefficient, repetitive survey procedures with a fully automated, voice interactive system for voice-activated data input. In pursuit of this goal, the NVID team developed a lightweight, wearable, voice-interactive prototype capable of capturing, storing, processing, and forwarding data to a server for easy retrieval by users. The voice interactive data input and output capability of NVID reduces obstacles to accurate and efficient data access and reduces the time required to complete inspections.

Results

JMO-T operations consisted of sending over 120 patient encounters via the TMCS LDET to the TMCS server located in the ECOC on the USS Coronado. Three nodes were operated by JMO-T personnel during ELB:

- HMMWV Mobile Node
- Area 41 Node
- Yuma Node

Two basic WaveLAN modes of operation were used. The first (and most commonly used) was the "standard" mode, which allowed the EUTs to communicate with the rest of the WaveLAN network via a WavePoint router connection, which translated the packets for use by the rest of the network. Because the power output of the individual WaveLAN card was only 25 milliwatts, the JMO-T EUT had to be located within 1,000 feet of a WavePoint in order to access the network. In practice, this range was extended to as much as 2,000 feet at Area 41, but this was due primarily to a high antenna mast (about 40 feet) for the Area 41 WavePoint antenna.

The other method of operation was called the "ad hoc demo" mode, which was accessed by selecting an option on the WaveLAN card "properties" window. When activated, this allowed the EUTs to communicate with each other (i.e., for training) without the need for a WavePoint.

Yuma Node

JMO-T participation at Yuma demonstrated far-forward message reach-back capability. JMO-T was assigned to operate from a WavePoint assigned to a Naval Research Lab (NRL) mobile commercial SATCOM system mounted in a HMMWV. This SATCOM link provided a 2-Mbps relay directly back to Area 41 at Camp Pendleton. EUT operational modification only required an IP change. As in Area 41, all JMO-T messaging was handled by a 1/5 Marines corpsman. The system was operated from the back of a vehicle within 200 feet of the NRL SATCOM HMMWV. Individual patient encounter messages were transmitted within 5-10 seconds. The ECOC TMCS server was able to be browsed to confirm delivery. Five additional images, including two 1.35-MB images, were transmitted via File Transfer Protocol (FTP). Small files were transmitted in 10-20 seconds, and large files took 2:20 each. The only operational problem noted was a tendency for the Global Positioning System unit to stop sending position information when requested. This was traced to a loose cable on the Quatech serial port card; however, the cable was tightened, and the system returned to normal operation.

ELB technology provided a number of excellent options for medical communications. When the network was not overwhelmed by the demands of Video conferencing, it provided an excellent method of collecting medical data—both TMCS textual data and images. During these times, Engineering Integrated Product Team (E-IPT) personnel working with data senders reported that TMCS data was sent in milliseconds, and the large files were transmitted in no more than 5 seconds. Data senders were able to use the handheld computers with ease. JMO-T participated in the longest leg of the exercise network by successfully sending TMCS and large data files from Yuma, Arizona.

The Libretto systems running Windows NT using 64 MB RAM performed satisfactorily; however, when the LDET, TCAS, Serial TCAS, and Medical Messaging Service (MMS) server were all running on one computer, the operation slowed significantly. One solution was to allow TCAS to speak TMCS (or Wave or any other medical software) in its native mode as a C++ object. Based on this experience, a more effective device for Echelon I use is a Windows CE computer, which weighs less than one pound, can easily fit into a Battle Dress Utilities (BDU) pocket, and provides resident software, a user-friendly screen, and a long-life, inexpensive battery.

Shipboard Node

The NVID node-established criteria for developing a lightweight, wearable, voice-interactive computer capable of capturing, storing, processing, and forwarding data to a server for retrieval by users. The prototype met many of these expectations. However, limitations in the current state of voice-recognition technologies create challenges for training and user interface. Integration of existing technologies, rather than development of new technology, was the intent of the design. A state-of the-art review of existing technologies indicated that commercial, off-the-shelf products cannot yet provide simultaneous walk-around capability and accurate speech recognition in the shipboard environment. Adaptations of existing technology involved trade-offs between speech recognition capabilities and wearability. Specific challenges and problems of the NVID prototype system that were examined and tested included:

- Shipboard operation in tight spaces
- Operation in high-noise environments
- Data gathering and checklist navigation
- Report generation
- Access to reference materials
- Comment capture capability
- Access to task schedule and survey data
- User system training
- Prototype effectiveness

Shipboard operation in tight spaces. Space and resource constraints on Navy ships make it necessary to complete surveys in enclosed, tight spaces. Ease of use, portability, and wearability of the NVID unit when maneuvering through these areas were validated based on surveys of military users. A study of the ergonomics associated with the use of an NVID computer was also performed. The human factors evaluated included, but were not limited to, the following parameters:

- Safety equipment compatibility
  - Work clothing, including gloves, glasses, and hard hats
  - Sound suppressors/hearing protection
  - Respirators
- Data input comparison and user acceptance (voice command vs. touchscreen) based on the opinions of Navy personnel aboard ship
- User interface evaluation (ease of use)
  - User comfort
  - User adjustability
  - Subcomponent connection procedure
  - Assessment of mean time to proficiency

Data gathering and checklist navigation. NVID prototype system users were capable of navigating through survey checklists by using voice commands, as well as other computer navigational tools, such as a mouse, touch pad, and stylus. The data collected were then automatically stored in an on-system database. To determine whether the system could successfully open each checklist and allow entry and storage of the required data, a script was developed that thoroughly tested the functionality of the hardware and software.

Report generation. The ability to generate reports and save them as files for downloading or printing was verified. Tests were performed to verify that the data were captured during inspection procedures and properly rendered into a usable working report.

Access to reference materials. Users may require access to survey reference materials, schedules, previous survey results, or discrepancies found during the survey process. Tests were performed to verify that the application software enabled access to designated reference material as specified within each checklist.

Comment capture capability. The NVID application provides the ability to document the inspector's notes via handwriting recognition, voice dictation, and a touchscreen keyboard. Verification of all three methods of data capture was performed using a predefined script of repeatable voice commands.

Access to task schedule and survey data. The NVID application software provides the ability to schedule tasks and review past reports. Verification of the software was performed using both voice command and touchscreen technologies.

User system training. To evaluate the effectiveness of user system training, the amount of training time required to achieve the desired level of voice recognition accuracy was first determined. Minimum training for the voice recognition software was conducted, and the length of time required to complete the training was documented. The system was then operated using a scripted, repeatable set of voice commands, and the number of errors was recorded. This process was repeated with additional training until the desired level of voice recognition accuracy was achieved.

Recommendations

Based on achievement of the stated objectives, the following recommendations are provided for continued wireless networking development:

- WaveLAN technology appears sufficiently mature to warrant use as a replacement for wired networking at field MTFs.
- A prototype network configuration to support an SC should be devised and prepared for testing.

The following recommendations are provided for continued TCAS software development:

- As demonstrated in the exercise, TCAS was based on a C++ Windows 95/98/ NT executable program. Operational experience with the Libretto NT system at Echelon I showed the need for a smaller, lighter computing system to support this highly mobile group. The Windows CE operating environment appears most suited to this requirement. Port TCAS software into the CE environment is recommended.

- The greatest asset (and liability) of the TCAS/J software is its flexibility.

Programming the various communications servers, forwarding rules, and message formats is similar to programming a full-featured network router. This implies that a TCAS operator must be both computer literate and network knowledgeable. Simplification of the user interface, perhaps with more graphical network connection screens, appears necessary. In addition, the software should feature some type of "system lock" that will keep all settings under a password-controlled environment so that an inexperienced operator cannot change them by accident.

- Continued developmental work is needed to incorporate the full range of medical database-specific messages into TCAS. Message delivery in the exercise was achieved via a complicated process involving multiple serial port data exchange and encoding. This process can be streamlined by the provision of a medical system communications server to the TCAS software developers so that they can test their message servers directly. The following recommendations are provided for continued NVID continuous voice recognition software development:
- Although users reported positive responses to the NVID prototype tested, the device exhibited the limitations of current speech recognition technologies. The processors in lightweight, wearable devices were not fast enough to process speech adequately. Yet, larger processors added unwelcome weight to the device, and inspectors objected to the 3.5 pounds during the walk-around surveys. In addition, throat microphones used in the prototype to limit interference from background noise also limited speech recognition. These microphones pick up primarily guttural utterances, and thus tended to miss those sounds created primarily with the lips, or by women's higher voice ranges. Heavier necks also impeded the accuracy of throat microphones.

DISCUSSION

The following are several of the lessons learned from each of the components of the BPR model:

PEOPLE LESSON

Military commanders, such as the CINC, should be knowledgeable and interact with the operations of JMO-T ACTD. The T&E members believe that all of the functional areas have a hands-on approach to IT. JMO-T ACTD used IT to redefine its business processes and adopt mobile e-commerce principles to telemedicine. They found that it is much easier to teach the CINC and top military commanders the fundamentals of technology than it is to teach technology people about strategic management of the battlespace. In addition, JMO-T ACTD also serves the medical needs of the military's internal personnel. Despite current limitations in speech recognition technology, the NVID prototype was successful in reducing the time needed to complete inspections, in supporting localreporting requirements, and in enhancing command-level intelligence. Attitudes of the end users toward the device were favorable, despite these restrictions. Users believed that the prototype would save time and improve the quality of reports.

ENVIRONMENT LESSON

If business processes are dependent on timely, accurate, and complete information, business re-engineering should be approached with a strategy that includes information re-engineering. In the contemporary military environment, information is especially important because it is very information intensive; hence, T&E choose a dual strategy of business and information re-engineering as JMO-T ACTD's ideological approach to BPR.

METHODOLOGY LESSON

BPR should be adopted based on a military need and not because "everyone else is doing it." T&E chose to redesign JMO-T ACTD processes because they were concerned about JMO-T ACTD's reputation with the top military commanders and the Senate Committees that fund their operations. Before re-engineering, no method for tracking soldiers' medical care in the field or during medical evacuation existed. The costs associated with medical evacuations was prohibitive, both in terms of lives and money. Reengineering the methodologies utilized to improve the processes through which the JMO-T ACTDs adopted mobile e-commerce technologies allows for on-site medical treatment without the fear of helicopter medical evacuations under enemy fire and/or during poor weather conditions.

Medical evacuations lead to a long cycle time from receipt of the wounded in the field until they could reach proper medical care. Long cycle times translate into increased mortality and morbidity for military personnel. Because JMO-T ACTD allows "real-time" treatment, T&E feels that telemedicine and mobile e-commerce technologies provide an edge for treating casualities. T&E believes that BPR has given the military that edge by decreasing cycle times and improving information sharing.

INFORMATION LESSON

T&E believes that it is important to choose a project that must work, so that its success can be sold to the rest of the company. Success is hard to resist. If a project is very successful, it will be much easier to get other departments and operational areas involved in BPR. Because the JMO-T ACTD project worked, it allowed the military to decentralize its information processing. Medical information processing was taking too long and negatively impacting soldier well-being; therefore, T&E took action and decided to embark on a major BPR project to rethink the existing medical information and inventory system and to decentralize medical treatment in the battlespace. This was a critical process and a risky venture, but the military had no choice. The JMO-T ACTD project succeeded because the potential for excellent results far outweighed the risk.

T&E believes that JMO-T ACTD must develop an independent JTF capability package in order to lead the IT re-engineering effort. JMO-T ACTD clients are the entire military. Because the IT capability package manages information flow throughout the military battlespace, it must be able to work with military commanders and end users to "show them the way." In other words, IT people in the JTF give a data view of the entire military organization. They know how the information is distributed to all departments and operational areas and are in an ideal position to work with information users as changes in business processes occur. This is a full-time job that requires individuals who are dedicated to carrying out this mission.

VISION LESSON

BPR projects require support from top commanders and those involved along the process path to succeed. If top military management does not formulate a vision for success and visibly support the BPR effort of JMO-T ACTD, politics will destroy the project. Most people are afraid of change, and given the opportunity to resist change, many will do just that. Moreover, changing the way that business is conducted will not be tolerated without top-level approval because top military officials are in charge. T&E believes that if those involved in the process are not part of the project, they will resist changes and most likely sabotage the BPR effort. After all, they are the ones who will most likely be affected by these changes.

T&E found that very few military personnel or officers know the overall military operational process; however, T&E believes that the JTF capability package must support an innovative approach to telemedicine improvement projects if it to serve all members of the military. T&E concluded, therefore, that top military management should form a JTF department and help it to gain knowledge about the military operations that it serves. The best strategy is to assign top military officers into the JTF department to add operational and strategic knowledge and experience.

OVERALL MODEL EVALUATION

The re-engineering project with mobile telemedicine provided many insights into how the military actually deals with BPR on an enterprise-wide basis. The project uncovered the ideological methodologies used to guide BPR efforts and the technologies used to help implement mobile e-commerce applications for telemedicine. The military radically redesigned T&E processes to improve overall performance. At the same time, they used technology and data warehouse methods to decentralize data management for increased information sharing, to more easily access data by those who need it, and to deliver data, products, and services in a more timely manner. Thus, their BPR strategy uses an approach to process improvement with information technology and mobile e-commerce applications as a complementary support mechanism.

It is realized that JMO-T ACTD must continue to provide telemedicine service to its military personnel, improve strategic awareness of the battlespace, and provide excellent information services to commanders and end users during times of both peace and war. The literature in BPR has not helped in this regard. In addition, it provides little insight into the complexities of dealing with military re-engineering and information re-engineering simultaneously. Each branch of the armed forces has a different set of problems to deal with. Books and periodicals can only provide basic ideas; therefore, I believe that a new methodology must be developed for dealing with change and process improvement.

CONCLUSIONS

Testing and evaluation of the JMO-T ACTD have produced tangible evidence for the military utility of mobile telemedicine. Results from Demonstration I indicate that the essential data collection and dissemination requirements of JMO-T can be met consistently, reliably, and cost effectively. This research focused on developing a holistic model of transformation. The model synthesizes current thinking on transformation into a holistic model and also explains the integrative influence of vision on the other four components of the model. The model was tested by T&E on the JMO-T ACTD. JMO-T ACTD has developed a very successful training program and is very aware of the importance of planned change. Top military officials are actively involved in change and are committed to people development through learning. The model served an applied purpose by allowing us to see how well the military organization fit current theory. The model also fit a theoretical purpose by organizing a holistic, comprehensive framework. Accordingly, we have organized and synthesized the literature into five interrelated components that act as a fundamental guide for research. The model also helped us to identify a theoretical link and apply it to the internal operations of mobile e-commerce and telemedicine in the military.

The ACTD promises the potential to demonstrate technology-enhanced data collection and dissemination of information, through the use of quality software, and robust communications infrastructure. Through its efforts, the JMO-T ACTD has developed a consistent pattern of progression. From an initial state of uncoordinated, service-unique solutions to the building of an overall architectural framework, this architectural solution is being developed and refined by several different concepts. Accuracy of speech recognition depended on the time a user committed to training the device to recognize his or her speech and changes in voice quality due to environmental or physical conditions. Accuracy rates varied from 85-98% depending on the amount of time users took to train the software. Optimal training time appeared to be 1 hour for Dragon Naturally Speaking software and 1 hour for NVID software. In addition, current software interprets utterances in the context of an entire sentence, so users had to form complete utterances mentally before speaking for accurate recognition. As speech recognition technologies evolve, many of these limitations should be addressed.

* * *

The views expressed in this paper are those of the author and do not reflect the official policy or position of the Department of the Army, Department of the Navy, Department of Defense, or the U.S. Government.

REFERENCES

Albers, J. (2000). Successful speech applications in high noise environments. *SpeechTEK Proceedings*, 147-154.

Amabile, T. M. (1997). Motivating creativity in organizations: On doing what you love and loving what you do. *California Management Review,* 1, 39–58.

Andrea, D. (2000). Improving the user interface: Digital far-field microphone technology. *SpeechTEK Proceedings,* 155-160.

Bangert, D., Doktor, R., & Warren, J. (1998). Introducing telemedicine as a strategic intent. *Proceedings of the 31st Hawaii International Conference on System Sciences* (HICSS-31), Maui, Hawaii.

Barney, J. B. (1991). Firm resources and sustained competitive advantage. *Journal of Management,* 99–120.

Batt, R. (1996). Outcomes of self-directed work groups in telecommunications services. In Paula B. Voos (ed.), *Proceedings of the Forty-Eighth Annual Meeting of the Industrial Relations Research Association* (p. 340). Madison, WI: Industrial Relations Research Association.

Bikel, D., Miller, S., Schwartz, R., & Weischedel, R. (1997). Nimble: A high performance learning name finder. *Proceedings of the Fifth Conference on Applied Natural Language Processing, Association for Computational Linguistics,* Washington, DC, 194-201.

Bokulich, F. (2000). JSF [Joint Strike Fighter] voice recognition. Aerospace Engineering Online [online]. Available at http://www.sae.org/aeromag/techupdate_5-00/03.htm.

Bourgeois, S. (2000). Speech-empowered mobile computing. *SpeechTEK Proceedings,* 223-228.

Caron, J. R., Jarvenpaa, S. L., and Stoddard, D. B. (1994). Business re-engineering at CIGNA Corporation: Experiences and lessons learned from the first five years. *MIS Quarterly,* September, 233–250.

Charry, M., Pimentel, H., & Camargo, R. (2000). User reactions in continuous speech recognition systems. *AVIOS Proceedings of the Speech Technology & Applications Expo,* 113-130.

Christ K. A. (1984). Literature review of voice recognition and generation technology for Army helicopter applications (Army Report No. HEL-TN-11-84). Aberdeen Proving Ground, MD: Human Engineering Lab.

Clark, C. E., Cavanaugh, N. C., Brown, C. V., and Sambamurthy, V. (1997). Building change-readiness capabilities in the IS organization: Insights from the Bell Atlantic experience. *MIS Quarterly,* 4, 21, 425–454.

Cooper, R., and Markus, M. L. (1995). Human Re-engineering. *Sloan Management Review,* 4, 39–50.

Cummings, A., and Oldham, G. R. (1997). Managing work contexts for the high-potential employee. *California Management Review,* 1, 22–38.

Dardelet, B. (1998). Breaking the wall: The rise of telemedicine as the new collaborative interface. *Proceedings of the 31st Hawaii International Conference on System Sciences* (HICSS-31), Maui, Hawaii.

Davenport, T. H. (1993). *Process Innovation: Re-engineering Work Through Information Technology*. Boston, MA: Harvard Business Press.

Davenport, T. H., and Short, J. E. (1990). The new industrial engineering: Information technology and business process redesign. *Sloan Management Review* (Summer), 11–27.

Davenport, T. H., and Stoddard, D. B. (1994). Re-engineering: Business change of mythic proportions. *MIS Quarterly, 18* (2),+121–127.

Department of Health and Human Services (1989). International classification of diseases (9th revision), clinical modification (3rd ed.). Washington, DC: Government Printing Office.

Doddington, G. (1999). Topic detection and tracking: TDT2 overview and evaluation results. *Proceedings of DARPA Broadcast News Workshop.* Herndon, VA.

Drucker, P. F. (1989). What businesses can learn from non-profits. *Harvard Business Review,* July-August, p. 89.

Ellis, D. (2000). Improved recognition by combining different features and different systems. AVIOS *Proceedings of the Speech Technology & Applications Expo.*, 236-242.

Erten, G., Paoletti, D., & Salam, F. (2000). Speech recognition accuracy improvement in noisy environments using clear voice capture (CVC) technology. AVIOS *Proceedings of the Speech Technology & Applications Expo*, 193-198.

Fiscus, J. G. (1997). A post-processing system to yield reduced word error rates: Recognizer output voting error reduction (ROVER). Proceedings, *1997 IEEE Workshop on Automatic Speech Recognition and Speech.*

Fiscus, J. G., Fisher, W. M., Martin, A. F., Przybocki, M. A., & Pallet, D. S. (2000). 2000 NIST evaluation of conversational speech recognition over the telephone: English and Mandarin performance results. *Proceedings of DARPA Broadcast News Workshop*, February-March.

Floro, F. C., Nelson, R., & Garshnek, V. (1998). An overview of the AKAMAI telemedicine project: A Pacific perspective. *Proceedings of the 31st Hawaii International Conference on System Sciences* (HICSS-31), Maui, Hawaii.

Gaddy, L. (2000a). The future of speech I/O in mobile phones. *SpeechTEK Proceedings*, 249-260.

Gaddy, L. (2000b). Command and control solutions for automotive applications. *SpeechTEK Proceedings,* 187-192.

Gagnon, L. (2000). Speaker recognition solutions to secure and personalize speech portal applications. *SpeechTEK Proceedings*, 135-142.

Garshnek, V., & Burkle, F. M. (1998). Telemedicine applied to disaster medicine and humanitarian response: History and future. *HICSS, 10* (6).

Goodliffe, C. (2000). The telephone and the Internet. AVIOS *Proceedings of the Speech Technology & Applications Expo*, 149-151.

Gorham, A., & Graham, J. (2000). Full automation of directory enquiries. A live customer trial in the United Kingdom. AVIOS *Proceedings of the Speech Technology & Applications Expo*, 1-8.

Greenberg, S., Chang, S., & Hollenback, J. (2000). An introduction to the diagnostic evaluation of switchboard-corpus automatic speech recognition systems. *Proceedings of DARPA Broadcast News Workshop*, February-March.

Gunn, R. (2000). "Voice": The ultimate in user-friendly computing. *SpeechTEK Proceedings,* 161-178.

Gupta, B., Nemati, H. R., and Harvey, J. D. (1999). Organizational factors affecting successful implementation of decision support systems: The case of fuel management systems at Delta Airlines. *Journal of Information Technology Cases and Applications, 1* (3), 4-25.

Hackman, J. R., Oldham, G., Janson, R., and Purdy, K. (1975). A new strategy for job enrichment. *California Management Review*, pp. 4, 17, 57–71.

Hammer, M. (1990). Re-engineering work: Don't automate, obliterate. *Harvard Business Review* (July-August), 18–25.

Hammer, M., and Champy, J. (1993). *Re-engineering the Corporation.* New York, NY: Harper Collins Books.

Harkness, W. L., Kettinger, W. J., and Segars, A. H. (1996). Sustaining process improvement and innovation in the information services function: Lessons learned at the Bose Corporation. *MIS Quarterly*, pp. 3, 20, 349–368.

Haynes, T. (2000). Conversational IVR: The future of speech recognition for the enterprise. AVIOS *Proceedings of the Speech Technology & Applications Expo*, 15-32.

Head, W. (2000). Breaking down the barriers with speech. *SpeechTEK Proceedings,* 93-100.

Herb, G., & Schmidt, M. (1994). Text independent speaker identification. *Signal Processing Magazine,* October, 18-32.

Hermansen, L. A., & Pugh, W. M. (1996). Conceptual design of an expert system for planning afloat industrial hygiene surveys. Technical Report No. 96-5E. San Diego, CA: Naval Health Research Center.

Hertz, S., Younes, R., & Hoskins, S. (2000). Space, speed, quality, and flexibility: Advantages of rule-based speech synthesis. AVIOS *Proceedings of the Speech Technology & Applications Expo*, 217-228.

Holtzman, T. (2000). Improving patient care through a speech-controlled emergency medical information system. AVIOS *Proceedings of the Speech Technology & Applications Expo*, 73-81.

Hutzell, K. (2000). Voice Interactive Display (VID) Contract Summary Report: Apr 98-May 2000. Johnstown, PA: MTS Technologies, Inc.

Ingram, A. L. (1991). Report of potential applications of voice technology to armor Training. Final Report: Sep 84-Mar 86. Cambridge, MA: Scientific Systems, Inc.

Institute of Medicine. (1996). *Telemedicine: A Guide to Assessing Telecommunications in Health Care.* National Academy Press: Washington, D.C.

Ives, B., & Jarvenpaa, S. L. (1993). Competing with information: Empowering knowledge networks with information technology. The Knowledge Economy Institute for Information Studies, 53-87.

Kangas, K. (1999). Competency and capabilities-based competition and the role of information technology: The case of trading by a Finland-based firm to Russia. *Journal of Information Technology Cases and Applications, 1* (2), 4-22.

Karam, G., & Ramming, J. (2000). Telephone access to information and services using VoiceXML. AVIOS *Proceedings of the Speech Technology & Applications Expo*, 159-162.

Kettinger, W. J., Weng, J. T. C., and Guha, S. (1998). Business process change: A study of methodologies, techniques, and tools. *MIS Quarterly,* 1, 21, 55–81.

Komissarchik, E., & Komissarchik, J. (2000). Application of knowledge-based speech analysis to suprasegmental pronunciation training. AVIOS *Proceedings of the Speech Technology & Applications Expo*, 243-248.

Kotter, J. P. (1995). Leading change: Why transformation efforts fail. *Harvard*

Business Review (March–April), 59–67.
Krause, B. (2000). Internationalizing speech applications. AVIOS *Proceedings of the Speech Technology & Applications Expo*, 10-14.
Kubala, F. (1999). Broadcast news is good news. *Proceedings of DARPA Broadcast News Workshop*, Herndon, VA, Feb.
Kubala, F., Colbath, S., Liu, D., Srivastava, A., & Makhoul, J. (2000) Integrated technologies for indexing spoken language. *Communications of the ACM, 43* (2), 48-56.
Kundupoglu, Y. (2000). Fundamentals for building a successful patent portfolio in the new millennium. AVIOS *Proceedings of the Speech Technology & Applications Expo,* 229-234.
Lai, J. (2000). Comprehension of longer messages with synthetic speech. AVIOS *Proceedings of the Speech Technology & Applications Expo,* 207-216.
Larson, J. (2000). W3C voice browser working group activities. AVIOS *Proceedings of the Speech Technology & Applications Expo*, 163-174.
Leitch, D., & Bain, K. (2000). Improving access for persons with disabilities in higher education using speech recognition technology. AVIOS *Proceedings of the Speech Technology & Applications Expo,* 83-86.
Macker J. P., & Adamson R. B. (1996). IVOX—The Interactive VOice eXchange application. Report No. NRL/FR/5520—96-980. Washington, DC: Naval Research Lab.
Mann, S. (1997). Wearable computing. *Computer, 30* (2), 25-32.
Markowitz, J. (2000). The value of combining technologies. AVIOS *Proceedings of the Speech Technology & Applications Expo*, 199-206.
Markus, M. L., and Benjamin, R. I. (1996). Change Agentry—The next IS frontier. *MIS Quarterly*, pp. 4, 20, 385–407.
Mata, F., Fuerst, W. F., Barney, J. B. (1995). Information technology and sustained competitiveness advantage: A resource-based analysis. *MIS Quarterly,* pp. 487-506.
McCarty, D. (2000). Building the business case for speech in call centers: Balancing customer experience and cost. *SpeechTEK Proceedings*, 15-26.
McGlashan, S. (2000). Business opportunities enabled by integrating speech and the Web. *SpeechTEK Proceedings*, 281-292.
McKenney, J. L., Mason, R. O., and Copeland, D. G. (1997). Bank of America: The crest and trough of technological leadership. *MIS Quarterly*, pp. 3, 21, 321-353.
Miller, R. (2000). The speech-enabled call center. *SpeechTEK Proceedings,* 41-52.
Mogford, R. M., Rosiles, A., Wagner, D., & Allendoerfer, K. R. (1997). Voice technology study report. Report No. DOT/FAA/CT-TN97/2. Atlantic City, NJ: FAA Technical Center.
Molina, E. A. (1991). Continued performance assessment methodology (PAM)

research (VORPET). Refinement and implementation of the JWGD3 MILPERF-NAMRL Multidisciplinary Performance Test Battery (NMPTB). Final Report: 1 Oct. 89-30 Sept. 91. Pensacola, FL: Naval Aerospace Medical Research Laboratory.

Morton, S. (1991). *The Corporation of the 1990s: Information Technology and Organizational Transformation.* New York: Oxford University Press.

Mott, D. (1972). Characteristics of effective organizations. San Francisco: Harper Collins, as reported by H. L. Tosi, Jr., and J. W. Slocum, Jr., Contingency theory: Some suggested directions. Journal of Management, (Spring, 1984). p. 11.

Newman, D. (2000). Speech interfaces that require less human memory. AVIOS *Proceedings of the Speech Technology & Applications Expo,* 65-69.

Oliver, R., Sheng, L., Paul, J. H., & Chih, P. W. (1999). Organizational management of telemedicine technology: Conquering time and space boundaries in health care services. *IEEE Transactions on Engineering Management, 46* (3), 279-288.

O'toole, J. (1999). Lead change effectively. *Executive Excellence,* p. 18.

Pallett, D. S., Garofolo, J. S., & Fiscus, J. G. (1999). 1998 broadcast news benchmark test results: English and non-English word error rate performance measure. *Proceedings of DARPA Broadcast News Workshop*, Herndon, VA, Feb.

Pallett, D. S., Garofolo, J. S., & Fiscus, J. G. (2000). Measurements in support of research accomplishments. *Communications of the ACM, 43* (2), 75-79.

Pan, J. (2000). Speech recognition and the wireless Web. *SpeechTEK Proceedings,* 229-232.

Paper, D. (1999). The enterprise transformation paradigm: The case of Honeywell's Industrial Automation and Control Unit. *Journal of Information Technology Cases and Applications, 1* (1), 4–23.

Paper, D., Rodger, J., and Pendharkar, P. (2000). Development and initial testing of a theoretical model of transformation. *HICCS Conference*, pp. 325-333.

Paul, D. L., Pearlson, K. E., & McDaniel, R. R. (1999). Assessing technological barriers to telemedicine: Technology-management implications. *IEEE Transactions on Engineering Management, 46* (3), 279-288.

Pearce, D. (2000). Enabling new speech-driven services for mobile devices: An overview of the ETSI standards activities for distributed speech recognition front-ends. AVIOS *Proceedings of the Speech Technology & Applications Expo*, 175-186.

Perednia, D. A., & Allen, A. (1995). Telemedicine technology and clinical applications. *Journal of the American Medical Association, 273* (6), 383-388.

Pfeffer, J. (1998). Seven principles of successful organizations. *California Management Review*, pp. 2, 40, 96-124.

Prizer, B., Thomas, D., & Suhm, B. (2000). The business case for speech.

SpeechTEK Proceedings, 15-26.

Przybocki, M. (1999). 1998 broadcast news evaluation information extraction named entities. *Proceedings of DARPA Broadcast News Workshop,* Herndon, VA, Feb.

Rasberry, M. S. (1998). The theater telemedicine prototype project: Multimedia e-mail in the Pacific. *Proceedings of the 31st Hawaii International Conference on System Sciences* (HICSS-31), Maui, Hawaii.

Rodger, J. A., & Pendharkar, P. C. (2000). Telemedicine and the Department of Defense. *Communications of the ACM, 43* (2).

Rolandi, W. (2000). Speech recognition applications and user satisfaction in the imperfect world. AVIOS *Proceedings of the Speech Technology & Applications Expo,* 153-158.

Schalk, T. (2000). Design considerations for ASR telephony applications. AVIOS *Proceedings of the Speech Technology & Applications Expo,* 103-112.

Scherr, A. L. (1993). A new approach to business processes. *IBM Systems Journal, 32* (1), 80–98.

Schneberger, S., and Krajewski, A. (1999). Common systems implementation at General Motors of Canada. *Journal of Information Technology Cases and Applications, 1* (3), 45-60.

Scholz, K. (2000). Localization of spoken language applications. AVIOS *Proceedings of the Speech Technology & Applications Expo,* 87-102.

Shepard, D. (2000). Human user interface HUI. *SpeechTEK Proceedings,* 262-270.

Sones, R. (2000). Improving voice application performance in real-world environments. *SpeechTEK Proceedings,* 179-210.

Soule, E. (2000). Selecting the best embedded speech recognition solution. *SpeechTEK Proceedings,* 239-248.

Sternberg, R. J., O'Hara, L. A., and Lubart, T. I. (1997). Creativity as investment. *California Management Review,* p. 40.

Stromberg, A. (2000). Professional markets for speech recognition. *SpeechTEK Proceedings*, 101-124.

Talwar, R. (1993). Business re-engineering—A strategy-driven approach. *Long Range Planning, 26* (6), 22-40.

Tanriverdi, H., & Venkatraman, N. (1998). Creation of professional networks: An emergent model using telemedicine as a case. *Proceedings of the 31st Hawaii International Conference on System Sciences* (HICSS-31), Maui, Hawaii.

Taylor, S. (2000). Voice enabling the Web_ modification or new construction. *SpeechTEK Proceedings,* 69-76.

Teng, T. C., Jeong, S. R., and Grover, V. (1998). Profiling successful re-engineering projects. *Communications of the ACM,* pp. 6, 41, 96-102.

Thompson, D., & Hibel, J. (2000). The business of voice hosting with VoiceXML.

AVIOS *Proceedings of the Speech Technology & Applications Expo,* 142-148.

Topchik, G. S. (1998). Attacking the negativity virus. *Management Review*, pp. 8, 61-64, 87.

URL: www.actd.tatrc.org

URL: www.actd.tatrc.org

URL: www.odusa-or.army.mil/TEMA/ref.htm).

URL: www.odusa-or.army.mil/TEMA/ref.htm).

Wenger, M. (2000). Noise rejection. The essence of good speech recognition. AVIOS *Proceedings of the Speech Technology & Applications Expo,* 51-63.

Wickstrom, T. (2000). Microphone voice recognition performance in noise: A proposed testing standard. *SpeechTEK Proceedings,* 211-219.

Wong, W.Y.L. (1998). A holistic perspective on quality quests and quality gains: The role of environment. *Total Quality Management,* pp. 4, 9, 241-245.

Woo, D. (2000). Desktop speech technology: A MacOS perspective. AVIOS *Proceedings of the Speech Technology & Applications Expo,* 39-50.

Yan, Y. (2000). An introduction to speech activities at Intel China Research Center. *SpeechTEK Proceedings,* 79-82.

Zajtchuk, R. S. (1995). Battlefield trauma care. *Military Medicine,* 160, 1-7.

Zapata, M. (2000). LIPSinc: We have ways of making you talk. *SpeechTEK Proceedings,* 273-280.

APPENDIX—MOBILE COMMERCE ACRONYMS

ACTD Advanced Concept Technology Demonstration
AE Aeromedical Evacuation
AEF Air Expeditionary Force
AELT Aeromedical Evacuation Liaison Team
AFB Air Force Base
AFFOR Air Force Forces
ALE Automatic Link Establishment
AMAL Authorized Medical Allowance List
AMC Air Mobility Command
AoA Analysis of Alternatives
AOR Area of Responsibility
ARFOR Army Forces
ASMB Area Support Medical Battalion
ASTS Air Medical Staging Squadron
ATH Air Transportable Hospital
BAS Battalion Aid Station
BDU Battle Dress Utilities
BMET Biomedical Engineering Technology
BUMED Bureau of Medicine
CG-1 KB Prime
CIA Care in the Air
CINC Commander-in-Chief
CJTF Commander Joint Task Force
CNSP Commander Naval Surface Force Pacific
COA Course of Action
COE Common Operating Environment
COI Critical Operational Issue
COIC Critical Operational Issues Criteria
COMPHIBGRU Commander, Amphibious Group
CONOPS Concept of Operations
COP Common Operating Picture
COTS Commercial-Off-the-Shelf
CPG-3 Commander, Amphibious Group-3
CPX Command Post Exercise
CRTS Casualty Receiving Treatment Ships
CSH Combat Support Hospital
CSI Command System, Incorporated
CSS Combat Service Support
C2 Command and Control
C4I Command, Control, Communications, Computers and Intelligence
DAMA Demand Assigned Multiple Access

DC Direct Current
DC II Desert Care II
DEPMEDS Deployable Medical Systems
DNBI Diseases and Non-Battle Injuries
DoD Department of Defense
DII Defense Information Infrastructure
ECOC Enhanced Combat Operations Center
E-IPT Engineering Integrated Product Team
ELB Extending the Littoral Battlespace
EMEDS Expeditionary Medical Service
EMT Emergency Medical Treatment
ENT Ear, Nose, and Throat
EUT End User Terminal
FBE-E Fleet Battle Experiment—Echo
FH Fleet/Field Hospital
FMC Fleet Medical Clinic
FMSS Field Medical Surveillance System
FSMC Forward Support Medical Company
FST Forward Surgical Team
FTP File Transfer Protocol
FTX Field Training Exercise
FY Fiscal Year
GICOD Good Idea Cut-Off Date
GMO General Medical Officer
GOTS Government-off-the-Shelf
GPS Global Positioning System
GUI Graphical User Interface
HAZMAT Hazardous Material
HF High Frequency
HM Hospital Corpsman
HM3 Hospitalman Third Class
HMC Hospitalman Chief
HMCS Hospitalman Chief Senior
HMMWV High-Mobility Multipurpose Wheeled Vehicle
HQ Headquarters
ICU Intensive Care Unit
IM Information Management
ICD-9 International Classification of Diseases 9th Revision
IDC Independent Duty Corpsman
INMARSAT International Maritime Satellite
IPT Integrated Product Team
ISDN Integrated Services Digital Network

JHSS Joint Health Service Support
JINC Joint Internet Controller
JMedSAF Joint Medical Semi-Automated Forces
JMeWS Joint Medical Workstation
JMO-T Joint Medical Operations-Telemedicine
JSIMS Joint Simulation Systems Acquisition Program
JTF Joint Task Force
JTTP Joint Tactics, Techniques, and Procedures
JV 2010 Joint Vision 2010
kb kilobyte
Kbps Kilobytes per second
KB 99 Kernel Blitz 99
LAN Local Area Network
LDET Lightweight Data Entry Tool
MARFOR Marine Forces
MASF Mobile Air Staging Facility
MB Megabyte
Mbps Million bits per second
MCM Military Command
MEDEVAC Medical Evacuation
MEF Marine Expeditionary Force
MES Medical Equipment Sets
MeWS Medical Workstation
MHS Military Health System
MHz Megahertz
MIEP Medical Information Engineering Prototype
MILNET Military Network
mm millimeters
MMS Medical Messaging Service
MOE Measures of Effectiveness
MOP Measures of Performance
MTF Medical Treatment Facility
M&S Modeling and Simulation
MSEL Master Scenario Events List
M3B Mobile Medical Monitor (B)
NAVFOR Navy Forces
NAVHOSP Navy Hospital
NEC Navy Enlisted Classification
NEPMU-5 Navy Environmental Preventative Medicine Unit-5
NIPRNET Unclassified Internet Protocol Router Network
NRL Naval Research Lab
NSHS Naval School of Health Sciences

NVID Navy Voice Identification Device
OPNAV Operations Navy
OR Operating Room
OTH Over-the-Horizon
PACOM Pacific Command
PC Personal Computer
PCB Poly Chlorinated Biphenyls, Pentachlorobenzole
PCMCIA Personal Computer Memory Card International Association
PIC Personal Information Carrier
P-IPT Performance Integrated Product Team
PPE Personal Protective Equipment
PROFIS Professional Filler System
PM 99 Patriot Medstar 99
PW 99 Pacific Warrior 99
RF Radio Frequency
RFTA Reserve Forces Training Area
RSO Regional Supply Officer
SAIC Science Applications International Corporation
SATCOM Satellite Communications
SC Surgical Company
SINCGARS Single-Channel Ground and Airborne Radio System
SPAWAR Space and Warfare
SSC SD Systems Center, San Diego
STP Shock Trauma Platoon
SURFPAC Surface Pacific
TAMC Tripler Army Medical Center
TCAS Team Care Automation System
TCEA Theater Clinical Encounter Application
TCP/IP Transmission Control Protocol/Internet Protocol
T&E Test and Evaluation
T&E-IPT Test and Evaluation Integrated Product Team
TMCS Theater Medical Core Services
TMIP Theater Medical Information Program
TTT Theater Telemedicine Team
T2P2 Theater Telemedicine Prototype Program
UHF Ultra-High Frequency
USAF U.S. Air Force
USFK U.S. Forces, Korea
UW Urban Warrior
VHF Very High Frequency
VTC Video Teleconference

Chapter XIII

Mobile Applications for Adaptive Supply Chains: A Landscape Analysis

Ravi Kalakota, Marcia Robinson, and Pavan Gundepudi
E-Business Strategies, Inc.

ABSTRACT

Streamlining supply chains is a high priority for corporations. In a volatile economy, customer satisfaction, market share and revenue growth become dependent on getting the right product to the right place at the right time. As a result, the notion of adaptive supply chains is emerging as the next competitive battlefield. Fulfillment velocity, inventory visibility, and supplier coordination versatility form the three pillars of adaptive supply chains. To support these business objectives, traditional tethered computing models are inadequate. Untethered models, enabled by mobile computing, facilitate the improvement, management and re-design of next generation supply chains. In this chapter, we examine the different ways mobility is morphing supply chain applications. Specifically, we show how mobile technology and infrastructure is transforming the key areas of procurement, supply execution, supply chain visibility and after-sales service management.

INTRODUCTION

Business is going mobile. In a fast-moving world, customers increasingly expect companies, suppliers and distributors to provide:

- Real-time information about their orders,
- The location of the product as it moves through the logistics and transportation network, and
- Accurate just-in-time delivery commitments.

These escalating customer demands are forcing corporations to adopt mobile technology to optimize and streamline the information, product and financial flows in the supply chain.

Clearly, as traditional supply chains become faster and faster, management issues like real-time visibility, control and flexibility become germane. Supporting these management objectives through the traditional tethered computing model is very difficult (if not impossible). As a result, leading companies are looking at innovative mobile solutions to enable their next generation supply chains. In this chapter, we illustrate the emerging landscape of mobile supply chain applications.

However, before we get into the specifics of mobile supply chain solutions, it is necessary to provide the context by illustrating the broader economic trends that are shaping next-generation supply chains.

Adaptive Supply Chains—Enabling Visibility, Velocity and Versatility

Streamlined supply chains create business value in the form of increased efficiency, low prices, and broad product selection. However, uneven demand, more frequent and shorter order-to-shipment times, and stricter customer compliance requirements are key variables that are shaping twenty-first-century business practices. Thus, modern supply chains must create value through their flexibility (Kalakota and Robinson, 2001). Their design must accommodate a customer changing his or her mind after placing the order. This is the essence of an emerging concept called adaptive supply chains.

As a result, a major trend today is the use of Internet and enterprise technology to increase supply chain velocity—the fast and accurate collection and manipulation of information—while maximizing service levels. The basics of supply chains such as purchase orders, invoices, advanced ship notices (ASNs), and bills of material are still the same. They are just being handled more quickly, more cheaply, and more accurately through the Internet.

Another trend is the replacement of inventory with information. The real-time transfer of inventory information helps companies create a more realistic demand picture. Under today's practices, order information is sent to warehouses, distribution centers, and retail stores in batches by telephone, fax, e-mail, or EDI. Since order information isn't matched with demand in real time, manufacturers often get caught in the "bullwhip effect," making too little product, then too much, in an attempt to keep up with the fluctuating market. The business challenge is to monitor demand effectively to able to synchronize actions along the supply chain.

The need for faster and more accurate fulfillment is also causing supply chain coordination problems. For example, as e-commerce became popular, JCPenney decided to minimize shipping time. So it asked manufacturers to ship direct to the customer instead of its central warehouse. Consequently, each manufacturer's function had to begin shipping to several thousands customer locations instead of a few warehouses. Also, this meant that the manufacturer was looking at item picks, small package shipments, and priority handling—all new processes they did not have to deal with before. Clearly, faster end-to-end order processing is dramatically affecting product handling and logistics.

The three trends—fulfillment velocity, inventory visibility, and coordination versatility—are laying the foundation for adaptive supply chains. Mobile applications are playing a central role in the enabling of that foundation. Supply chain processes in companies today are desktop-centric. That is, processes from the receiving dock, production, and shipping areas slowly make their way to a desktop before resuming their natural flow. Mobile devices are deconstructing desktop-centric workflows. Figure 1 illustrates the application of different mobile device capabilities for different supply processes.

Figure 1: Mobile device capabilities

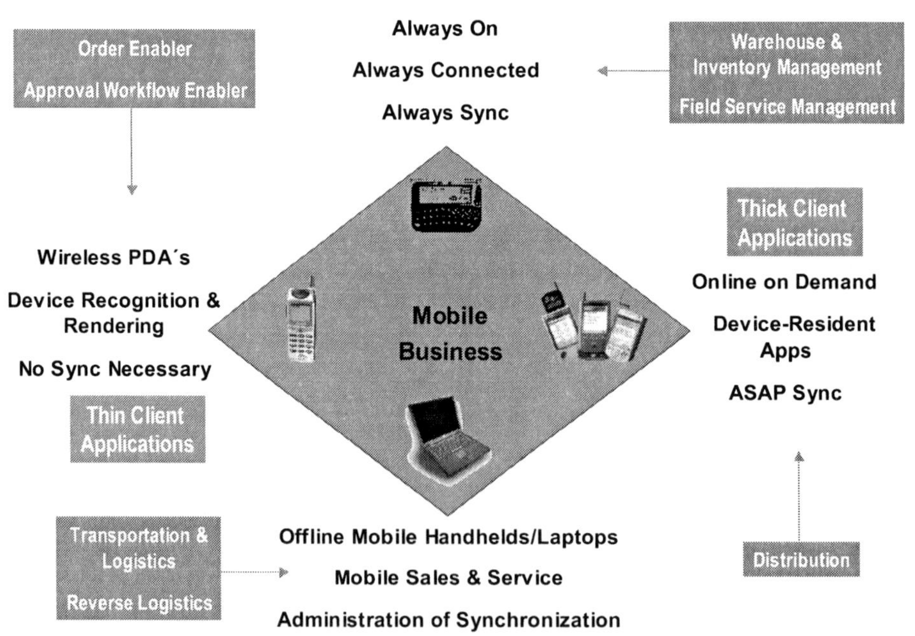

Trends Driving Mobile Applications in Supply Chains

Customers are impatient. The order-to-delivery time is often longer than the customer is prepared to wait. Several market trends are driving mobile investments to shorten lead-time.

- **Streamlining the order-to-cash process.** Customer orders drive supply chains. The complexity of the ordering process and the need for order information upstream to support supply chain planning has resulted in unusual pressure to streamline the entire order-to-cash process.
- **Coordinating the order-fulfillment process.** Customer demand for faster and more customized delivery has disrupted traditional inventory management policies and transportation models. Successful order fulfillment requires local, regional, national, and global coordination, which itself requires sophisticated synchronization of multiple distribution channels.
- **Better asset tracking and utilization.** Customers are demanding greater visibility into the supply chain execution processes for real-time order status information. As a result, companies are investing more in real-time asset tracking to achieve inventory reductions, eliminate sources of order-fulfillment variance, reduce leakage, and cause fewer returns.
- **More responsive service management.** Lower profit margins to maintain and increase market share are today's mandate. In response, companies are redesigning post-sale service and support functions to eliminate unnecessary tasks and reduce process complexity. The benefits include reducing delays, errors, and cost. For example, reverse logistics is an area of increasing focus.

Figure 2: Supply chain framework

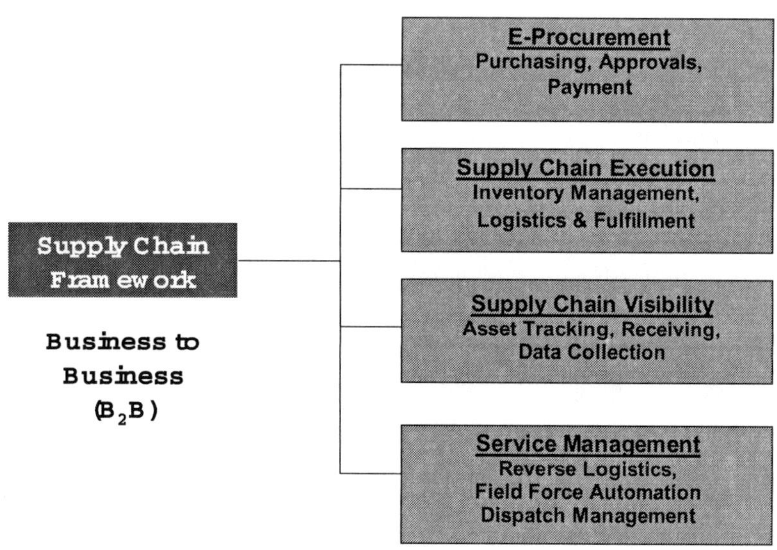

Figure 2 shows the four major segments where mobile solutions will significantly impact supply chains.

Mobile Procurement

Traditional paper- and fax-based processes do not provide real-time visibility into the order-to-cash cycle. In retail, for example, the data on what was on the store shelves takes days or even weeks to make its way back to the manufacturers. This makes it difficult to match production schedule to actual demand. Three types of mobile applications are enabling the order-to-cash cycle: order and transaction enablers, approval workflow enablers, and receiving and payment enablers.

Order and Transaction Enablers

Supply chains are increasingly market-driven, capable of reading and responding to real customer demand. In support of this trend, new mobile applications are changing the face of order taking.

Kroger, the largest grocery chain in the United States, with the help of PocketScript has built an electronic-prescription interchange (PR Newswire, Jan. 12, 2001). Using a handheld device, physicians send e-prescriptions directly to any Kroger-owned pharmacy. The system is expected to reduce the number of illegible handwritten prescriptions and thus decrease the time spent by the pharmacies in making callbacks to physicians to verify prescriptions' contents.

Approval Workflow Enablers

Workflow applications such as document approval, expense reporting, payment, and purchase orders are another area to which mobile innovations are bringing considerable benefit.

In particular, keeping track of orders being placed is a priority at many companies. But current e-procurement applications require the approver to be at the office and by the desktop. Mobile applications relax the tethered constraint and allow the approving manager to authorize a requisition from anywhere, at anytime.

Critical purchase orders, such as those impacting important customers or projects, no longer must wait in a queue until the manager returns or be expedited as an exception because the approval process is slow. Exception purchases often prove expensive because the purchase is rushed and a lower price or more favorable shipping terms can't be negotiated with the supplier.

Receiving and Payment Workflow Enablers

While many companies are focusing their efforts on implementing mobile solutions to improve their order workflow, they are ignoring the need to bring the same types of solutions to two processes: receiving and accounts payable. Most companies still work with 60- to 120-day payment cycles. Errors made during the receipt of products can easily dissipate the efficiency gains from improvements to the front end of the procurement process.

With the compression of the supply chain, receiving and inventory put-away processes are being reexamined to save time, cut costs, reduce inventory, and speed up service to customers. To improve efficiency, corporate receiving departments are using mobile scanners. Over the past twenty years, handheld technology has advanced well beyond simple scanning. For example, today's handhelds can download the original PO at the time the shipment is received, allowing the checker to display each item and confirm its receipt. Verifying the receipt of all purchase-order items initiates the accounts payable process.

Mobile Supply Chain Execution

Supply chain execution—the process of fulfilling customer-specific needs for goods and value-added services in a timely, efficient, and cost-effective manner—is a key differentiator in increasingly competitive markets.

Traditionally, mobile applications were used extensively in warehouse and transportation operations. The focus is now shifting to integrating these operations with corporate planning systems and other enterprise software. The result of this trend is new applications of mobility in three key business functions: warehousing and inventory management, inbound logistics and transportation, and outbound finished-products distribution.

Mobility in Warehousing and Inventory Management

Warehousing and inventory management seek to minimize inventory risk and out-of-stock situations, and maximize inventory turns. These strategies track and manage the movement of product from the supplier to the customer. They focus on providing exceptional customer service and faster, more accurate order delivery.

Mobile solutions increase the efficiency of the warehouse operations. Bar code scanners are used to track pallets of inventory that are shipped and received, providing an accurate, real-time count of on-hand amounts. For instance, when product comes into the receiving bay at the warehouse, the bar code tag containing product description, quantity, lot number, and pallet number is scanned. The information is then instantly transmitted to a warehouse PC and corporate server. The PC processes the data and generates a put-away list, indicating the warehouse storage location. This put-away task is automatically downloaded to the appropriate forklift operator, who retrieves the pallet of product and places it in the required location.

Inventory management also occurs at the store level. It involves the counting of items on shelves twice a year. The cost of this exercise can be staggering for a grocery retailer like Kroger with more than 1,500 stores. To speed up the process and also make it more accurate, industrial handheld computers (HHC) with laser scanners are used. Important information is collected instantaneously, eliminating manual rekeying and volumes of paperwork.

Mobility in Transportation and Logistics

Widely dispersed assets, tight delivery schedules, and minimal room for process error characterize today's logistics and transportation industries. Mobile applications provide customers with the ability to track their shipments across a network of multimodal transportation.

The USPS moves millions of parcels across international borders. Unlike UPS or FedEx, each of which has its own fleet of jets, it relies on a variety of third-party logistics (3PL) providers to move mail. USPS developed a performance-tracking mobile system to monitor the airline carrier and shipment delivery performance. Using handheld scanners, ramp clerks automatically enter data for each mail container and synchronize, via an ethernet connection, with corporate databases, eliminating manual data entry and the need to interpret handwriting. The system enables the tracking of any mail-handling irregularity, such as an airline's failure to load a bag of mail onto a plane.

Mobility in Distribution "Last Mile"

Distribution management oversees the end-to-end process of transporting goods from manufacturers to distribution centers to the consumer. Mobile applications give distributors faster access to shipping, tracking, and delivery data.

Kraft Foods uses handheld computers to capture delivery quantities for the products it sells. When delivering an order to the back door receiving area of a local supermarket, a Kraft truck driver unloads the shipment and then records any changes to the order in the handheld. In cases where the supermarket chain supports EDI back door receiving, the driver connects the handheld computer directly to the supermarket's computer system and transfers the delivery data. When the driver returns to the distribution center, the daily transaction data is uploaded from the handheld into the company's back-end systems and any updated delivery information for the next day's route is downloaded.

Global Positioning Satellites (GPS) and Automatic Vehicle Location (AVL) mobile technologies hold significant potential for the delivery industry. They will allow firms to have continuous contact with their drivers and to map delivery routes. For example, Roadnet and Descartes routing systems use a satellite geocode to plot customers on a longitude and latitude grid and calculate the most logical delivery route. This tool also enables truck dispatchers to know where drivers are at all times.

Supply Chain Visibility

To better monitor and optimize asset utilization, customers want visibility into inventory in motion and inventory at rest, and a real-time view of their assets.

Customers also want immediate notification when supply chain performance fails to meet delivery terms and service agreements.

Supply chain visibility requires the technological ability to match a unique customer transaction with that customer's products as they flow through the supply chain. This matching process is often done manually or through visual inspection and increases the potential for error. Today's companies are seeking technologies that, using either a customer serial number or pallet ID, enable the tracking of products from the original product components to the product's receipt by the customer.

Bar Code

The need to improve supermarket efficiency has led to the bar code's creation in 1970.[1] The Uniform Product Code (UPC) system assigns each type of food or grocery product a unique code represented by a small rectangle of black and white bars. Automated checkout stands contain technology capable of processing the code. The system thus provides a standard product identifier for all packaged foodstuffs, regardless of where the products were manufactured or sold.

The bar code has enabled the creation of important new applications ranging from tracking customer buying habits to managing inventory. For example, bar code technology was key to breakthroughs such as Efficient Consumer Response (ECR) and Quick Response (QR). These technologies capture on-demand data using point-of-sale or point-of-use devices. ECR and QR transformed the retailer's ability to hear the voice of the market and to respond directly to it. Bar codes are a case in which technology sought markets, markets found unappreciated properties, and retailers exerted themselves to translate properties into supply chain capabilities that resulted in tremendous value.

RFID Tags and Auto-ID

In recent years, new applications have been developed that complement bar-code functionality. These applications permit the tagging and tracking of physical goods. For example, Radio Frequency ID (RFID) tags along with tag readers, known as "interrogators," allow companies to track products and materials. RFID systems use low-power wireless signals to read and update information on a "tagged" item or component. Unlike bar codes, the signals don't require line of sight or manual scanning.

Burlington Northern Santa Fe Corp., a railway line, used to employ an army of some 500 clerks armed with pencils and clipboards to walk up and down the tracks at its depots and switching stations to read numbers painted on the sides of its railway cars. The information was then handed to data entry personnel who would key it in, so the company's mainframe systems could track cars.

Today, all of the company's rail cars are tagged, and it has 443 readers positioned at key junctures along 33,500 miles of track in 28 states and two Canadian provinces. As a result, the company eliminated all of its trackside clerks in 1997.

Now, the system has paid for itself, and it reads 100,000 tags a day with virtually no errors. Burlington can provide customers with more accurate data about where their shipments are. The system has also dramatically reduced track delays. In the old days, when a car was out of place, people had to spend hours trying to figure out where it should be, and that caused delays throughout the system. Today, that's almost never a problem (Roberti, 2002).

A recent extension of the RFID method is the Auto-ID technology. MIT's Auto-ID Center has developed a numbering scheme—the "Electronic Product Code"—that is embedded in a microchip with wireless antennas. The microchip transmits a unique item-level product code signal when exposed to energy from a reader. This code identifies products individually, not just by product type, as today's UPC codes do. When the Auto-ID readers lining warehouse or retail store shelves intercept a tag's radio signal, which contains the product code, they use the code to identify the product and track it as it flows through the supply chain.[2] Proctor & Gamble in collaboration with Wal-Mart is piloting this technology to track products from factory floor to store shelves and prevent sudden out-of-stock shortages (Roberti, 2002).

The bar code, RFID, and Auto-ID technology extend visibility more deeply into supply chain performance. However, despite today's sophisticated technology, successfully managing supply chain visibility and performance hinges on the accuracy of transaction data captured at the line level.

Supply Chain Event Management

Asset utilization and quality improvements are the focus of many supply process efforts to achieve sustainable growth. To reduce supply process defects, companies are increasingly focusing on Supply Chain Event Management (SCEM) tools that enable better visibility through more real-time event monitoring and data collection.

Savi Technology, a SCEM vendor, for example, enables every supply chain event across land, sea, and air to be tracked and monitored in real time. The objective is to enable immediate responses to unexpected events, preventing time-critical, expensive catastrophes from occurring. The Savi platform collects data and translates it into meaningful supply chain information about assets, inventory, shipments, and orders. The data collection is done by leveraging existing EDI data and legacy interfaces, or tracking in real time via RFID or GPS/cellular locating systems.

In 1990, the U.S. DoD was preparing for war. The Army shipped 40,000 containers to the Persian Gulf for Operation Desert Shield (later called Desert Storm), then had to open up 25,000 of them to see what was inside. The Army estimated that if an effective RFID way of tracking the location and content of the cargo containers had existed at the time, the DoD would have saved roughly $2 billion.[3]

Mobile Service Management

Service delivery is the final task in a finished-goods supply chain. A variety of new mobile applications are currently being deployed to increase asset utilization and speed of service delivery functions.

Sears Roebuck, for example, wanted to maximize the time employees spent with the customer. It is giving its stockroom staff and salesclerks handhelds for inventory tracking, shipping, receiving, and price checks, thus reducing the number of their trips to the stockroom for inventory checks. Also, the handhelds' simple interface helps compress new hire training time (Wagner, 2001), which is important because workforce turnover in the retail industry is high.

Service companies are positioning themselves to help meet customers' demand levels of product quality and service performance by offering solutions in three primary service areas: reverse logistics, field force management, and mobile dispatch.

Reverse Logistics

Reverse logistics manages the movement of goods back through the supply chain from the customer to their point of origin in the most efficient and cost-effective way possible. Generous product warranties have resulted in a growing trend of customers returning products. However, supply chains are designed for products and services to flow in one direction only—downstream to the customer.

As a result, product returns, whether due to a product defect, spoilage, or a customer changing his or her mind, present significant challenges for a business process optimized to move products forward. In addition to the strain they place on the supply chain, returns also impact company financials since they must be reconciled with the company books as the purchase transaction is reversed.

Webvan, the now defunct home delivery service, tried to deliver a high-quality customer experience by using handhelds to track deliveries, client accounts, and other information. When couriers went into the distribution center to pick up their deliveries for the day, all the information they needed was downloaded into a mobile handheld. When the delivery person arrived at the customer's home with the order, the mobile handheld displayed the customer's name, address, phone number, and the items included in the order. During the drop-off, if the customer wished to return certain products, the delivery person entered the information into the device and generated an instant credit that is applied against the new bill. The driver also had a small printer, which printed out an updated customer receipt (Gonzales, 2001). The entire returns process was easy and convenient. Result: high customer satisfaction.

Field Service and Service Parts Management

Field service companies, from small repair businesses to large corporations, are seeking to automate the costly, unproductive and paper-intensive processes associated with performing field service. The improvement efforts focus on every major field service process, including communicating with company headquarters, receiving work assignments, completing work orders, submitting billing information to accounting, and ordering parts.

The primary goal of these efforts is to increase field technician productivity by removing mundane and time-consuming data-entry tasks. Immediate invoicing updates also greatly improve cash flow and enable billing to be accomplished in hours or days rather than weeks. Billing clerks no longer waste time entering job information. This reduces data-entry errors, further speeding up the billing process.

The American Automobile Association (AAA) is one of the largest providers of roadside assistance, with a membership of more than 45 million people. In the past, AAA used a two-way radio system to dispatch its field units. However, only limited amounts of information could be transmitted to the company's fleet vehicles and field contractors. Using two-way radios also resulted in recurrent miscommunications between dispatchers and drivers, ultimately increasing customer waiting time.

AAA initiated a mobile strategy to reduce customer wait time by up to fifteen minutes per call in some regions.[4] Under the new mobile system, customer assistance calls are wirelessly routed over the Cingular Network to the nearest call center. The call center verifies the caller's membership and obtains other incident information. Incident details are then forwarded to the RIM 950 Wireless Handheld of the closest available driver. An estimated 2,100 AAA fleet vehicles are equipped with the service.

Field service processes can also be improved through the deployment of telemetry applications. These applications use wireless modems installed in equipment assets from computers to elevators at the customer's location to replace expensive dial-up lines or private data networks. Many organizations can use telemetry to monitor remote equipment such as vending machines, security systems, and meter readers. Telemetry applications proactively monitor the status of equipment assets by sending an alarm when certain preset threshold conditions are reached. The result is automatic dispatching in near real time, reduced field service management costs, and faster service.

Vehicle Dispatch Solutions

Vehicle dispatch and identification applications complement field service solutions. Mobile applications in civilian emergency, such as police, fire, and ambulance, allow officers to obtain real-time information from local, state, and national databases. Onboard devices give law-enforcement officers remote access to filed incident reports and other paperwork. Ambulances are enabled with applications used to transmit patient data before the vehicle arrives at the hospital, which reduces time and thus saves lives.

Mobile solutions for non-emergency service such as taxi, shuttle, private fleets, and couriers are also growing. BostonCoach, provider of high-quality luxury vehicle service with 935,000 rides a year in 450 cities, is entirely reservation-based. The company uses the same computerized reservation systems that travel agents use and also takes reservations via its website. BostonCoach passengers—typically corporate employees with a travel account—simply sign a voucher at the end of their ride. No payment is accepted or tipping allowed in a BostonCoach vehicle.

In its early years, BostonCoach used traditional methods—voice radio—for linking drivers and dispatchers. But by 1993, there were simply too many drivers vying for the limited radio bandwidth space. Drivers trying to get an open channel to call dispatchers often experienced delays, and deciphering directions garbled by static radio transmissions was frustrating. The problem became a major concern for the company's reputation as a reliable, high-quality service.

To solve the problem, BostonCoach upgraded its communications technology by installing mobile data terminals (MDTs) in each vehicle. The MDTs allowed drivers to communicate directly with a host computer system, freeing dispatchers to devote more time to ensuring efficient use of the fleet. But they proved expensive, limited in functionality, and used proprietary operating systems and hardware. The bulky terminals were not portable, which was a major disadvantage at locations, particularly airports, where meet-and-greet drivers needed to leave their cars.

To support the company's growth and improve efficiency, BostonCoach developed a new system using handhelds running Windows CE.[5] The system is small, mobile, and easily removed and operated, allowing drivers to carry the unit with them when they are out of the vehicle greeting customers. It also functions on a ubiquitous wireless network, not constrained by one wireless provider or mobile service with limited service coverage.

At the end of each ride, the trip's data is sent via a wireless modem to the dispatch office over the wireless network. It takes only a few seconds from the time the driver presses the "send" button until the system acknowledges receipt of the data. Ride billing data is transferred directly from the driver's data terminal into the back-office systems, which ensures that customer billing is done quickly and accurately.

CONCLUSION

In today's hyper-competitive environment, management focus is shifting to streamlining supply chains. As market share and revenue growth become dependent on getting the right product to the right place at the right time, the notion of adaptive supply chains is emerging as the next competitive battlefield.

Adaptive supply chains represent the ability of supply chains to rapidly react and readjust to environmental changes and supply and demand fluctuations. Fulfillment velocity, inventory visibility, and coordination versatility form the three principles of adaptive supply chains. Mobility represents a key enabler for designing and implementing real-time supply chains built on these principles. Automating customer and supplier relationships means upgrading existing systems architecture, infrastructure, and applications from their current PC-centric orientation to one that supports multidevice capabilities like handhelds, RFID devices and Wireless LANs.

However, research on the implications of mobile applications for supply chains today is nascent, almost nonexistent. We conclude by highlighting a few important research questions that need to be addressed:

- Do emerging mobile technology innovations impact existing supply chain operational strategies such as Vendor Managed Inventory (VMI), Continuous Replenishment, Cross-Docking, etc.?
- How can mobile technology reduce or eliminate the "bullwhip effect" through better visibility?
- What is the impact of mobile technology on inventory management? Will inventory buffers be significantly affected by the availability of information from mobile devices?

In summary, we expect that companies in many industries must evolve their supply chain strategies to accommodate increasingly complex issues, including guaranteed delivery of products on a daily basis, the challenge of globalization, and the ability to respond quickly to changes in market demand and competition. It seems intuitively obvious that increasing the quality and speed of information along the supply chain is a strategy that most companies will adopt.

REFERENCES

Adshead, A. (2001). E-supply chain savings are more than just dotcom hype. *Computer Weekly,* May, p. 20.

Brown, S. A. (1997). *Revolution at the Checkout Counter: The Explosion of the Bar Code.* Harvard University Press.

Gonzales, A. (2001). Electronics keep delivery services on right track. *San Francisco Business Times, 15* (33), 29.

Kalakota, R., and Robinson, M. (2001). *M-Business: The Race to Mobility.* McGraw-Hill.

Leibowitz, E. (1999). Bar codes: Reading between the lines. *Smithsonian,* 29, 130.

PR Newswire, Kroger and PocketScript to launch electronic prescription interchange service for in-store pharmacies, January 12, 2001.

Roberti, M. (2002). RFID: From just-in-time to real-time, *CIOInsight,* April 12.

Wagner, M. (2001). Handhelds nudge PCs in the enterprise. *InternetWeek,* May 14.

ENDNOTES

1 For a comprehensive history of bar codes see Leibowitz, 1999, and Brown, 1997.

[2] For more information on Auto-ID, see the website: auto-id.mit.edu. For more on tagging technology see the website: www.aimglobal.org.

[3] Source: www.savi.com/index.html

[4] For more specific details on the implementation see case study: AAA by Cingular Interactive Data Services (www.bellsouthwd.com).

[5] For more technical information on the Boston Coach solution see Dynamic Mobile Data (http://www.dmdsys.com/).

About the Authors

Ee-Peng Lim is an associate professor in the School of Computer Engineering, Nanyang Technological University (NTU), Singapore. He received his Ph.D. degree in computer science from the University of Minnesota, Minneapolis, in 1994. His research interests include Web warehousing, electronic commerce, database integration and digital libraries. He has published more than 100 refereed journal and conference articles. He has also chaired several conferences and workshops, including the Fifth International Conference on Asian Digital Libraries (ICADL 2002) and the ACM Workshop on Web Information and Data Management (WIDM 2001 and 2002). He has served in the program committee of numerous international conferences. Dr. Lim is currently the director of the Centre for Advanced Information Systems at NTU. He is a senior member of IEEE and a member of ACM.

Keng Siau is an associate professor of management information systems (MIS) at the University of Nebraska, Lincoln (UNL). He received his Ph.D. degree from the University of British Columbia (UBC) where he majored in management information systems and minored in cognitive psychology. He has published more than 40 refereed journal articles, and these articles have appeared in journals such as *Management Information Systems Quarterly, Communications of the ACM, IEEE Computer, Information Systems, ACM's Data Base, Journal of Database Management, Journal of Information Technology, International Journal of Human-Computer Studies, Transactions on Information and Systems, Quarterly Journal of ECommerce,* and many others. In addition, he has published over 60 refereed conference papers in proceedings such as ICIS, ECIS, WITS, and HICSS. He served as the organizing and program chairs for the International Workshop on Evaluation of Modeling Methods in Systems Analysis and Design (EMMSAD) (1996–2002). For more information about him, please refer to his personal website at http://www.ait.unl.edu/siau/.

* * *

Maristella Agosti is a professor of computer science in the Department of Electronics and Computer Science and faculty of Humanities, University of Padua, Italy. She is the leader of a research group in the department focusing on database systems, digital libraries, and information retrieval research. Her research areas of interest include design and implementation of digital libraries applications; distributed multichannel access to information and data stored in digital libraries of distributed collections of semi-structured text and multimedia digital documents; information retrieval on the Web; and multilingual information retrieval. She has published more than 100 refereed articles on journals and conference proceedings and authored or coauthored books and journal issues on hypertext and information retrieval, database design, and automatic construction of hypertexts. She is a coordinator of research activities in the context of national and European research projects. She served as the Program Chair for the 6th European Conference on Digital Libraries—ECDL 2002, Rome, and the director of First DELOS International Summer School on Digital Library Technologies—ISDL 2001, Pisa. She also served as the program committee member of several international conferences, including ACM-SIGIR, CIKM, and ACM-DL. She is a member of the editorial board of *Information Processing & Management,* the subject area editor for *Hypermedia of Information Retrieval.* She is also a member of IEEE-CS and ACM.

Stuart J. Barnes is associate professor of electronic commerce at the School of Information Management, Victoria University of Wellington, New Zealand. He has been teaching and researching in the information systems field for over a decade. His academic background includes a first-class degree in economics from University College London and a Ph.D. in business administration from Manchester Business School. His current research interests include evaluating Website and e-commerce quality, e-commerce strategy, information systems implementation, knowledge management systems, and business applications of wireless information technologies. He has published and presented more than forty articles in leading conferences, academic journals, professional outlets, and edited books. He has published three books: *E-Commerce and V-Business* in 2001, a bestseller for Butterworth Heinemann, *Knowledge Management Systems* in 2002, and *M-Business* in 2003. Recently, consulting assignments have included those for the UK Inland Revenue, UK Customs and Excise and the OECD.

Peter Bertok received his master's of engineering degree from the Technical University of Budapest, Hungary, and his Ph.D. from the University of Tokyo, Japan. Peter has worked in the area of computing and automation as a senior researcher and a senior software developer. Recently, he has been teaching Computer Science at the Royal Melbourne Institute of Technology. His research interests include distributed and networked systems and mobile computing. Peter is a member of IEEE, ACM and IFIP.

Petter Bae Brandtzæg received a master's degree in work and organizational psychology at the Norwegian University of Technical Science, Trondheim, Norway in 2000. He is currently working at the Institute of Telecom and Informatics, SINTEF, Oslo, Norway. His research interests are concerned with human-computer interaction.

Aslihan Celik is an assistant professor of the operations & management information systems in the Leavey School of Business and Administration at Santa Clara University. She received the B.S. and M.S. degrees in industrial engineering from Bilkent University, Ankara, Turkey, in 1992 and 1994, respectively. She later received her Ph.D. in management information systems from the University of Arizona. Her research interests include designing and testing wireless protocols for data and Web page delivery. She has published in the *ACM Transactions on Database Systems* and various conferences.

Susy S. Chan is an associate professor and the director of the Center for E-Commerce Research in the School of Computer Science, Telecommunications and Information Systems at DePaul University. Her research focuses on e-business strategies, e-commerce and IS curriculum, and usability of m-commerce.

Anindya Datta is the CEO and founder of a venture-backed software company called Chutney Technologies and is regarded as an industry authority on Web infrastructure issues. Chutney is defining and leading an emerging category of solutions called Web Application Optimization, which allows enterprise Web applications to scale to support significantly higher user loads and to deliver faster performance and QoS levels. As a renowned "thought leader," Anindya is frequently invited to speak at industry events. Recently, he delivered presentations at Supercomm 2001, Networld+Interop Fall 2001, and the SunTrust Internet Acceleration Conference. Prior to founding Chutney, Anindya built extensive experience leading teams in the development and implementation of large-scale database systems, including a large commercial project for the USDA. He has served as a consultant for AT&T, US West, IBM and the Israeli government in the fields of data warehousing, data mining and e-commerce. A substantial contributor to several innovations, Anindya holds numerous patents for a variety of data management and Internet technologies. One of his most recent contributions was in developing technologies similar to those incorporated by IBM in the DB2 product for the AS/400 platform. He has also worked on broadcast technologies for mobile users and the access security of subscription-based broadcast information services. In addition to his leadership of Chutney, Anindya is an associate professor at the Georgia Institute of Technology and founder of the iXL Center for Electronic Commerce. Previously, he was an assistant professor at the University of Arizona, after finishing his doctoral studies at the University of Maryland, College Park. Anindya's undergraduate education was completed at the Indian Institute of Technology, Kharagpur. His primary research interests lie in studying technologies that have the potential to significantly impact the automated processing of organizational information. Examples of such technologies include electronic commerce, data warehousing/OLAP, and workflow systems. He has published over 50 papers in prestigious refereed journals such as *ACM Transactions on Database Systems, IEEE Transactions on Knowledge and Data Engineering, INFORMS Journal of Computing,* and the *VLDB Journal,* and in reputed conferences such as ACM SIGMOD, and VLDB. He has also chaired as well as served on the program committees of reputed international conferences and workshops.

Do van Thanh obtained his MSc in electronic and computer sciences from the Norwegian University of Science and Technology in 1984 and his Ph.D. in informatics from the University of Oslo in 1997. In 1991 he joined Ericsson R&D Department in Oslo after 7 years of R&D at Norsk Data, a minicomputer manufacturer in Oslo. In 2000 he joined Telenor R&D and is now in charge of PANDA (Personal Area Network & Data Applications) research activities with a focus on SIP, XML and next generation mobile applications. He holds also a professor position at the Institutt for Telematikk at the Norwegian University of Science and Technology in Trondheim. He is the author of numerous publications and an inventor of a dozen patents.

Xiaowen Fang is an assistant professor in the School of Computer Science, Telecommunications and Information Systems at DePaul University. He is the associate director for the Center for E-Commerce Research. His research focuses on user-centered design of Web search tools, usability of e-business, and usability of mobile commerce.

Nicola Ferro has been a Ph.D. student in computer science in the Department of Electronics and Computer Science at the University of Padua, Italy since January 2002. He received the Laurea degree in Telecommunications Engineering at the University of Padua in April 2001 with a thesis focused on *online information access through handheld devices*. His main research interests are the design and implementation of applications for searching and retrieving documents through mobile clients; examples of such applications are digital libraries and search engines.

Pavan Gundepudi is a senior research analyst at e-Business Strategies. He specializes in using economic and mathematical modeling to address management problems. Pavan is completing a Ph.D. in business administration at the University of Rochester. With background in process control, management science, and operations management, Pavan is part of the research group on Next Generation Supply Chains. He is a member of INFORMS.

Jan Heim received a Ph.D. degree from University of Trondheim in 1982. From 1971 to 1985, he was an assistant professor at the Department of Psychology, University of Trondheim. After that he joined the Norwegian Computing Center as a research scientist. In 1992, he joined SINTEF. He is now the senior research scientist with SINTEF. His main fields of research are cognitive psychology and human-computer interaction, evaluation and design of user interfaces. He has a special interest in individual prerequisites for the use of complex software, user interfaces and usability methods for elderly and disabled, fitness-for-purpose of communication technology, and user interface and experience of interactive TV.

Ravi Kalakota is cofounder and CEO of e-Business Strategies, a technology research and consulting practice based in Atlanta, Georgia. His current research and consulting focuses on multichannel e-business, mobile business strategies, brick-and-click process models, and design of e-service platforms. Ravi holds the distinction of coauthoring several bestselling books on e-commerce, e-business and now m-business. His new book *M-business: The Race to Mobility* (McGraw-Hill, 2001), coauthored with Marcia Robinson, looks at the evolution of enterprise applications into the mobile economy. Ravi received a Ph.D. in e-commerce from the University of Texas at Austin.

John Krogstie has a Ph.D. (1995) and a M.Sc. (1991) in information systems, both from the University of Trondheim (NTNU). He is currently a senior research scientist at SINTEF and a group leader for the Group for Cooperative Information Systems. He is also a (part-time) professor at NTNU. He was employed as a manager in Accenture 1991–2000. John Krogstie is the Norwegian Representative for IFIP TC8 and a member of IFIP WG 8.1, where he is the initiator and leader of the task group for Mobile Information Systems. He has published approximately 40 refereed papers in journals, books and archival proceedings since 1991.

Sai Ho Kwok received a BEng (Hons) in electronic and communications engineering (1992) from the University of North London. He received his Diploma of Imperial College (DIC) (1997) from the Imperial College of Science, Technology and Medicine, and his Ph.D. in digital image processing (1997) from the University of London. He is currently assistant professor of the Department of Information and Systems Management at the Hong Kong University of Science and Technology

(HKUST). He was visiting scholar and a research assistant in the Department of Electronic and Information Engineering at the Hong Kong Polytechnic University (1994–1995). His research interests include digital watermarking, digital rights management, copyright and intellectual property protection, knowledge management, and electronic commerce applications.

Hong Va Leong received his Ph.D. from the University of California at Santa Barbara and is currently an associate professor at the Hong Kong Polytechnic University. He has served on the program committees and organization committees for many international conferences. He has also served as the program co-chair of several international conferences and is a reviewer for a number of international journals, including *ACM Transactions, IEEE Transactions,* and *Information Systems.* His research interests mainly lie in mobile computing, internet computing, distributed systems, distributed databases, and digital libraries. He is a member of the ACM and IEEE Computer Society.

Andreas L. Opdahl is professor of information science in the Department of Information Science at the University of Bergen in Norway. He received his Ph.D. from the Norwegian University of Science and Technology in 1992. He is the author, coauthor or coeditor of more than thirty journal articles, book chapters, refereed archival conference papers and books within multi-perspective enterprise modelling, object-oriented modelling, requirements engineering, software performance engineering, and other areas. Professor Opdahl is a member of IFIP WG8.1 on Design and Evaluation of Information Systems. He serves regularly as a reviewer for internationally recognised journals and on the program committees of several international conferences and workshops.

Marcia Robinson is cofounder and president of e-Business Strategies. Marcia has extensive background in service delivery and customer side of e-Business. She was responsible for multi-channel delivery initiatives such as single sign-on, industry trend examination, customer analysis and integration strategy development. Her book *e-Business: Roadmap for Success* (Addison-Wesley, 1999) and *e-Business 2.0: Roadmap for Success* (Addison-Wesley, 2001), coauthored with Ravi, is an international bestseller and was ranked #3 on Amazon.com's Business Best Seller list. This was the first book on e-Business that looked at the organization and changes necessary in order to compete in the digital economy.

James A. Rodger received his doctorate (1997) in Management Information Systems from Southern Illinois University at Carbondale. He earned an MBA from IUP (1990) and a B.S. in biology (1970) from the University of Pittsburgh. Dr. Rodger joined the MIS and Decision Sciences Department at IUP as an associate professor in 1999, teaching e-commerce, networking, system architecture, and introduction to MIS at both the graduate and undergraduate level. Previously, Dr. Rodger worked as an assistant professor in the business department at the University of Pittsburgh at Johnstown. Dr. Rodger has an extensive and diversified portfolio of scholarly activities. He has published his work in 33 journals, five book chapters, and 50 conferences. His work has appeared in *Annals of Operations Research, Communications of the ACM, Expert Systems with Applications, Journal of Database Management*, and several other journals and proceedings of international, national, and regional conferences. Dr. Rodger has consulted with the US Department of Defense, the Navy Health Research Center, Army Medical Board, and several local health care companies. Dr. Rodger has served as a journal editor for the *Pennsylvania Journal of Business and Economics.* He is on the advisory board of several journals and Info-Science Online, with Idea Group Publishing Company. In addition, he has served as a reviewer for many academic journals and conferences. Dr. Rodger received a research grant from the Central Research Development Fund, Small Grants Program, University of Pittsburgh, Pittsburgh, Pennsylvania, in order to fund several of his research projects in health care. Dr. Rodger is also active in free consulting to small businesses in the region through the Small Business Institute. He received the Outstanding MIS and Decision Sciences Faculty award in academic year 2000–2001. He is also a member of the board of directors for Goodwill Industries of the Laurel Highlands, Inc.

Zixing Shen is a graduate student in Department of Management at University of Nebraska—Lincoln. She earned her bachelor's degree from Sichuan University, Chengdu, China, and worked in China Tobacco Import and Export Sichuan Corporation for three years.

Zahir Tari is an associate professor at RMIT University. He received his bachelor's degree in mathematics at University of Algiers (Algeria), master's degree in operational research at University of Grenoble (France), and a Ph.D. degree in computer science at the University of Grenoble. He is currently the leader of Distributed and Networking Systems activity unit at School of Computer Science and Information Technology. His research is mainly focused on middleware and Web services, in particular dealing with mobility, interoperability, security, and performance (such as caching and load balancing). Dr Tari has been the general cochair and PC cochair of several international conferences, including CoopIS/DOA/ODASE 2002, DOA 2001/2000/1999, IFIP WG 11.3 on Database Security 2000 and IFIP WG 2.6 on Data Semantics 1998. He has coauthored several books, most recently, *Fundamentals of Distributed Object Systems* (John Wiley, 2001). He is a senior member of IEEE. More details about Dr. Tari can be found at http://www.cs.rmit.edu.au/~zahirt.

Jari Veijalainen received a B.Sc. degree in mathematics in 1978 and a M.Sc. degree in computer science in 1983 from the University of Helsinki, Finland, and a

Ph.D. in computer science (Informatik) in 1989 from the TU-Berlin, Germany. During 1989–1997 he worked as a senior research scientist and as a group manager at the Technical Research Center of Finland (VTT). Since 1996 he has been a full professor in computer science at the University of Jyväskylä, Finland. His research interests include heterogeneous transaction management, CSCW transaction models and mechanisms, electronic commerce systems, formal modeling, mobile computing, and multimedia data management. He is a member of ACM and IEEE CS, and a member of the editorial board of *ACM WINET* and *VLDB Journal*. See www.cs.jyu.fi/~veijalai for more details.

Abhinav Vora is a Ph.D. candidate at RMIT University, Melbourne, Australia, and a member of the Distributed and Networking Systems Activity Unit at the School of Computer Science and Information Technology. His research interests are in the area of mobile and pervasive computing. His research supervisors are Associate Professor Zahir Tari and Dr. Peter Bertok. He received his B.S. (Honours) in computer science from RMIT University in 2001.

Mathias Weske received a doctoral degree from the University of Koblenz in 1993 and a habilitation degree from the University of Muenster in 2000. Since 2001 he has been a professor of software systems technology at the Hasso Plattner Institute for Software Systems Engineering at the University of Potsdam, Germany, where he leads a business process technology research group. His current research interests include various topics in workflow management, Web services technology, and enterprise application integration. He is a member of the GI, vice chair of the executive committee of GI SIG EMISA, and a member of IEEE and ACM.

Index

L

M

N

O

P

R

Y